# Studies in Computational Intelligence

Volume 1182

The series "Studies in Computational Intelligence" (SCI) publishes new developments and advances in the various areas of computational intelligence—quickly and with a high quality. The intent is to cover the theory, applications, and design methods of computational intelligence, as embedded in the fields of engineering, computer science, physics and life sciences, as well as the methodologies behind them. The series contains monographs, lecture notes and edited volumes in computational intelligence spanning the areas of neural networks, connectionist systems, genetic algorithms, evolutionary computation, artificial intelligence, cellular automata, self-organizing systems, soft computing, fuzzy systems, and hybrid intelligent systems. Of particular value to both the contributors and the readership are the short publication timeframe and the world-wide distribution, which enable both wide and rapid dissemination of research output.

Indexed by SCOPUS, DBLP, WTI AG (Switzerland), zbMATH, SCImago.

All books published in the series are submitted for consideration in Web of Science.

Thipendra P. Singh · Chandan Jyoti Kumar ·
Ajith Abraham · Kingsley Theophilus Igulu
Editors

# Revolutionizing Healthcare: Impact of Artificial Intelligence on Diagnosis, Treatment, and Patient Care

Springer

*Editors*
Thipendra P. Singh
School of Computer Science Engineering and Technology
Bennett University
Greater Noida, Uttar Pradesh, India

Chandan Jyoti Kumar
Department of Computer Science and Information Technology
Cotton University
Guwahati, Assam, India

Ajith Abraham
Bennett University
Greater Noida, Uttar Pradesh, India

Machine Intelligence Research Labs (MIR Labs)
Auburn, AL, USA

Kingsley Theophilus Igulu
Department of Computer Science
Kenule Beeson Saro-Wiwa Polytechnic Bori
Bori, Nigeria

ISSN 1860-949X ISSN 1860-9503 (electronic)
Studies in Computational Intelligence
ISBN 978-3-031-80812-8 ISBN 978-3-031-80813-5 (eBook)
https://doi.org/10.1007/978-3-031-80813-5



This Springer imprint is published by the registered company Springer Nature Switzerland AG
The registered company address is: Gewerbestrasse 11, 6330 Cham, Switzerland

# Preface

As the healthcare landscape continues to evolve, the integration of artificial intelligence (AI) is becoming a transformative force in this domain. Collection of articles in the current book comes at a critical time in the discourse on intelligent techniques in healthcare. AI is no more a futuristic idea; rather it has become an essential component of modern healthcare ecosystem. AI has transformed everything from in this domain; starting from disease diagnosis, drug design to patient engagement.

This book volume aims to shed light on the profound impact of AI in various areas of healthcare. The articles incorporated here presents a comprehensive overview of state-of-the-art technologies and methods, ranging from the neuronal dynamics to modern object detection techniques like YOLO. Traditionally, the use of AI in healthcare has been driven by the aspiration to overcome the limitations of conventional methods. The use of machine learning (ML) and deep learning (DL) technologies in this area has opened new possibilities for understanding complex medical conditions. For example, in one article, the authors explored how ML algorithms can analyze clinical parameters to predict health risks. This demonstrates a proactive approach to patient care with the help of AI Technology. Few of the articles have reviewed the progress of automatic diagnosis of breast cancer and diabetic retinopathy. It is emphasized that DL can help identify diseases early and effectively. This in turn might help to reduce the death rate. Thus the articles illustrate how ML/DL integrated diagnosis systems may bring down the cost and increase the accuracy of diagnosis. Furthermore, AI is also extensively used in diagnosis and treatment of various mental health conditions. The way AI have transformed the field of Mental health disorders like Major depressive disorder, Schizophrenia, Addiction etc. are extensively discussed in various chapters of this book.

Nevertheless, as we embrace the advancements of AI technology, it is also extremely important to steer the ethical considerations inherent in AI's integration into healthcare domain. Various chapters in this volume present the challenges associated with it. Some of the challenges include maintaining patient privacy, ensuring equity in AI applications, and the moral responsibilities of healthcare providers etc. As we move forward in this direction, these discussions will be very critical in guiding the integration of AI in healthcare industry.

The authors also discuss the hurdles related to integrating AI technology into existing healthcare systems. Strategies for successful implementation are critical for maximizing the benefits of these innovations to minimize potential risks. As the volume discusses, collaboration among stakeholders—from healthcare professionals to technology developers—will be essential to foster an environment conducive to AI integration.

In summary, this collection of articles presents the state of the art of the AI technology in various domains of healthcare. In addition, the book also looks ahead to the potential of AI by providing valuable insights into how smart techniques can revolutionize patient diagnosis, treatment, and care. As we stand on the edge of a new era in healthcare, we must continue to explore these technologies with a critical eye to ensure that they will be of greater benefit to the society. This volume invites readers to engage with the possibilities and challenges of AI in healthcare, providing a platform for further discussion and exploration in this dynamic field.

Greater Noida, India Prof. Thipendra P. Singh

# Introduction

The Artificial intelligence (AI) technology is changing the world in all the fields a significant manner, and healthcare is one such domain AI is transforming the most. With the growth in population, the demand for precise, efficient and accessible healthcare services in increasing day by day. In such a situation, AI is presenting unparalleled opportunities for reshaping the diagnostic methods, drug design, treatment plans, and patient care. This book, *Revolutionizing Healthcare: Impact of Artificial Intelligence on Diagnosis, Treatment, and Patient Care*, presents an in-depth investigation of various impacts the AI technology is having on today's healthcare industry. Through a structured, chapter-by-chapter examination, this work presents how data driven AI technology is enhancing and potential to improve medical accuracy, early detection further, and may assist healthcare professionals in making more informed and timely decisions.

The book covers a variety of ML/DL applications in healthcare, starting from disease diagnosis to the management of complex medical information. Furthermore, it discusses ethical implications and practical challenges in integration of AI technology in healthcare domain. Every one of the chapters focuses on a unique dimension of healthcare where AI is either already making a difference or holds transformative potential and tries to give a unique perspective in that direction.

The opening chapters of the book discusses the use of Artificial Intelligence (AI) technology in the domain of healthcare is transforming the diagnosis of disease, genomics, drug discovery, population health management, and patient care. One of the chapters discusses the possible New Effects of Neuronal Dynamics, while another discusses the transformative potential of YOLO in healthcare industry. The use of machine learning (ML) and deep learning (DL) techniques for diagnosis of cardiovascular diseases, breast cancer, and diabetic retinopathy is thoroughly investigated in different chapters of the volume. This book highlights the crucial role interdisciplinary research plays in generating high-quality data that can empower real-time AI applications. Especially, the collaboration between medical professional and ML/DL researchers is essential for development of real time solution in this domain. However, as the reliance on data grows, so does the need for rigorous data privacy and security protocols. This book advocates for a responsible approach to AI in healthcare, one that balances innovation with the ethical duty to protect patient information.

AI's potential in mental health conditions and disorders is another domain of significant interest. Multiple chapters in this book investigate into how ML/DL models can be useful in identifying mental health conditions. These models can be effective in diagnosis and treatment of various disorders like ASD, ADHD, and SUD etc. The book also navigates some of the logistical and ethical hurdles AI faces within healthcare. The integration of AI-driven solutions into electronic health records (EHRs), the automation of healthcare delivery, and the introduction of surgical robots like the da Vinci system exemplify how AI can optimize operational efficiencies and enhance patient care. Additionally, by examining AI's impact on ethical principles such as beneficence and autonomy, this book provides readers with a framework for understanding the moral responsibilities that accompany AI deployment in healthcare.

By focusing on both the promises and the challenges of AI in healthcare, this volume highlights the importance of a balanced, ethically grounded approach to innovation in this field. As the chapters unfold, readers are invited on a journey through the evolving intersection of AI and healthcare, highlighting the transformative possibilities and research challenges that lie ahead.

# Contents

# Contributors

**Kirti Aggarwal** Department of Computer Science and Engineering and Information Technology, Jaypee Institute of Information Technology, Noida, India

**Taibat Onome Asunogie** Department of Computer Science, Edo State University, Uzairue, Edo State, Nigeria

**Dhritiraj Barman** Department of Computer Applications, Assam Don Bosco University, Azara, Assam, India

**Arun Kr. Baruah** Department of Mathematics, Dibrugarh University, Dibrugarh, India

**Sangita Baruah** Cotton University, Guwahati, Assam, India

**Kanak Chandra Bora** Department of Computer Science, University of Science and Technology, Baridua, Meghalaya, India

**Reviricha Bora** Cotton University, Assam, India

**Dhruba Jyoti Borah** Majuli University of Culture, Chitadarchuck, Majuli, Assam, India

**Hemashree Bordoloi** Department of Electronics and Communication Engineering, Assam Don Bosco University, Guwahati, Assam, India

**Sangeeta Borkakoty** Department of Computer Science, University of Science and Technology, Baridua, Meghalaya, India

**Lal Omega Boro** Assam Don Bosco University, Azara, India

**Subrat Chetia** Pandit Deendayal Upadhyaya Adarsha Mahavidyalaya, Dalgon, Assam, India

**Levi Adawari S. Damingo** Department of Computer Science, Ignatius Ajuru University of Education, Rumuolumeni, Port Harcourt, Rivers State, Nigeria

**Kaushik Das** Department of Computer Science and Engineering, Dibrugarh University, Dibrugarh, India

**Pranab Das** Department of Computer Applications, Assam Don Bosco University, Azara, Assam, India;
Computer Applications, Assam Don Bosco University, Guwahati, Assam, India

**Priti Rekha Das** Department of Psychology, Gauhati University, Guwahati, Assam, India

**Carl James Debono** Faculty of ICT, University of Malta, Msida, Malta

**Balendra Kr. Dev Choudhury** Department of Physics, Pub Kamrup College, Baihata Chariali, India

**Kirtan Dua** School of Computer Science, University of Petroleum and Energy Studies, Dehradun, India

**Okpu Okpomo Eterigho** Delta State University of Science and Technology, Ozoro, Nigeria

**Lalit Garg** Faculty of ICT, University of Malta, Msida, Malta

**Deba Kanta Gogoi** Department of CSE, Tezpur University, Tezpur, Assam, India

**Minakshi Gogoi** Girijananda Chowdhury University, Assam, Guwahati, India

**Prashant K. Gupta** School of Computer Science Engineering and Technology, Bennett University, Greater Noida, India

**Irani Hazarika** Department of Computer Science, Gauhati University, Guwahati, Assam, India

**Manasi Hazarika** Department of Computer Science and Engineering, Assam Don Bosco University, Azara, Assam, India;
Computer Science and Engineering, Assam Don Bosco University, Guwahati, India

**Kingsley Theophilus Igulu** Department of Computer Science, Kenule Beesor Saro-Wiwa Polytechnic, Bori, Rivers State, Nigeria

**Atowar Ul Islam** Department of Computer Science, University of Science and Technology, Baridua, Meghalaya, India

**Sanjib Kr. Kalita** Department of Computer Science, Gauhati University, Guwahati, India

**Kunal Kartikeya** Department of Computer Science and Engineering and Information Technology, Jaypee Institute of Information Technology, Noida, India

**Eluemunor Kizito** Department of Information Systems and Technology, Delta State University of Science and Technology, Ozoro, Nigeria

**Elliot Nkem Kizzy** Department of Computer Science, Informatics and Cybersecurity, Federal University Otuoke, Otuoke, Bayelsa State, Nigeria

**Ankit Kumar** University of Petroleum and Energy Studies, Dehradun, India

**Chandan Jyoti Kumar** Department of Computer Science and IT, Cotton University, Guwahati, Assam, India

**Yotam Lurie** Ben-Gurion University of the Negev, Beersheba, Israel

**Anjana Kakoti Mahanta** Department of Computer Science, Gauhati University, Guwahati, Assam, India

**Lipi B. Mahanta** Mathematical and Computational Sciences Division, Institute of Advanced Study in Science and Technology, Guwahati, Assam, India

**Shlomo Mark** Shamoon College of Engineering, Ashdod, Israel

**Bireshwar Dass Mazumdar** School of Computer Science Engineering and Technology, Bennett University, Greater Noida, India

**Manmohan Mishra** Department of Master of Computer Application, United Institute of Management, Prayagraj, Uttar Pradesh, India

**Vinaytosh Mishra** Thumbay Institute for AI in Healthcare, Gulf Medical University, Ajman, UAE

**Gypsy Nandi** Department of Computer Applications, Assam Don Bosco University, Guwahati, Assam, India;
Assam Don Bosco University, Azara, India

**Bapynshngainlang Nongkynrih** Department of Computer Applications, Assam Don Bosco University, Guwahati, Assam, India

**Ubrurhe Ogheneochuko** Delta State University of Science and Technology, Ozoro, Nigeria

**Vijay Prakash** Faculty of ICT, University of Malta, Msida, Malta

**Debashis Saikia** Department of Instrumentation and USIC, Gauhati University, Guwahati, Assam, India

**Samar Jyoti Saikia** Department of Electronics and Communication Engineering, Assam Don Bosco University, Guwahati, Assam, India

**Arup Sarmah** Department of Computer Application, Assam Don Bosco University, Guwahati, India

**Sonia Sarmah** Department of Computer Applications, Assam Don Bosco University, Azara, Assam, India;
Computer Applications, Assam Don Bosco University, Guwahati, Assam, India

**Amal Satheesh** Department of Computer Applications, Assam Don Bosco University, Azara, Assam, India

**Nne Rena Saturday** Department of Computer Science, Rivers State University, Port Harcourt, Rivers State, Nigeria

**Rosy Sharma** Department of CSE, Tezpur University, Tezpur, Assam, India

**Thipendra P. Singh** School of Computer Science Engineering & Technology, Bennett University, Greater Noida, India

**Vikalp Srivastava** Department of Computer Science and Engineering and Information Technology, Jaypee Institute of Information Technology, Noida, India

**Jyotismita Talukdar** Department of CSE, Tezpur University, Tezpur, Assam, India

**Rita Rani Talukdar** Department of Psychology, Gauhati University, Guwahati, Assam, India

**Abhishek Tripathi** Department of Accounting and Information Systems, School of Business, The College of New Jersey, Ewing Township, NJ, USA

# Abbreviations

| | |
|---|---|
| ACC | Accuracy |
| AcidLSD | Lysergic Acid Diethylamide |
| ADHD | Attention-Deficit Hyperactivity Disorder |
| AI | Artificial Intelligence |
| AMD | Age-related Macular Degeneration |
| ANN | Artificial Neural Network |
| AP | Action Potential |
| API | Application Programming Interface |
| ASD | Autism Spectrum Disorder |
| ASPP | Atrous Spatial Pyramid Pooling |
| AUC | Area Under the Curve |
| AUC_PR | Area Under the Precision & Recall Curve |
| BCR-UNet | Bi-directional ConvLSTM Residual U-Net |
| BI-RADS | Breast Imaging Reporting and Data System |
| CADx | Computer Aided Diagnostic System |
| CapsNet | Capsule Network |
| CBIS-DDSM | Curated Breast Imaging Subset of the DDSM |
| CBT | Cognitive Behavioural Therapy |
| CC | Craniocaudal |
| CER | Character Error Rate |
| CLAHE | Contrast Limited Adaptive Histogram Equalization |
| CNN | Convolutional Neural Network |
| CNS | Central Nervous System |
| CT | Computed Tomography |
| CTPN | Connectionist Text Proposal Network |
| CVD | Cardiovascular Disease |
| DBN | Deep Belief Network |
| DCNN | Deep Convolutional Neural Network |
| DDSM | Digital Database for Screening Mammography |
| DICOM | Digital Imaging and Communications in Medicine |
| DL | Deep learning |

| | |
|---|---|
| DM | Diabetes Mellitus |
| DME | Diabetic Macular Oedema |
| DMHI | Digital Mental Health Interventions |
| DQAD | Digital Q learning Auxiliary Diagnostics |
| DR | Diabetic Retinopathy |
| DSM | Diagnostic and Statistical Manual of Mental Health Disorders |
| DTC | Decision Tree Classifier |
| ECG | Echocardiogram |
| EHR | Electronic Health Record |
| EMR | Electronic Medical Record |
| EU | European Union |
| EX | Exudates |
| FA | Firefly Algorithm |
| fCNN | Fully Convolutional Neural Networks |
| FFDM | Full-Field Digital Mammography |
| FN | False Negative |
| FP | False Positive |
| FPN | Feature Pyramid Network |
| FPPI | False Positives Per Image |
| GA | Genetic Algorithms |
| GABA | Gamma Amino Butyric |
| GAM | Generalized Additive Model |
| GAN | Generative Adversarial Network |
| GDLS | Gray Level Difference Statistics |
| GDPR | General Data Protection Regulation |
| GFFNN | Generalized Feedforward Neural Network |
| GIoU | Generalized Intersection over Union |
| GLCM | Gray-Level Co-occurrence Matrix |
| GRU | Gated Recurrent Unit |
| HIPAA | Health Insurance Portability and Accountability Act |
| HM | Haemorrhages |
| HOS | Higher Order Statistics |
| HRF | High-Resolution Fundus |
| HSI | Hyperspectral Imaging |
| IARC | International Agency for Research on Cancer |
| ICD-10 | International Classification of Diseases-10 |
| ICGC | International Cancer Genome Consortium |
| ICT | Information and Communication Technology |
| IoT | Internet of Things |
| IoU | Intersection over Union |
| IRMAs | Intraretinal Microvascular Abnormalities |
| KMCM | Kardia Mobile Cardiac Monitor |
| KNN | K-Nearest Neighbor |
| LBP | Local Binary Pattern |
| LDA | Latent Dirichlet Allocation |

| | |
|---|---|
| LDA | Linear Discriminant Analysis |
| LIME | Local Interpretable Model-agnostic Explanations |
| LIWC | Linguistic Inquiry and Word Count |
| LLMs | Large Language Models |
| LR | Logistic Regression |
| LSTM | Long Short-Term Memory |
| MA | Microaneurysms |
| mAP | Mean Average Precision |
| mAUC_PR | Mean Area Under the Precision & Recall Curve |
| MCC | Matthews Correlation Coefficient |
| MDD | Major Depressive Disorder |
| MDMA | 3,4-methylenedioxymethamphetamine |
| MDSCC | Moderately Differentiated Squamous Cell Carcinoma |
| MEWS | Modified Early Warning Score |
| MI | Mutual Information |
| MIAS | Mini-Mammographic Image Analysis Society Digital Mammogram Database |
| ML | Machine Learning |
| MLO | Mediolateral Oblique |
| MNN | Modular Neural Network |
| MP | Multilayer Perceptron |
| MRI | Magnetic Resonance Imaging |
| MRT-SegNet | Maximum Response Texton-Based SegNet |
| Mult-Attn-U-Net | Multiscale Attention U-Net |
| NASSS | Non-adoption, Abandonment, Scale-up, Spread, Sustainability |
| NB | Naive Bayes |
| NBCF | National Breast Cancer Foundation |
| NGS | Next-Generation Sequencing |
| NHS | National Health Service |
| NLP | Natural Language Processing |
| NMIFS | Normalized Mutual Information Feature Selection |
| NMS | Non-Maximum Suppression |
| NN | Neural Network |
| NPDR | Non-Proliferative Diabetic Retinopathy |
| OCR | Optical Character Recognition |
| OCT | Optical Coherence Tomography |
| OECD | Organization for Economic Cooperation and Development |
| OSCC | Oral Squamous Cell Carcinoma |
| PCG | Phonocardiogram |
| PD | Parkinson's disease |
| PDR | Proliferative Diabetic Retinopathy |
| PHI | Protected Health Information |
| PHQ | Patient-Health Questionnaire |
| PNN | Probabilistic Neural Network |
| PSO | Particle Swarm Optimization |

| | |
|---|---|
| PTSD | Post-Traumatic Stress Disorder |
| QAI | Quantum AI |
| QPOP | Quadratic Phenotypic Optimisation Platform |
| r/Depression | Reddit community for Depression |
| r/SuicideWatch | Reddit community for Suicide Watch |
| RBES | Rule Based Expert Systems |
| R-CNN | Region-Based Convolutional Neural Network |
| RF | Random Forest |
| RFE | Recursive Feature Elimination |
| RMSE | Root Mean Squared Error |
| RNN | Recurrent Neural Network |
| ROC | Retinopathy Online Challenge |
| ROI | Regions of Interest |
| RPA | Robotic Process Automation |
| RPM | Remote Patient Monitoring |
| SAA | Shock Advisory Algorithm |
| SCC | Squamous Cell Carcinoma |
| SDG | Sustainable Development Goals |
| SDRB | Structured Dropout Residual Block |
| SD-UNet | Structured Dropout U-Net |
| SE | Soft Exudates |
| SHAP | Shapley Additive Explanations |
| SMANet | Segmentation Multiscale Attention Network |
| SMOTE | Synthetic Minority Over-sampling Technique |
| SOFM | Self-Organizing Feature Map |
| SUD | Substance Use Disorder |
| SVM | Support Vector Machine |
| TCGA | The Cancer Genome Atlas |
| TDC | Typically Developing Children |
| THC | Tetra Hydro Cannabinol |
| TN | True Negative |
| TP | True Positive |
| VA | Virtual Assistants |
| VHA | Virtual Health Assistants |
| VR | Virtual Reality |
| WBC | Wisconsin Breast Cancer |
| WCRF | World Cancer Research Fund |
| WDSCC | Well Differentiated Squamous Cell Carcinoma |
| WER | Word Error Rate |
| WHO | World Health Organization |
| WIOU | Weighted Intersection Over Union |
| XAI | Explainable AI |
| XGBoost | Extreme Gradient Boosting |
| YOLO | You Only Look Once |

# Introduction to Intelligent Techniques in Healthcare

**Chandan Jyoti Kumar and Thipendra P. Singh**

**Abstract** The use of Artificial Intelligence (AI) technology in the domain of healthcare is transforming the diagnosis of disease, genomics, drug discovery, population health management, and patient care in a rapid pace. AI, mainly Machine Learning (ML) and Deep Learning (DL) based techniques, has displayed significant potential in enhancing the accuracy and efficiency of diagnosis. Various fields like early detection of cancer with ML/DL techniques have emerged as most dominant research topic. Furthermore, ML/DL techniques have been applied extensively in research related to chronic diseases such as diabetes and cardiovascular conditions, for providing advanced diagnostic capabilities and decision support systems. In addition, in field like genomics, AI plays a crucial role in analyzing genetic variations and thereby contributing towards development of personalized medicine. By leveraging next-generation sequencing (NGS) and high-throughput genomic technologies, AI facilitates the identification of disease-related mutations and supports the development of targeted therapies. AI also assists in the study of non-coding regions of the genome, which are important for understanding of various genetic disorders. ML/DL based techniques through its predictive power may enhance the ability to forecast health risks of a population thereby may help in population health management and contribute towards optimize interventions, and manage resources effectively in a society. These ML/DL based systems has the potential to transform patient care service by providing virtual health assistants, to improve accessibility as well as support mental health, and offer continuous monitoring of the patient. In spite of all these advantages, there are various challenges associated with it. As the AI based approaches are data driven, the reliability of data is of prime concern. Collaborative interdisciplinary research has to be done for generating these data which can facilitate the development of such real time application. Additionally, data privacy is another important aspect that needs to be handled well. This chapter underscores AI's

C. J. Kumar (✉)
Department of Computer Science and IT, Cotton University, Guwahati, Assam, India
e-mail: chandan14944@gmail.com

T. P. Singh
School of Computer Science Engineering & Technology, Bennett University, Greater Noida, India
e-mail: thipendra@gmail.com

 
T. P. Singh et al. (eds.), *Revolutionizing Healthcare: Impact of Artificial Intelligence on Diagnosis, Treatment, and Patient Care*, Studies in Computational Intelligence 1182,
https://doi.org/10.1007/978-3-031-80813-5_1

transformative potential in healthcare, advocating for responsible use and continuous innovation to enhance patient outcomes and care efficiency.

**Keywords** Artificial intelligence · Disease diagnostics · Genomics · Drug discovery

## 1 Introduction

The availability of enormous amounts of data in today's era is a key factor driving the growth of data-driven technologies like AI. The field of AI/ML aims to develop machines which are effective in performing tasks that usually require human intelligence. This field includes various techniques such as ML/DL, Digital image processing and natural language processing (NLP). Recent advancement to this field are the Large Language Models (LLMs) which are Transformer learning based AI algorithm that uses massively large data sets to understand, summarize, generate, and predict new text-based content (Suleimenov et al. 2020).

With technological advancements and the evolution of scientific theories the field of medicine, especially biotechnology, is becoming increasingly digitized and data-driven (Bhadra et al. 2023). The growing availability of diverse data types—ranging from genomics and clinical information to economic, demographic, and phenotypic data—combined with advancements in mobile technology, the Internet of Things (IoT), computing power, and data security, signals a pivotal moment for the intersection of healthcare and technology. This shift has given rise to smart healthcare, which integrates cutting-edge information technologies (Kumar et al. 2019). However, smart healthcare represents more than just technological progress; it signifies a comprehensive, multi-dimensional transformation. This transformation is evident in several areas: changes in medical models (shifting from disease-centred to patient-centred care), advancements in informatization (expanding from clinical to regional medical information systems), new approaches to medical management (moving from general to personalized care), and evolving concepts of prevention and treatment (emphasizing preventive care over reactive disease treatment). These changes aim to meet the personalized needs of patients while enhancing healthcare efficiency, ultimately transforming the healthcare experience and guiding the future of modern medicine (Bajwa et al. 2021). This chapter will examine the current landscape of AI in healthcare, highlighting its potential benefits, limitations, and challenges. This is an attempt to provide insights into the future development of AI technologies, aiming to enhance understanding of AI's role in healthcare and support its effective integration into clinical practice.

## 2 Overview of Intelligent Techniques

The history of AI goes beyond simply replicating or replacing human cognitive functions with computers. AI is better described as the field dedicated to the study and application of scientific and engineering principles to create intelligent systems, particularly through the development of sophisticated computer programs. While AI involves using computer systems to understand aspects of human intelligence, it is not limited to methods that mirror physiological processes. Instead, AI focuses on leveraging computational systems to grasp and address the complexities of human intelligence. AI can be described as an interdisciplinary field that combines computer science with large datasets to solve complex problems. It encompasses various subfields, including Machine Learning (ML) and Deep Learning (DL), which are integral components of AI as shown in Fig. 1. These subfields focus on developing algorithms designed to create advanced systems that can make predictions or classifications based on the data they analyze (Bhadra et al. 2024).

Machine learning (ML) is broadly divided into three main subfields: supervised learning, unsupervised learning, and reinforcement learning. In supervised learning, algorithms are developed to perform prediction or classification tasks using data that is already labelled. ML algorithms require inputs (predictors) that correlate with an output (response). When the output consists of distinct categories, the task is classification. If the output is a continuous range of values, the task is regression. Supervised learning includes a variety of techniques such as linear and nonlinear regression, neural networks, random forests, and decision trees. Unsupervised learning, on the other hand, focuses on discovering patterns and trends within unlabeled data. The goal here is not to predict a specific outcome but to identify underlying similarities within the data through clustering algorithms and similar approaches. Reinforcement learning involves training and deploying learning agents within an environment with

**Fig. 1** Structural framework of AI, ML, and DL

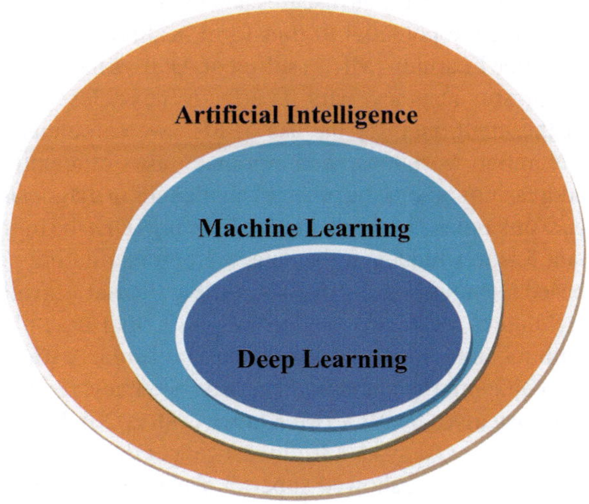

the goal of maximizing their rewards. A specialized subset of ML is DL techniques, which encompasses a variety of specialized neural network architectures designed to tackle different types of problems. Deep learning architectures such as CNNs, LSTMs, GANs, and Transformers have distinct strengths to different types of data and tasks. CNNs are designed for processing image data, while RNNs, including LSTMs, works with sequential data. GANs are adept at generating new data samples, and Transformers have transformed NLP by efficiently managing complex language tasks. Each of these architectures is essential for pushing forward the capabilities and applications of AI and machine learning. In addition to learning algorithms, Fuzzy logic is one field that allows systems to handle uncertainty and approximate reasoning by working with degrees of truth rather than binary values. This makes it particularly useful for control systems and decision-making processes where inputs and conditions are imprecise. Genetic algorithms (GAs) are another such widely used intelligent technique. It draws inspiration from natural evolution for solving optimization problems. Selection, crossover, and mutation are various mechanisms used by GAs. Techniques like Bayesian networks provide a probabilistic framework to draw relationships between variables. Together, these techniques offers support for developing automated systems that can tackle various problems related to the domain of healthcare.

## 3 Applications of AI in Healthcare

### *3.1 In Disease Diagnostics*

Addressing challenges related to healthcare industry remains a significant issue globally, particularly when it comes to developing early diagnostic tools amidst the complexity of various diseases and their overlapping symptoms (Kumar et al. 2022). AI holds the potential to transform healthcare, especially in the area of diagnosis. Machine Learning (ML), a subset of AI, uses the pattern hidden in the data to improve diagnostic accuracy, with its effectiveness closely tied to the quality and quantity of the input data. By the use of ML, we can enhance decision-making processes, streamline workflows, and automate tasks efficiently and cost-effectively, thereby addressing some of the inherent challenges in diagnosis and improving overall healthcare outcomes. Similarly in deep learning (DL) the initial layers can extract the important features hidden in the data, thereby facilitating proper classification. These DL based techniques are very useful if big dataset is available to train the model. These techniques are widely used in diagnosis of disease like Cancer. Diagnosis of Cancer with mammograms is a widely explored area. A study in the UK claimed to achieve a notable improvement in diagnostic accuracy, reducing false positives by 5.7% and false negatives by 9.4% (McKinney et al. 2020). In a South Korean study, researchers evaluated an AI system for breast cancer diagnosis against the performance of radiologists. The AI showed superior sensitivity, identifying breast cancer with a 90%

accuracy rate in cases involving masses, compared to 78% for radiologists. It also excelled in detecting early-stage breast cancer, with a sensitivity of 91%, while radiologists achieved 74% (Kim et al. 2020). In a separate study, deep learning methods were used to do the diagnosis of skin cancer (Haenssle et al. 2018). An AI system employing CNNs demonstrated high accuracy in diagnosing melanoma and recommending treatment options, outperforming dermatologists in diagnostic performance (Han et al. 2020). For the diagnosis of breast cancer, researchers utilized a backpropagation neural network and achieved an impressive accuracy of 99.27% on the WBC dataset, which contains nine features (Bhattacherjee et al. 2020). Similarly, Alshayeji et al. (2022) employed the WBCD and WDBI datasets to develop a shallow artificial neural network (ANN) model for breast cancer classification. Their model demonstrated a remarkable classification accuracy of up to 99.85% without the need for feature selection or algorithm tuning (Alshayeji et al. 2022). Similar investigation on the diagnosis of breast cancer with various artificial neural network (ANN) architectures on the WBC dataset was also carried out by (Sultana et al. 2021). Their study explored several models, including the multilayer perceptron (MLP), Jordan/Elman neural network, modular neural network (MNN), generalized feedforward neural network (GFFNN), self-organizing feature map (SOFM), SVM neural network, probabilistic neural network (PNN), and recurrent neural network (RNN). Among these, the PNN achieved the highest accuracy at 98.24%, surpassing the other neural network models employed in the study. However, similar to many other studies, this research lacks interpretability, as it does not identify which features are most influential during the prediction process. Another researcher employed deep learning (DL) techniques to detect breast cancer using the WBC dataset. The authors trained seven DL models: artificial neural network (ANN), convolutional neural network (CNN), gated recurrent unit (GRU), long short-term memory (LSTM), multilayer perceptron (MLP), probabilistic neural network (PNN), and recurrent neural network (RNN). Among these, LSTM and GRU achieved the highest performance, with accuracies of approximately 99% (Ghosh et al. 2021).

This technology has been also used in chronic disease such as diabetes, cardiovascular diseases etc. For instance, Ansari et al. developed an automated system for diagnosing coronary heart disease using neuro-fuzzy integrated systems, achieving an accuracy of approximately 89% (Ansari and Gupta 2011). However, a notable limitation of this study is the absence of a clear explanation regarding how their method performs in various scenarios, such as multiclass classification, big data analysis, and imbalanced class distribution. Additionally, there is no discussion on the credibility of the model's accuracy, which is increasingly important in medical fields to help users outside of the medical domain understand the effectiveness of the approach. Some researchers have employed deep CNN approaches to detect irregular cardiac sounds (Rubin et al. 2017). To enhance the sensitivity and specificity of their training dataset, the authors modified the loss function. Their model was evaluated in the 2016 PhysioNet computing competition, where it achieved a commendable second-place finish, with a specificity of 0.95 and a sensitivity of 0.73. A similar study developed a DL technique for diagnosing fetal health, focusing on multiclass morphological patterns (Miao and Miao 2018). Their model aims to distinguish and classify the morphological patterns associated with pregnancy complications. The preliminary results of their computations include an accuracy of 88.02%, a precision of 85.01%, and an F-score of 0.85. To tackle overfitting issues, the study employed various dropout strategies, which extended the training time—a tradeoff they acknowledged for achieving higher accuracy. Although ML based research have been extensively done for diagnosing heart disease, there has been a lack of research addressing the challenges associated with unbalanced data in multiclass classification scenarios (Ahsan et al. 2022) Numerous studies have extensively explored the use of ML systems for diagnosis of diabetes. For instance, a study compared various ML classifiers for classifying diabetes mellitus patients (Kandhasamy and Balamurali 2015). Their experiment, conducted using the UCI Diabetes dataset, found that the KNN ($K = 1$) and RF classifiers achieved accuracy of approx. 1. However, a notable limitation of this study is its use of a simplified Diabetes dataset with only eight binary-classified parameters, making it unsurprising to achieve 100% accuracy given the dataset's simplicity. Additionally, the study does not discuss how the algorithms impact the final predictions or how the results should be interpreted from a nontechnical perspective. In a similar study, researchers developed a Decision Support System to aid physicians in diagnosing diabetes, with the use of a range of ML techniques including support vector machines (SVM), random forest (RF), and deep CNN (Yahyaoui et al. 2019). In their computations, RF outperformed all other algorithms with an accuracy of 83.67%, while the deep learning (DL) and SVM approaches achieved accuracies of 76.81% and 65.38%, respectively. A study to diagnosis diabetic retinopathy showed its superiority over humans (Li et al. 2021). A similar study using various ML models showed the effectiveness of these techniques in predicting the need of appendix surgery. In addition to the diseases mentioned above ML/DL techniques are also extensively used in severe diseases such as kidney disease, COVID-19 as well as mental health related diseases/disorders like Parkinson's, Alzheimer's, Autism Spectrum Disorder (ASD) etc. (Kumar et al. 2021). The literature shows how the ML/DL model doesn't only help in disease diagnosis but also can work as a decision support system. AI tools can enhance diagnostic accuracy, lower costs, and expedite processes compared to traditional methods (Sharma et al. 2024a). They also help minimize human errors and deliver more precise results more quickly. Looking ahead, AI technology has the potential to support medical decisions by offering clinicians real-time assistance and valuable insights (Sharma et al. 2024b).

### 3.2 *In Genomics and Drug Discovery*

The integration of AI with genotype analysis offers tremendous potential for advancing disease surveillance, prediction, and personalized medicine (Maqsood et al. 2024).

A key objective of genomics is to uncover mutations or changes in DNA sequences that can disrupt cellular processes and lead to disease, typically through genome-wide association studies (GWAS). It's important to note that not all mutations are disease-causing (Grigorenko and Dozier 2013). Identifying which variations are linked to specific diseases can lead to more effective treatments, drugs, and potentially cures. A study highlighted that understanding genetic variation can enhance our knowledge of why individuals respond differently to the same medications (McGuire et al. 2020). This insight fuels the concept of personalized medicine, where pharmacogenomics is used to tailor treatments based on an individual's genetic profile. Cancer arises from alterations or mutations in the DNA sequences of genes that control cell survival, division, and other key processes. These genetic changes can disrupt normal cell growth mechanisms, leading to uncontrolled cell proliferation and the spread of abnormal cells. As these cells continue to accumulate mutations, they can form tumours. The specific mechanisms of DNA sequence alteration can differ among various types of cancer. Researchers have long sought to understand the biological foundations of cancers with variable clinical outcomes. Genomics offers valuable insights into this complex and heterogeneous disease. Through genomic studies, researchers can identify numerous mutated genes responsible for cancer, known as oncogenes. For example, the TP53 gene, frequently mutated across various cancers, serves as a notable instance (Teer 2014). Next-generation sequencing (NGS) technologies enable comprehensive genome profiling of cancer patients, facilitating the identification of clinically relevant genetic variations that can be targeted for potential therapies. Projects like the International Cancer Genome Consortium (ICGC) and The Cancer Genome Atlas (TCGA) have primarily concentrated on characterizing cancer genomes by examining genetic variations in the coding regions. While these efforts have uncovered numerous novel mutations, it's important to note that only 2% of the human genome consists of coding regions, with the remaining 98% being non-coding. Consequently, there is limited information on how variations in the non-coding regions influence cancer development. Another study have reported that although mutations in non-coding regions are abundant, their effects are not yet well understood (Elliott and Larsson 2021). Recent studies, however, indicate that variations in non-coding regulatory regions of the genome are strongly associated with disease conditions. Identifying and understanding these gene regulatory regions is crucial not only for grasping genome function but also for deepening our understanding of disease mechanisms and providing a more comprehensive overview of disease states. For instance, mutations in the regulatory region of the RB1 gene have been linked to the development of glioblastoma, a type of brain cancer (Arabzadeh et al. 2021). Genes feature multiple non-coding regulatory elements, including promoters, activators, enhancers, and silencers, which play critical roles in gene regulation. Accurate prediction of enhancers is crucial for understanding the role of non-coding genome regions in disease contexts. A rule-based explainable AI (XAI) model was developed by researchers for identifying enhancers in Drosophila melanogaster cell lines (Wolfe et al. 2021). This model was trained using histone modification ChIP-seq data and STARR-seq data. For assessment of its performance, the XAI model developed was compared with conventional ML models for enhancer prediction and annotation. Thus the newly developed ML model, trained on the same histone modification data, demonstrated the ability to accurately predict enhancer locations and generalize to other cell lines without further adjustment. The project had the following objectives: (1) Train the XAI model using histone modification ChIP-seq data, (2) Define, interpret, and implement rules for enhancer prediction with the XAI model, and (3) Apply the model to predict changes in enhancers across different developmental, physiological, or disease contexts. In oncology, categorizing cancers into clinically relevant molecular subtypes through transcriptomic profiling has become a powerful tool (Shlyueva et al. 2014). This approach, initially developed for breast cancer and later applied to other cancers such as colorectal, ovarian, and sarcomas, has significant implications for diagnosis, prognosis, and treatment selection (Shrikumar et al. 2017). Traditional computational methods for cancer subtyping, like support vector machines (SVMs) or k-nearest neighbours, can be prone to errors due to batch effects and may focus narrowly on a limited set of signature genes, potentially overlooking critical biological information (Sizilio et al. 2012). However, the rise of high-throughput genomic sequencing technologies, combined with advancements in AI and machine learning, is rapidly advancing the field of personalized medicine and drug discovery (Ström et al. 2020).

### *3.3 AI Assistance in Population Health Management*

Revolutionizing healthcare: the role of artificial intelligence in clinical practice

In population health management, predictive analytics is becoming increasingly crucial for designing effective health initiatives. ML/DL algorithms are instrumental in analyzing diverse data sets to create models that improve patient outcomes and reduce healthcare costs. For example, these analytics can identify individuals at risk for chronic conditions like endocrine or cardiac diseases by evaluating data such as medical history, demographics, and lifestyle factors. This enables targeted prevention and treatment interventions. Similarly, predictive analytics can anticipate hospital readmissions by examining patient demographics, medical history, and social health factors, allowing for preventive measures that lower readmission rates. This not only aims to cut costs but also enhances patient care and outcomes (Alowais et al. 2023).

AI algorithms are particularly adept at processing large volumes of data to reveal patterns and relationships that might be missed by human analysts. This boosts the accuracy of predictive models and ensures that patients receive appropriate interventions. Additionally, AI can streamline public health management tasks such as patient outreach and care coordination, further reducing costs and improving care by enabling timely and effective interventions.

Nonetheless, the success of predictive analytics in public health management relies heavily on the quality of the data and the strength of the technological infrastructure supporting these models. Human oversight is also critical to ensure that interventions for at-risk patients are appropriate and effective. In summary, while predictive analytics is increasingly vital in population health, its success depends on a blend of high-quality data, advanced technology, and careful human supervision.

## 3.4 AI-Powered Patient Care

As global demand for healthcare services rises and resources remain constrained, innovative solutions are essential to meet these challenges. AI-powered virtual health assistants are emerging as a transformative technology in this realm, designed to support healthcare professionals and improve patient care. These digital tools simulate human interaction to deliver personalized support based on patient inputs. They use AI technologies, including chat-bots, voice interfaces, and other advanced systems, to perform a range of functions such as diagnosing symptoms, providing medical advice, sending medication reminders, scheduling appointments, and monitoring vital signs. Virtual health assistants can also collect daily health information from patients and communicate it to their healthcare providers, alleviating the workload on medical staff and enhancing patient outcomes. Their round-the-clock availability ensures easier access to healthcare services, making it more convenient for patients to receive timely assistance. A notable application of AI in healthcare is the use of mobile apps for triage, which evaluate the urgency of symptoms reported by users. For example, the National Health Service (NHS) in north London has implemented an app that serves around 1.2 million users, who prefer this AI-driven chat-bot over traditional non-emergency contact methods. Additionally, intelligent speakers offer significant benefits for elderly or chronically ill patients who may find smart phone apps challenging, thus expanding the reach of virtual health services.

Another major domain where AI has the potential to revolutionize is mental health support, particularly in providing personalized and accessible care options. Research has already shown that web-based or internet-based cognitive-behavioural therapy (CBT) can be an effective psychotherapeutic intervention (Das et al. 2024). While traditional psychiatric care relies on direct patient interaction and behavioural observation, AI-powered tools can significantly enhance these approaches in several ways (Ahsan et al. 2021). AI-driven mental health applications are capable of aiding in the early detection and diagnosis of mental health conditions, offering tailored treatment, and providing ongoing support. These tools can deliver 24/7 assistance, reducing the need for in-person visits and minimizing wait times. They also facilitate continuous monitoring of patient progress and medication adherence, offering valuable insights into the effectiveness of treatments. Current research highlights depression as the most frequently studied mental health disorder within AI applications. Additionally, AI-powered apps have shown promise in managing substance use disorders. Nevertheless, there are important limitations to consider. A major concern is the potential for bias in AI data and algorithms. If the data used to train these systems is not diverse, it may lead to biased and inaccurate diagnoses. Moreover, AI may struggle to fully capture the complexity of mental health conditions, which can vary widely among individuals. There is also a risk that AI tools might lack the personalization and empathy that are crucial for effective mental health care (Choudhury et al. 2022).

Furthermore, AI can play a crucial role in improving patient education and reducing healthcare provider burnout. A key application is the use of AI-powered chat-bots, which are being introduced in various healthcare contexts, including for diet advice, smoking cessation, and CBT. Patient education is essential for helping individuals grasp their medical conditions, explore treatment options, and adopt preventive measures. When patients are well-informed, they are more likely to adhere to treatment plans and achieve better health outcomes. For this, AI has the potential to greatly enhance patient understanding by providing personalized, interactive information and guidance.

However, integrating AI into patient education presents several challenges. Ensuring the accuracy, reliability, and transparency of AI-generated information is paramount. Additionally, safeguarding patient privacy and preserving the human element in communication are crucial for delivering compassionate care. As AI technology continues to evolve, it could significantly enhance patient education by offering more effective and personalized support. Ongoing advancements in AI tools may pave the way for innovative approaches to help patients better manage their health.

## 4 Conclusion

The integration of AI technology in the domain of healthcare offers great potential for revolutionizing patient care and improving treatment outcomes. This data driven predictive analytics can significantly improve the accuracy, efficiency, and cost-effectiveness of disease diagnosis and clinical laboratory testing. Furthermore, AI can greatly impact population health management and guideline development by delivering real-time, precise information and optimizing medication choices. Its applications in virtual health and mental health support are also promising, with the potential to improve patient care and increase accessibility.

It is noteworthy note, for useful implementation of AI/ML technology in healthcare industry, it is crucial to address challenges such as biases and the lack of personalization for ensuring equitable use. Key measures for responsible implementation include enhancing cyber-security to safeguard patient data, fostering collaboration among healthcare organizations, AI researchers, and regulatory bodies to develop directives and standards, and investing in research and development to advance AI technologies customized to healthcare needs.

This chapter has presented how AI algorithms can analyze population demographics, disease prevalence, and geographical distribution to identify high-risk patients and predict potential healthcare events. This capability aids in proactive prevention and treatment, as well as in efficiently allocating resources like vaccines. Public perception of AI in healthcare is mixed; while many are open to use of AI technology for healthcare purposes, they still prefer human practitioners for complex issues. Building trust and making the patients aware about the technology are crucial for successful AI integration. Addressing challenges related to data quality, privacy, bias, and the need for human oversight is essential for responsible and effective AI use. Collaboration among stakeholders, ongoing research, and innovation are key to developing robust AI systems and ethical guidelines, ultimately leading to improved patient outcomes, greater efficiency, and enhanced access to personalized care.

## References


Ahsan, M. M., Luna, S. A., & Siddique, Z. (2022). Machine-learning-based disease diagnosis: A comprehensive review. Healthcare, 10, 541. https://doi.org/10.3390/healthcare10030541.

Ahsan, M. M., Mahmud, M., Saha, P. K., Gupta, K. D., & Siddique, Z. (2021). Effect of data scaling methods on machine learning algorithms and model performance. Technologies, 9, 52.

Alowais, S. A., Alghamdi, S. S., Alsuhebany, N., Alqahtani, T., Alshaya, A. I., Almohareb, S. N., Aldairem, A., Alrashed, M., Bin Saleh, K., Badreldin, H. A., Al Yami, M. S., Al Harbi, S., & Albekairy, A. M. (2023). Revolutionizing healthcare: The role of artificial intelligence in clinical practice. BMC Medical Education, 23(1), 689. https://doi.org/10.1186/s12909-023-04698-z.

Alshayeji, M. H., Ellethy, H., & Gupta, R. (2022). Computer-aided detection of breast cancer on the Wisconsin dataset: An artificial neural networks approach. Biomedical Signal Processing and Control, 71, 103141.

Ansari, A. Q., & Gupta, N. K. (2011). Automated diagnosis of coronary heart disease using neuro-fuzzy integrated system. In Proceedings of the 2011 World Congress on Information and Communication Technologies (pp. 1379–1384). Mumbai, India.

Arabzadeh, A., Mortezazadeh, T., Aryafar, T., Gharepapagh, E., Majdaeen, M., & Farhood, B. (2021). Therapeutic potentials of resveratrol in combination with radiotherapy and chemotherapy during glioblastoma treatment: A mechanistic review. Cancer Cell International, 21(1), 1–15.

Bajwa, J., Munir, U., Nori, A., & Williams, B. (2021). Artificial intelligence in healthcare: Transforming the practice of medicine. Future Healthcare Journal, 8(2), e188–e194. https://doi.org/10.7861/fhj.2021-0095.

Bhadra, S., & Kumar, C. J. (2023). Enhancing the efficacy of depression detection system using optimal feature selection from EHR, Computer Methods in Biomechanics and Biomedical Engineering, Taylor and francis.

Bhadra, S., Kumar, C. J., & Bhattacharyya, D. K. (2024). Multiview EEG signal analysis for diagnosis of schizophrenia: An optimized deep learning approach. Multimedia Tools and Applications.

Bhattacherjee, A., Roy, S., Paul, S., Roy, P., Kausar, N., & Dey, N. (2020). Classification approach for breast cancer detection using back propagation neural network: A study. In Deep Learning and Neural Networks: Concepts, Methodologies, Tools, and Applications (pp. 1410–1421). IGI Global.

Choudhury, A., Biswas, A., Singh, T. P., & Ghosh, S. K. (Eds.). (2022). Smart agriculture automation using advanced technologies: Data analytics and machine learning, cloud architecture, automation and IoT. Springer Nature.

Das, P. R., Talukdar, R. R., & Kumar, C. J. (2024). Exploring the interplay of abstinence self-efficacy, locus of control, and perceived social support in substance use disorder recovery. *Current Medical Research and Opinion, 40*(9), 1625–1635. https://doi.org/10.1080/03007995.2024.2390046.

Elliott, K., & Larsson, E. (2021). Non-coding driver mutations in human cancer. Nature Reviews Cancer, 21(8), 500–509.

Grigorenko, E. L., & Dozier, M. (2013). Introduction to the special section on genomics. Child Development, 84(1), 6–16.

Ghosh, P., Azam, S., Hasib, K. M., Karim, A., Jonkman, M., & Anwar, A. (2021). A performance-based study on deep learning algorithms in the effective prediction of breast cancer. In Proceedings of the 2021 International Joint Conference on Neural Networks (IJCNN) (pp. 1–8). Online.

Haenssle, H. A., Fink, C., Schneiderbauer, R., Toberer, F., Buhl, T., Blum, A., et al. (2018). Man against machine: Diagnostic performance of a deep learning convolutional neural network for dermoscopic melanoma recognition in comparison to 58 dermatologists. Annals of Oncology, 29(8), 1836–1842. https://doi.org/10.1093/annonc/mdy166.

Han, S. S., Park, I., Eun Chang, S., Lim, W., Kim, M. S., Park, G. H., et al. (2020). Augmented intelligence dermatology: Deep neural networks empower medical professionals in diagnosing skin cancer and predicting treatment options for 134 skin disorders. Journal of Investigative Dermatology, 140(9), 1753–1761. https://doi.org/10.1016/j.jid.2020.01.019.

Kim, H.-E., Kim, H. H., Han, B.-K., Kim, K. H., Han, K., Nam, H., et al. (2020). Changes in cancer detection and false-positive recall in mammography using artificial intelligence: A retrospective, multireader study. Lancet Digital Health, 2(3). https://doi.org/10.1016/s2589-7500(20)30003-0.

Kandhasamy, J. P., & Balamurali, S. (2015). Performance analysis of classifier models to predict diabetes mellitus. Procedia Computer Science, 47, 45–51. [CrossRef].

Kumar, M., Singh, T. P., Choudhury, T., & Gupta, S. C. (2019). ICT—the smart education system in India. In 2019 International Conference on Contemporary Computing and Informatics (IC3I) (pp. 279–282). IEEE.

Kumar, C. J., & Das, P. R. (2021). The diagnosis of ASD using multiple machine learning techniques. International Journal of Developmental Disabilities, Taylor and francis.

Kumar, C. J., Das, P. R., & Hazarika, A. (2022). Autism spectrum disorder diagnosis and machine learning: a review. International Journal of Medical Engineering and Informatics, Inderscience, vol. 14(6).

Li, S., Zhao, R., & Zou, H. (2021). Artificial intelligence for diabetic retinopathy. Chinese Medical Journal, 135(3), 253–260. https://doi.org/10.1097/CM9.0000000000001816.

Maqsood, K., Hagras, H., & Zabet, N. R. (2024). An overview of artificial intelligence in the field of genomics. Discover Artificial Intelligence, 4, 9. https://doi.org/10.1007/s44163-024-00103-w.

McGuire, A. L., Gabriel, S., Tishkoff, S. A., Wonkam, A., Chakravarti, A., Furlong, E. E., et al. (2020). The road ahead in genetics and genomics. Nature Reviews Genetics, 21(10), 581–596.

McKinney, S. M., Sieniek, M., Godbole, V., Godwin, J., Antropova, N., Ashrafian, H., et al. (2020). International evaluation of an AI system for breast cancer screening. Nature, 577(7788), 89–94. https://doi.org/10.1038/s41586-019-1799-6.

Miao, J. H., & Miao, K. H. (2018). Cardiotocographic diagnosis of fetal health based on multiclass morphologic pattern predictions using deep learning classification. International Journal of Advanced Computer Science and Applications, 9, 1–11. [CrossRef].

Rubin, J., Abreu, R., Ganguli, A., Nelaturi, S., Matei, I., & Sricharan, K. (2017). Recognizing abnormal heart sounds using deep learning. arXiv:1707.04642.

Sharma, M., Kumar, C. J., & Bhattacharyya, D. K. (2024a). Machine/deep learning techniques for disease and nutrient deficiency disorder diagnosis in rice crops: A systematic review, Biosystems Engineering, Elsevier.

Sharma, M., Kumar, C. J., Singh, T. P., Talukdar, J., Sharma, R. K., & Ganguly, A. (2024b). Enhancing disease region segmentation in rice leaves using modified deep learning architectures. Archives of Phytopathology and Plant Protection, Taylor and Francis.

Shlyueva, D., Stampfel, G., & Stark, A. (2014). Transcriptional enhancers: From properties to genome-wide predictions. Nature Reviews Genetics, 15(4), 272–286.

Shrikumar, A., Greenside, P., & Kundaje, A. (2017). Learning important features through propagating activation differences. In International Conference on Machine Learning (pp. 3145–3153). PMLR.

Sizilio, G. R., Leite, C. R., Guerreiro, A. M., & Neto, A. D. D. (2012). Fuzzy method for pre-diagnosis of breast cancer from the fine needle aspirate analysis. Biomedical Engineering Online, 11(1), 1–21.

Sultana, Z., Khan, M. R., & Jahan, N. (2021). Early breast cancer detection utilizing artificial neural network. WSEAS Transactions on Biology and Biomedicine, 18, 32–42. [CrossRef].

Ström, P., Kartasalo, K., Olsson, H., Solorzano, L., Delahunt, B., Berney, D. M., et al. (2020). Artificial intelligence for diagnosis and grading of prostate cancer in biopsies: A population-based, diagnostic study. Lancet Oncology, 21(2), 222–232.

Suleimenov, I. E., Vitulyova, Y. S., Bakirov, A. S., & Gabrielyan, O. A. (2020). Artificial intelligence: What is it? In Proceedings of the 2020 6th International Conference on Computer Technology and Applications (pp. 22–25). https://doi.org/10.1145/3397125.3397141.

Teer, J. K. (2014). An improved understanding of cancer genomics through massively parallel sequencing. Translational Cancer Research, 3(3), 243.

Wolfe, J. C., Mikheeva, L. A., Hagras, H., & Zabet, N. R. (2021). An explainable artificial intelligence approach for decoding the enhancer histone modifications code and identification of novel enhancers in Drosophila. Genome Biology, 22, 308. https://doi.org/10.1186/s13059-021-02532-7.

Yahyaoui, A., Jamil, A., Rasheed, J., & Yesiltepe, M. (2019). A decision support system for diabetes prediction using machine learning and deep learning techniques. In Proceedings of the 2019 1st International Informatics and Software Engineering Conference (UBMYK) (pp. 1–4). Ankara, Turkey.

# Kuramoto Phase Model: Possible New Effects of Neuronal Dynamics

**Arup Sarmah, Balendra Kr. Dev Choudhury, and Sanjib Kr. Kalita**

**Abstract** The Kuramoto model is a model for the behavior of a large set of coupled oscillators. The propagation of the signal through Neuronal Network is being considered as one of the complicated problems in neuronal dynamics. In this study concept of a filter circuit is proposed incorporating inductance in neuronal coupling. Thereafter, the permissible impendence ratio is used as a vector to define the perturbating external impulse in Kuramoto Phase Model.

**Keywords** Kuramoto model · Neuronal system · Dynamic system

## 1 Introduction

In the physical world, it is the common-man experience that some happening may occur through precise timing only. A simple thought is sufficient to provide us some intriguing feeling. To be scientific, it demands some hidden principle as well as physical dynamics. In this area, some commonly quoted examples are:

(i) Suppose there was some long standing problem. Years after years different researchers pursued the problem. But nobody was successful. In some unexplained moment, say, Einstein resolves this mystery associated with equivalence principle. The resolving of the problem is generally quoted as accidental; but is it some long time tuning of still unexplained brain activities?

A. Sarmah (✉)
Department of Computer Application, Assam Don Bosco University, Guwahati, India
e-mail: arupsarmah21@gmail.com

B. Kr. Dev Choudhury
Department of Physics, Pub Kamrup College, Baihata Chariali, India

S. Kr. Kalita
Department of Computer Science, Gauhati University, Guwahati, India

 
T. P. Singh et al. (eds.), *Revolutionizing Healthcare: Impact of Artificial Intelligence on Diagnosis, Treatment, and Patient Care*, Studies in Computational Intelligence 1182,
https://doi.org/10.1007/978-3-031-80813-5_2

(ii) In some musical concert, there are different musical instruments used. When these are performed at some moment the concert becomes pleasant. Our hearts beat rhythmically. How does it happen?

(iii) Some low level living creatures show extremely coherent physiological actions among themselves, as for example flashing of fireflies during their mating (Mirollo and Strogatz 1990). How it may be possible?

All the above mentioned examples are easy to observe but difficult to explain. The simple explanatory words associated with these happenings are concentration, rhythm, tuned collective effort, luck, coincidence etc. Through step by step analysis, it is released that those *performances whether it is in individual or collective approach become possible only through* some synchronized ways. In this area, though the relevance of synchronization is firmly grounded, its dynamics is still not fully understood. Thus physics of synchronization whether it is in biological, chemical, physical or some social systems is intensely demanding the special attention from researcher community. The successful understanding of synchronization is supposed to lead towards the implicit general principle. So, it is understood why century long efforts are being followed from different angles. Winfree is actively involved of the research in the model of non-linear interaction, and thus he is in a position to remark that synchronization can be understood as a threshold process (Winfree 1967). In most of the same categorized biological processes, rhythmical activity is of the prime importance. The main advantage of rhythmicity is the formulation of phase variable (Sharma et al. 2024). The specific value of the phase variable or the relation among the phase variables of the different members of the network motivates to think about the coupling among the individual member of the assembly. The basic specificity of this dynamics is that it is a system generated dynamics; and it means that it is a collective process. Being collectively coherent, the physical properties of the system make threshold frequency (phase) possible. In this perspective the most successful approach was due to Kuramoto (1975). His model stands for the phase oscillators. Here its activity appears at arbitrary intrinsic frequencies. The coupling among the oscillators comes through the sine of their phase differences. It is also tempting enough to explore the cause of existing intrinsic frequencies.

In the present work Kuramoto model is used to understand the neuronal coupling and thereby attempting to understand 'Neuronal Dynamics' in terms of electrical properties of the system. The present work is organized as follows. The next section provides a brief preliminary of Kuramoto model and the short introduction of electrical properties of neuronal system (Bhadra et al. 2024). The present thought is also covered in this chapter followed by the discussion.

## 2 Kuramoto Model

Kuramoto model presents a successful analysis of phase oscillator dynamics. Here the system is supposed to be of consisting of a population of N phase oscillators. The oscillators are assumed to have arbitrary intrinsic properties. During the evolution they are coupled through the sine of their phase differences. The simple mathematical form of the model is

$$\dot{\phi}_i = \omega_i + \sum_{j=1}^{N} C_{ij} \sin(\phi_j - \phi_i) \tag{1}$$

where $i = 1 \cdots N$ indicates individual oscillator having intrinsic frequency $\phi_i$ and natural frequency of the system is $\omega_i$ which is distributed in the given probability density g(w) plane. Here the physical behaviour of the system is captured mathematically with the rotation of $\phi_i$ in probability density induced plane. The rotation $\phi_i$ is due to coupling among the phase oscillators appears as incoherent. But there is found appropriate coupling which provides a threshold frequency at which collective synchronization emerges spontaneously. There may be spiking or some other important physical activity may occur at this specific value of phase (frequency).

Since we are planning to study neuronal dynamics we present the Kuramoto model with this bias. Our presentation is directly based on the excellent book of Eugene M. Izhikevich (Izhikevich 2005). Appropriate addendum is followed from another excellent book by Strogatz (Strogatz 2015). A motivating quote included in Eugene book is “Information processing depends not only on the anatomical substances of synoptic circuits but also on the electro physical properties of neurons. Even if two neurons in different regions of the nervous system possess identical morphological features, they may respond to the same synaptic input in very different manners because of each cell’s intrinsic properties.” It is challenging whether this complex behaviour can be captured in Kuramoto model. One line summary of Eugene is that neuronal system is a nonlinear dynamic system. Interestingly resonant response of a neuron is an attractive feature to think about Kuramoto type model. Figure 1a, b shows Invitro recording of rebound spikes and Resonant response respectively.

A dynamical system in the state space, most simply can be expressed by state variable (x). So its dynamical equation is

$$\dot{x} = \frac{dx}{dt} = f(x) \tag{2}$$

Let us assume that the system state variable is increased by a constant amount ‘D’ in precise interval of time. Then the equation can be modified as

$$\dot{x} = f(x) + D\delta(t - t_s) \tag{3}$$

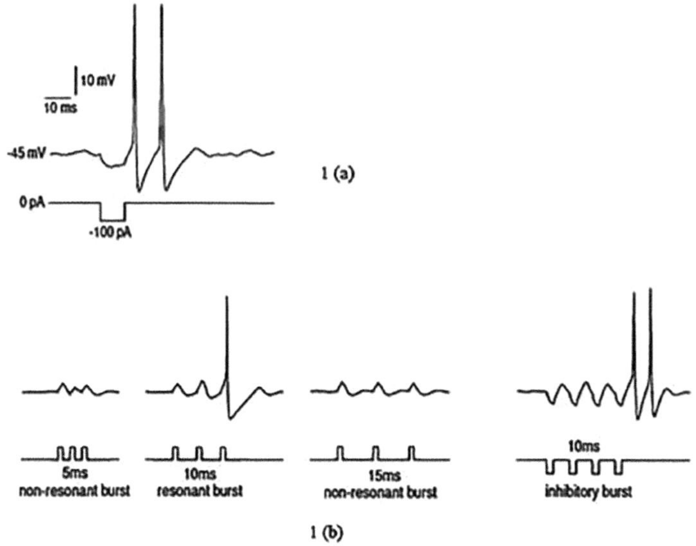


**Fig. 1** **a** Invitro recording of rebound spikes of rat's brainstem mesV neuron in response to a brief hyperpolarizing pulse of current. **b** Resonant response of the mesencephalic V neuron of rat brainstem to pulses of injected current having 10 ms period (in vitro) (Courtesy to Eugene M. Izhikevich's book "Dynamical Systems in Neuroscience: The Geometry of Excitability and Bursting, page, 5)

where δ(t) is the Dirac-delta function. The expression provides a stable limit cycle oscillation.

Now the question arises that every individual neuron has its own limit cycle i.e. its own phase leading incoherence, how the coherence may be established?

Let two neurons have their own separate state variable evolution, say y(t) and x(t) respectively. If $y(t) \rightarrow x(t)$ as $t \rightarrow \alpha$.

Then there may be one manifold where the dynamical solutions carry same phase. Thus it is the journey from incoherence to coherence. The physical concept is concretized through the mathematical terminology 'isochrones'.

For our case a weakly coupled dynamical system is considered. Its dynamical equation may be expressed as

$$\dot{x} = \frac{dx}{dt} = f(x) + \varepsilon p(t) \tag{4}$$

Here $\dot{x} = f(x)$ represents the unperturbed oscillator. Remembering the oscillator is in the network formed by the assembly of neurons, its perturbation-input is included through the term εp(t). Assuming ε << 1, the weak coupling is ensured.

Now, coming to the neuron, it is assumed that the neuron gives spike at some specific phase periodically. But every coupling due to other neurons in the network, though causes perturbation may not be capable for satisfying the condition for another spiking, it only changes the timing i.e. the phase of the subsequent spikes. So it is understood as the resetting of phase which gives response curve known as phase resetting curve (PRC). Understanding phase shift, synchronization is defined say for, two periodic pulses if they occur at the same time or at a constant phase shift.

Kuramoto's approach is a further simplification of Winfree's approach. It is isochrones when they are highly dense, then phase resetting curve (PRC) may offer linearization. So linear scaling of PRC with reference to the strength of the pulse may be expected to provide

$$\mathrm{PRC}(\phi, \mathrm{C}) \approx \mathrm{F}(\phi)\mathrm{C} \tag{5}$$

Where $\mathrm{F}(\phi) = \frac{\partial \mathrm{PRC}\,(\phi,\mathrm{e})}{\partial \mathrm{C}}$ $at\ C = 0$ is the linear response or sensitivity function. It gives the small variation of rate, or instantaneous frequency of oscillation due to a small oscillation. Using the concept of Poincare phase map, we can have

$$\phi(\mathrm{t}_{\mathrm{n}+1}) = \{\phi(\mathrm{t}_{\mathrm{n}}) + \mathrm{F}\,(\phi(\mathrm{t}_{\mathrm{n}}))\varepsilon \mathrm{p}(\mathrm{t}_{\mathrm{n}})\mathrm{h} + \mathrm{h}\}$$

Here h is a small inter-pulse interval and $\mathrm{C} = \varepsilon \mathrm{p}(\mathrm{t}_{\mathrm{n}})\mathrm{h}$. Hence

$$\frac{\phi(\mathrm{t}_{\mathrm{n}} + \mathrm{h}) - \phi(\mathrm{t}_{\mathrm{n}})}{\mathrm{h}} = \mathrm{F}(\phi(\mathrm{t}_{\mathrm{n}}))\varepsilon \mathrm{p}(\mathrm{t}_{\mathrm{n}}) + 1$$

Considering as continuous variable we can write

$$\phi = \frac{\mathrm{d}\dot{\phi}}{\mathrm{dt}} = 1 + \varepsilon \mathrm{F}(\phi)\mathrm{p}(\mathrm{t}) \tag{6}$$

Setting $\varepsilon = 0$, assuming $\boldsymbol{\phi}(\mathrm{x})$ to denote the phases of points in the vicinity of limit cycle attractor one can proceed

$$\frac{\mathrm{d}\phi(\mathrm{x})}{\mathrm{dt}} = \mathrm{grad}\boldsymbol{\phi}.\frac{\mathrm{dx}}{\mathrm{dt}} = \mathrm{grad}\boldsymbol{\phi}.\mathbf{f}(\mathbf{x}) \tag{7}$$

For $\varepsilon = 0$, the expression (6) shows

$$\frac{\mathrm{d}\phi(\mathrm{x})}{\mathrm{dt}} = 1$$

So from (7)

$$\mathrm{grad}\phi.\mathrm{f}(\mathrm{x}) = 1 \tag{8}$$

This is a very useful equality which indicates the mapping of isochrones to isochrones by the flow of vector field f(x).

Applying chain rule to the perturbed system represented by Eq. (4) Kuramoto finds

$$
\begin{aligned}
\frac{d\phi(x)}{dt} &= \mathrm{grad}\boldsymbol{\phi}.\frac{dx}{dt} \\
&= \mathrm{grad}\phi\{f(x) + \varepsilon p(t)\} \\
&= \mathrm{grad}\phi.\ f(x) + \varepsilon \mathrm{grad}\phi.p(t)
\end{aligned}
$$

And using (8) we have

$$
\dot{\phi} = 1 + \varepsilon \mathrm{grad}\phi.p(t) \tag{9}
$$

The Eq. (9) represents the Kuramoto phase model. Following different subsequent follow up and consideration of Fourier term Kuramoto obtains the form of the model as depicted in the Eq. (1).

## 3 Electrical Properties of Neuronal System

Biological neuronal system analysis in terms of electrical properties of the system components is not new one. Famous Hodgkin-Huxley model (H-H model) (Hodgkin and Huxley 1952) and various literatures show the confirmed electrical properties of biological neuronal system. Figure 2 shows the equivalent neural circuit of axons. Though mainstream research community is less interested to incorporate the role of inductance in neuronal dynamics, the electromyography (EMG) recording of cortical and pelvic electricalnerve stimulation shows that in general neural circuit resembles a parallel RLC circuit. From Hao Wang et al.'s publication, we can place some stimulating diagrammatic features:

Our aim is to study phase dynamics of biological neuronal system on the basis of Kuramoto's phase model using electrical properties. The synchronization is supposed to provide the resonance type condition which in practice triggers some peculiar physical effect.

In Kuramoto's phase model the concept of phase is basically a mathematical artifact. To have the concept of natural angle of rotation instead of phase rotation is crucial for experimental purpose. In this situation Josephson-Junction arrays analysis and its classical analog appears as a motivating factor. So we take a closer look on Josephson junction from Strogatz book, and then investigate its pendulum analog. It gives some idea how natural rotation can be linked to phase. Two superconductor samples may be coupled through weak coupling as shown in the Fig. 3.

Josephson Effect (1962) demands the passing of current between two superconductors in ground state even without potential difference through quantum mechanical tunneling. For us, main focus point is that constant phase difference $\phi = \phi_1 - \phi_2$ gives the dynamics and as a result the current relation is obtained as

$$
I = I_c \sin\phi \tag{10}
$$

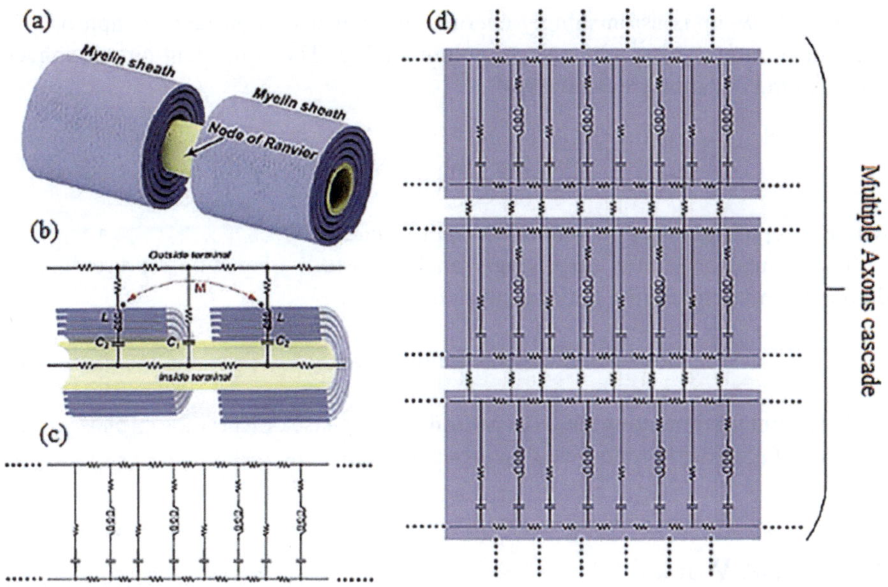


**Fig. 2** The equivalent neural circuit of axons by considering the myelin sheaths as inductors. **a** 3D illustration of the biological structure of a myelinated axon segment; **b** the cross-sectional view of the axon segment with the equivalent circuit model; **c** the equivalent circuit cascade of a myelinated axon; **d** the equivalent cascaded circuit network of multiple axons (Courtesy to Wang et al. for reference (Wang et al. 2018))

| | |
|---|---|
| $\Psi_1 e^{\Phi_1}$ | ←------------ superconductor1 |
| | ←------------- weak coupling |
| $\Psi_2 e^{\Phi_2}$ | ←-------------Superconductor2 |

**Fig. 3** Schematic figure for Josephson-Junction [Courtesy to Strogatz included in 2015]

So bias current $(I)$ increases as phase difference increases.

The dynamical change of phase difference through slipping of current provides Josephson voltage–phase relation.

$$V = \frac{h}{2e}\dot{\phi} \tag{11}$$

where V(t) is the instantaneous voltage across the junction h dressed Plank constant and e is the charge of the electron.

The follow–up is astonishingly encouraging when we consider the approach of equivalent circuit along with the pendulum analog. The equivalent circuit concept forwards the equation (Stewart 1968).

$$\frac{hc}{2e}\ddot{\phi} + \frac{h}{2eR}\dot{\phi} + I_c \sin \phi = I \tag{12}$$

where Josephson concept is intuitively appropriated.

The expression (12) is surprisingly analogous to the equation representing the damped pendulum motion under a constant torque.

$$mL^2\ddot{\theta} + b\dot{\theta} + mgL \sin \theta = \Gamma \tag{13}$$

This point is stimulating for us to naturalize the phase of Kuramoto's phase model in terms of electrical properties of neuronal system.

## 4 Present Work

In our work, the spiking, propagation of impulses and as a whole neuronal dynamics is attempted to understand in terms of electrical properties with the concept of RLC circuit. As the equation of dynamics, Kuramoto's phase model equation is followed. The motivation for naturalization of the phase model is followed from Josephson junction and pendulum analog. Coherence and timing towards a desired frequency have very close relation. The concept of synchronization is also rooted in such type of relation.

The fine tuning is not an acceptable concept physically in scientific pursuit. It is the simple realization that the drawing the system, as for example electrical and acoustic systems, at some desired frequency is practically difficult to achieve. To get rid of this difficulty the idea of filters is developed.

The variability of external stimuli against the neuron indicates the need of band pass filter type concept. Thus the need of fine- tuning may be avoided. Instead of mere synchronization, it seems more favorable to neuron-response towards external stimulus in band pass filter type structure. For an electrical analogy, a toy model type band pass filter circuit can be considered as shown in Fig. 4.

In band pass filter series arms have a series resonant circuit and the shunt arm has parallel resonant circuit. It can be shown when $\left(-4 < \frac{Z_1}{Z_2} < 0\right)$ i.e. only currents (electrical waves) of frequencies for which the impedance ratio lies between this internal $\left(-4 < \frac{Z_1}{Z_2} < 0\right)$ will pass through the network without attenuation (no loss of magnitude) while others are lost.

From any undergraduate book (Garcia et al. 2012), we have

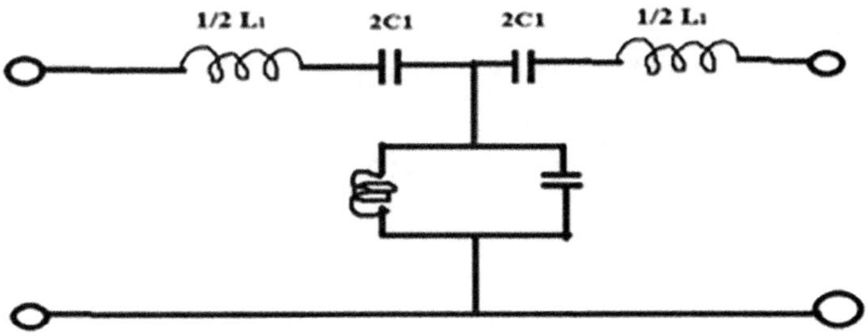


**Fig. 4** Simple band pass filter circuit

$$\frac{z_1}{z_2} = \frac{(\omega^2 L_1 C_1 - 1)(1 - \omega^2 L_2 C_2)}{\omega^2 C_1 L_2} \quad (14)$$

The $\frac{z_1}{z_2}$ determine current (electric waves) of frequencies which will pass through the network without attenuation (no loss of magnitude) while others are lost. In our case network is neuron network. Our assumption is naïve and hence it is toy model, but still it upholds the concept. Let it be a simplest one where the network represents a "two neurons-network". Understanding the response of neuron-network, we assume—

(i) So far the external stimulus is at resonance with the neuron network the network cannot provide effective spike as it is in energy accumulating state.
(ii) Again for off resonance condition there are two cut-off frequencies where effective spike can be hoped.

Now we propose to link $\frac{z_1}{z_2}$ with the perturbating factor p(t) in the dynamical equation. Here Poincare phase may which can be associated with two phase coupled oscillators may provide some idea. Considering phase locking, P:1 phase locking can be understood as p spikes per unit pulse. In the expression of dynamical system

$$\frac{dx}{dt} = f(x) + \varepsilon p(t)$$

$\varepsilon p(t)$ is the time dependent input which perturbs periodic oscillator $\frac{dx}{dt} = f(x)$. The perturbation is the cause for providing effective spike p. The effective spike is the collective synchronized spike where perturbation input propagation through coupling from neuron to neuron is a prerequisite. Thus it is understood p is to be implicitly related to $\frac{z_1}{z_2}$.

Let us understand p(t) and $\frac{z1}{z2}$ respectively as post synaptic current and the impedance of our toy-type neuron network. We propose a simple ohmic type relation between p(t) and $\frac{z_1}{z_2}$ in the state of neuron-network where effective spiking is caused as

$$p(t) = \frac{1}{k}\left(\frac{z_1}{z_2}\right) \quad (15)$$

where k is constant

Using (15) in Kuramoto's equation, we have

$$\frac{d\phi}{dt} = 1 + \varepsilon grad\boldsymbol{\phi}.\frac{1}{\boldsymbol{k}}\left(\frac{z_1}{z_2}\right) \tag{16}$$

So,

$$\frac{d\phi}{1 + \varepsilon grad\boldsymbol{\phi}.\frac{1}{\boldsymbol{k}}\left(\frac{z_1}{z_2}\right)} = dt \tag{17}$$

The mechanical properties of neuronal structure are a highly involved active research field. It is strongly pointed out (Ouyang et al. 2010) the need of importance to study the spatial gradient of elastic mechanical properties of axon. The research pursuit is also active showing mechanically coupled structures such as microfilaments, neurofilaments or tau proteins. Various literatures (Bernal et al. 2007) uphold that the consideration of Action Potential as purely electrical is traditional only; it therefore becomes reasonable to ask whether mechanical deformation may itself play a role in AP generation and vice versa.

With the spirit of welcoming mechanical properties, in our case we assume that input electrical pulse gives rises to shearing stress to the neuronal-network system. Let this shear be θ; so it is measurable by tanθ. Borrowing the concept of naturalization of phase ϕ, we replace θ by ϕ, that is tanθ by tanϕ. In Eq. (16) we assume

$$\varepsilon \text{grade}\phi.\frac{1}{\text{k}}\left(\frac{\text{z}_1}{\text{z}_2}\right) = \tan\phi \tag{18}$$

From (16) and (17), we have

$$\frac{\text{d}\phi}{1 + \tan\phi} = \text{dt}$$

On integration

$$\text{t} = \int \frac{\text{d}\phi}{1 + \tan\phi}$$

$$\left[\frac{\text{d}\phi}{1 + \tan\phi} = \frac{\cos \text{d}\phi}{\sin\phi + \cos\phi}\right]$$

Let $\cos\phi = \lambda.\frac{\text{d}}{\text{d}\phi}(\cos\phi + \sin\phi) + \mu(\cos\phi + \sin\phi)\cos\phi = \lambda.\frac{\text{d}}{\text{d}\phi}(\cos\phi + \sin\phi) + \mu(\cos\phi + \sin\phi)$

Then

$$\cos\phi = \lambda.(-\sin\phi + \cos\phi) + \mu(\cos\phi + \sin\phi)$$

Comparing the coefficient of sin ϕ and cos ϕ on both sides we get

$$-\lambda + \mu = 0 and \lambda + \mu = 1$$

Therefore $\lambda = \mu = \frac{1}{2}\frac{1}{2}$

$$\begin{aligned} t &= \frac{\cos\phi d\phi}{\cos\phi + \sin\phi} \\ &= \frac{\frac{1}{2(-\sin\phi+\cos\phi)} + 1/2(\sin\phi + \cos\phi)}{\cos\phi + \sin\phi} d\phi \\ &= \frac{1}{2}\int \frac{(-\sin\phi + \cos\phi)}{\cos\phi + \sin\phi} d\phi + \frac{1}{2}\int \frac{(\sin\phi + \cos\phi)}{\cos\phi + \sin\phi} d\phi \end{aligned}$$

Putting $y = \cos\phi + \sin\phi$, we have

$$\begin{aligned} t &= \frac{1}{2}\int \frac{dy}{y} + \frac{1}{2}\int d\phi \\ t &= \frac{1}{2}\log y + \frac{1}{2}\phi \\ t &= \frac{1}{2}\phi + \frac{1}{2}\log(\sin\phi + \cos\phi) \end{aligned} \tag{19}$$

Here the Eq. (18) gives a complicated time evolution against the change in ϕ

Again from (17) we get,

$$\phi = \tan^{-1}\left\{\varepsilon grade\#.\frac{1}{\mathbf{k}}\left(\frac{\mathbf{z}_1}{\mathbf{z}_2}\right)\right\} \tag{20}$$

## 5 Discussion

From Eq. (15) we have if $\frac{d\phi}{dt} = 0$

$$\varepsilon grade\phi.\frac{1}{k}\left(\frac{z_1}{z_2}\right) = -1$$

Using (19), then we have

$$\tan\boldsymbol{\phi} = -1 = \tan\frac{-\pi}{4}$$

So, $\phi = \frac{-\pi}{4}$

Thus this value of $\phi$ should have some special importance. Differentiating Eq. (18) with respect to time t, we get

$$\frac{1}{2}\frac{\mathrm{d}\phi}{\mathrm{dt}}\left(1+\frac{\cos\phi-\sin\phi}{\sin\phi+\cos\phi}\right)=1$$
$$\frac{d\phi}{dt}=\frac{2}{1+\frac{\cos\phi-sin\phi}{\sin\phi+\cos\phi}} \tag{21}$$

The Eq. (20) shows when $\phi=\frac{-\pi}{4}$, $\frac{d\phi}{dt}=0$. Hence the instantaneous non-dynamical state in Kuramoto's equation is captured by the specific value of $\varphi$. It seems physically most important. Remembering the band pass filter condition, $-4<\frac{z_1}{z_2}<0$, so to have passage of signal at $\phi=\frac{-\pi}{4}$, we assume signal passing condition is satisfied at $\frac{z_1}{z_2}=-1$. Then

$$\frac{\varepsilon}{\mathrm{k}}\mathrm{grad}\phi=1$$

$$\frac{\mathrm{d}\phi}{\mathrm{dx}}=\frac{\mathrm{k}}{\varepsilon}$$

$$\phi=\frac{\mathrm{k}}{\varepsilon}.\mathrm{x}$$

a simple linear relation; but when this linearity is being maintained, $\phi$ loses its dynamical behavior. Mathematically it is only possible if $\frac{dx}{dt}=0$ since $\frac{d\phi}{dt}=\frac{d\phi}{dx}.\frac{dx}{dt}$. It means there state variable (x) loses its time dependency, more correctly f(x) is non-existent. As per the Eq. (4), for a while the neuron even ceases to be unperturbed oscillator. Naively, this situation can be compared to simple pendulum. At equilibrium i.e. at $\theta=0$, the pendulum should come to rest(i.e. non-dynamical) but it remains dynamic due to a physical property, inertia. At $\theta=0$, again a new oscillation begins only due to inertia of motion. In the present case, at every $|\phi|=\frac{\pi}{4}$, a new effective spiking starts, after spiking the neuron regains its intrinsic oscillating character. So in one full cycle there should be eight effective spiking sites. The question arises, like inertia, what is the physical property here to make $\phi$ dynamic.

## 6 Mechanical Property of Axon

Pondering on non-dynamicity of Kuramoto equation at $\varphi=\pi/4$, crudely twisting couple of the axon can be represented. Suppose the axon is of the length l and its radius is a. Then the twisting couple of the axon is given as $M=\frac{\pi^2\sigma a^4}{8l}$ where torsional rigidity is $\frac{\pi\sigma a^4}{2l}$. If experimental set-up can be carried on, expression for rigidity modulus $\eta$ is simple, that is, $=\frac{8\pi Il}{T^2a^4}$. $\boldsymbol{\sigma}=\frac{\boldsymbol{Il}}{2\pi^3\boldsymbol{T}^2\boldsymbol{a}^4}$.

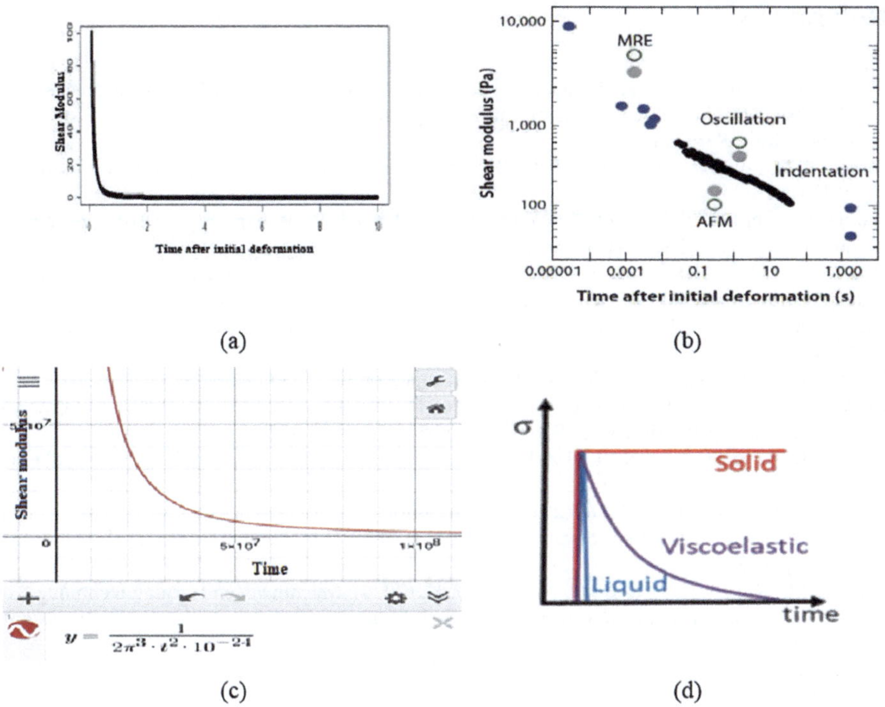


**Fig. 5** **a** $1/T^2$ versus T (Constants are standardized as 1) Data points on axes: 99,001 [randomized], Increments: i + 0.00001 secs, R plot **b** [Rheology] Shear modulus versus time after initial deformation. *Source* Mechanics in Neuronal Development and Repair **c** DESMOS PLOT **d** [Rheology] Shear modulus versus time source: LS instruments, viscoelasticity

Where **l** length of axon, **I** is moment of Inertia, T is time, **a** is radius of axon.

The compared graphical interpretation, though pin—pointedly not supporting the experimental data but viscoelastic character of the neuron is upheld in the graphs Fig. 5 when electrical property is coupled with the mechanical property.

But whatever may be our conjecture, first we need the appropriate biology; the quantification (physics and mathematics) will come later on. We feel that the statement made by Bernal et al. "the few references studying the stiffness of axons are unfortunately generally not taking into account like spatial gradient of mechanical properties" may be a stepping stone to start freshly. First, we need the appropriate biology; the quantification (physics and mathematics) will come later on. We feel that the statement made by Bernal et al. "the few references studying the stiffness of axons are unfortunately generally not taking into account like spatial gradient of mechanical properties" may be a stepping stone to start freshly.

## References


Bhadra, S., Kumar, C. J., & Bhattacharyya, D. K. (2024). Multiview EEG signal analysis for diagnosis of schizophrenia: An optimized deep learning approach. Multimedia Tools and Applications, Springer.

Bernal, R. P., Pullarkat, P., & Melo, F. (2007). Mechanical properties of axon. Phys. Rev. Lett., 99, 018301.

Garcia, J. A., et al. (2012). A model of the spatially dependent mechanical properties of the axon during its growth. CMES, 1(4), 1–22.

Hodgkin, A. L., & Huxley, A. F. (1952). A quantitative description of membrane current and its application to conduction and excitation in nerve. The Journal of Physiology, 117(4), 500–544.

Izhikevich, E. (2005). Dynamical Systems in Neuroscience: The Geometry of Excitability and Bursting. The MIT Press.

Josephson, B. D. (1962). Possible new effects in superconductive tunneling. Phys. Lett., 1, 251.

Kuramoto, Y. (1975). International Symposium on Mathematical Problems in Theoretical Physics, Lecture Notes in Physics, Vol. 39. Springer, New York, 420–422.

Mirollo, R. E., & Strogatz, S. H. (1990). Synchronization of pulse-coupled biological oscillators. SIAM J. Appl. Math., 50, 1645.

Ouyang, H. E., Nauman, & Shi, R. (2010). Contribution of cytoskeletal elements to the mechanical property of axon. Micr/Nano Symposium (UGIM), 2010, 18th Biennial University/Government/ Industry, 1–5.

Sharma, M., Kumar, C. J., & Bhattacharyya, D. K. (2024). Machine/deep learning techniques for disease and nutrient deficiency disorder diagnosis in rice crops: A systematic review. Biosystems Engineering, 244, 77–92.

Stewart, W. C. (1968). Current-voltage characteristics of Josephson junctions. Appl. Phys. Lett., 12, 277.

Strogatz, S. H. (2015). Nonlinear Dynamics and Chaos with Applications to Physics, Biology, Chemistry, and Engineering, Second Edition. Westview Press.

Wang, H., et al. (2018). Inductance in neural systems. https://doi.org/10.1101/343905, bioRxiv.

Winfree, A. T. (1967). J. Theor. Biol., 16, 15.

# Object Detection in the Healthcare Industry Using Yolo: A Better Way Than CNN

**Kaushik Das and Arun Kr. Baruah**

**Abstract** In the recent study of Object Detection techniques, Deep Learning has an impact on development of the methodologies in the areas of Healthcare Industry such as cancer diagnosis, Medicine, Diabetic, Cardiac Imaging, Tumour Detection and many more. The recent study of object detection has made a comparison on utilisation of several Machine Learning models in the areas of healthcare. In this study, an attempt has been made to focus on the comparison between results of object detection from medical images using both YOLO (You Only Look Once) and CNN (Convolutional Neural Network). Classification techniques in computational demands and interpretability are reviewed in this study. This review allows the readers to understand the strengths and challenges across methodologies, thereby finding the way for informed decisions in real time applications. The study also attempts in addressing challenges like computational resources of image processing techniques. As the author of this paper, I appreciate the constructive feedback received through the double-blind peer review process, which significantly enhanced the clarity and depth of the research.

**Keywords** Object detection · YOLO (You Only Look Once) · Convolutional Neural Network (CNN) · Image segmentation

## 1 Introduction

It has been observed that the technique of using Electronic Medical Record (EMR) is gaining popularity eventually. The EMR allows to obtain medical knowledge from the clinical experience of Human which are recorded in text. The tool used to scan those clinical reports are available in open source. However, there is a scope of reducing risk

K. Das (✉)
Department of Computer Science and Engineering, Dibrugarh University, Dibrugarh, India
e-mail: kd.duiet@dibru.ac.in

A. Kr. Baruah
Department of Mathematics, Dibrugarh University, Dibrugarh, India



T. P. Singh et al. (eds.), *Revolutionizing Healthcare: Impact of Artificial Intelligence on Diagnosis, Treatment, and Patient Care*, Studies in Computational Intelligence 1182,
https://doi.org/10.1007/978-3-031-80813-5_3

factor and maintaining confidentiality at the same time. The Human interpretation is kept on loop while verification of the labelled data required to compare with specific drug name(s) against the concerned disease(s). Zhao and Li (2020), created an integrated framework for annotating and extracting information from clinical texts. They utilized the HwaMei-1200 corpus for modelling and designed a thorough operational protocol for a medical information system. A similar technique has been attempted through this research work which defines different entity types for further characterization in the design. Their definitions are listed below:

(i) **Disease**: A condition or state that hinders regular functioning, usually characterized by specific symptoms and signs. The internal or external cause of disease and description about the damages in organs or any disinfections, disturbances or deficiencies are to be categorized in this stage.
(ii) **Symptom**: An observation or indicator that is frequently brought on by and suggests the existence of an illness. It usually reflects the subjective discomfort that the patient has described. One example of such an anomaly is excessive blood pressure.
(iii) **Test**: A technique employed to detect signs of diseases or their causes, which may include physical examinations, laboratory tests, imaging studies, and similar procedures. A routine test can refer to an individual test or a set of tests, depending on the level of human intervention involved.
(iv) **Treatment**: A technique for managing diseases or reducing symptoms. This stage may include surgery, chemotherapy, radiotherapy, anticoagulant, anti-shock and general nursing.
(v) **Drug**: A medication employed to manage diseases or relieve symptoms. Here, the drug's name, its specifications, and the prescribed dosage or quantities may be detailed.
(vi) **Body**: A body part or substance. It specifies an organ or area of the human body and describes the contents of that particular body part. For example, stomach contents.
(vii) **Personal History**: The patient's medical history activities. For example, smoking or consuming alcohol history.
(viii) **Equipment**: Medical tools, instruments, and devices. For example, an electric knife.
(ix) **Department**: The name of a medical specialty. For example, cardiology.

Initially, we will employ a simple regex with maximum recall to capture samples with more prevalent positive instances. Subsequently, we have to annotate the data, train the model, and leverage its predictions to expedite the annotation of the remaining data. We may apply this approach to refine a model pre-trained on unlabeled medical data for extracting cancer stages from free text, achieving notable results. In this work, the area of image processing has played vital role by detecting the texts present on the medical record which are sometimes very difficult for human identification. The YOLOv3 and CNN are compared between themselves in order to have a well oriented discussion for paving a way out to develop a particular one stop solution model in near future which will help the medical workers. High-level entity relations are an essential semantic component, and ideally, accurate annotation of medical relations should be grounded in precise text expression and alignment with external knowledge. However, reports by analysts reveal that real clinical texts often contain numerous typos, grammatical errors, and content inaccuracies. These issues can make it challenging to interpret relations and may lead to inconsistencies between the text's intended meaning and external knowledge. Therefore, annotations should prioritize the literal interpretation of the text over external knowledge. If a medical relation is clearly indicated by the text, the annotation should reflect that. If not, analysts can use their medical expertise to infer implicit relations from the text.

## 2 Methods

The methodology will resemble the traditional way available on a work-sheet. The Fig. 1, illustrate the medical relationships between entities such as Status, Information and Intervention. The working procedure may be explained as below: The status of Disease, Symptom and Personal History will require the Intervention through treatment or drug dose which will again bring some update on the status of the disease. At the same time, the status will be dependent upon the information cause to get a response as suggestion. During this process, the Information entity will require the intervention from treatment and drug quantity.

This is a simple way to implement the text retrieval system on medical record(s). In our proposed framework, the idea is to identify the texts present on the clinical reports and automatically the result will be obtained as texts (Fig. 2).

Image segmentation is a core component of digital image processing that involves dividing an image into separate regions based on their features and characteristics. The main goal of image segmentation is to simplify image analysis by breaking the image into areas with similar properties, referred to as image objects. This process is a foundational step in image interpretation and plays a vital role in computer vision applications. Without effective segmentation, comprehensive image analysis

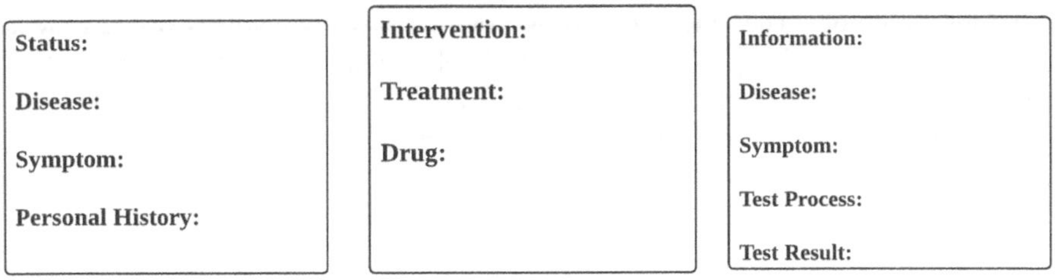


(a) Fig. 1.1 (b) Fig. 1.2 (c) Fig. 1.3

**Fig. 1** Illustration of the relationships between status, information, and intervention in a medical context

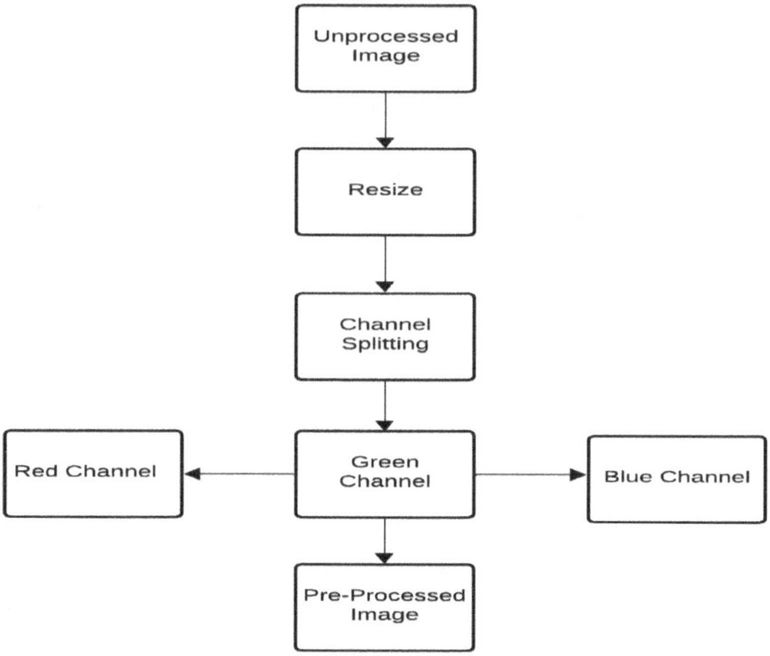


**Fig. 2** Flowchart of pre-processing

becomes challenging. Segmentation techniques allow for the isolation and classification of individual pixels within an image, facilitating the assignment of values and organization of pixels based on their characteristics. This process facilitates the outlining of boundaries, detection of lines, and identification of specific objects of interest amid unrelated elements in the image. In artificial intelligence, the information obtained from image segmentation aids both supervised and unsupervised learning, allowing for the exploration of related issues. Object detection is a crucial aspect of computer vision that focuses on identifying objects within digital images and videos. These objects can include categories such as people, animals, vehicles, buildings, and more. The object detection process generally involves two key steps: localization and classification. Localization involves pinpointing the exact spatial lo cation of objects using bounding boxes, while classification assigns semantic labels to these objects, such as "person", "car", "dog" and so forth.

**Algorithm 1** Pre-processing of Input Image

1. Start
2. Define PreprocessImage(inputImage)
3. Apply Module 1: Gaussian Smoothing Filter
4. Define a function Gaussian Smoothing Filter (input Image)
5. Initialize sigma = 1.0
6. Calculate s = 2.0 * sigma * sigma
7. Initialize GKernel as a 5x5 array
8. Initialize sum = 0.0
9. Iterate over rows from -2 to 2:
10. Iterate over columns from -2 to 2:
11. Calculate r = sqrt(x^2 + y^2)
12. Compute GKernel[x + 2][y + 2] = (exp(-(r^2) / s)) / (π * s)
13. Increment sum by GKernel[x + 2][y + 2]
14. Normalize the kernel:
15. Iterate over rows from 0 to 4:
16. Iterate over columns from 0 to 4:
17. Divide GKernel[i][j] by sum
18. Convolve the input image with GKernel to apply Gaussian smoothing
19. Return the smoothed image
20. Apply Module 2: CLAHE Contrast Enhancement Method:
21. Define a function Apply CLAHE(input Image) to apply Contrast Limited Adaptive Histogram Equalization (CLAHE):
22. Read the input Image
23. Apply Contrast Limited Adaptive Histogram Equalization (CLAHE) to the input image:
    - Initialize clip Limit and tile Grid Size parameters
    - Apply Contrast Limited Adaptive Histogram Equalization (CLAHE) to the input image using the specified parameters.
24. Return the contrast-enhanced image
25. Apply Module 3: Image Resizing, Normalizing, and Batching:
26. Resize the pre-processed image to the desired dimensions
27. Normalize the pixel values of the resized image
28. Convert the image to an appropriate format for adding into the neural network
29. Return the processed image batch
30. End

**Importance and Applications**:

- **Search Engines**: Object detection is vital for search engines as it helps in retrieving images that contain specific objects or scenes.
- **Surveillance Systems**: It facilitates real-time monitoring and tracking of objects or individuals.
- **Autonomous Vehicles**: Object detection is crucial for recognizing and interpreting the surrounding environment to ensure safe and accurate navigation.

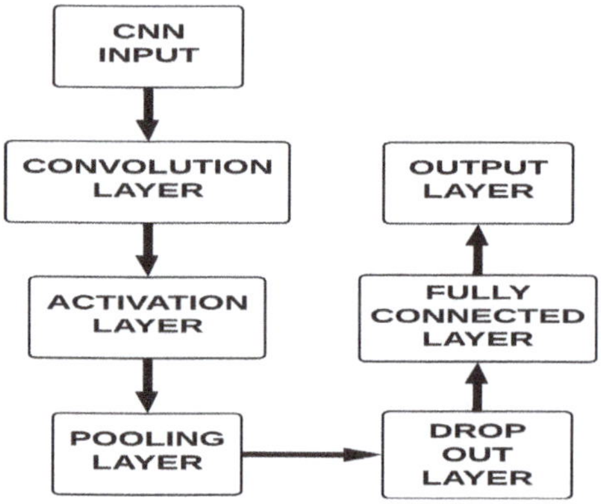


**Fig. 3** Different layers of convolution neural network

- **Robotics**: It is used for tasks such as object manipulation, navigation, and interacting with the environment (Fig. 3).

In recent years, object detection has advanced significantly, primarily due to deep learning techniques, especially Convolutional Neural Networks (CNNs) (Kumar et al. 2022). Unlike traditional methods that depended on manually crafted features and shallow machine learning models, CNN-based object detectors can automatically extract relevant features from data. This results in improved performance and better adaptability across different tasks and datasets.

Figure 4 illustrates the proposed methodology for detailed image analysis, covering both object detection and text extraction. The process starts by reading the input image and then advances through pre-processing steps to improve its quality and make it more suitable for analysis. Object detection and image feature extraction are performed using the advanced YOLO V3 model, which utilizes its hybrid SwishReLU activation function to accelerate training.

This stage involves a Fully Convolutional Network (FCN) combined with Feature Pyramid Networks (FPN) and Atrous Spatial Pyramid Pooling (ASPP) to identify potential text regions and boundaries, ensuring high recall rates. Additionally, a refinement network, modelled after the Connectionist Text Proposal Network

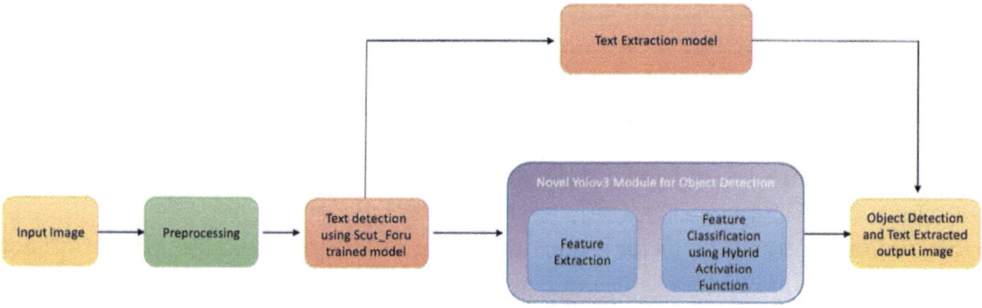


**Fig. 4** Proposed framework of object detection and text extraction

(CTPN), improves text detection accuracy. Non-Maximum Suppression (NMS) is then applied to consolidate text lines.

The integration of Faster R-CNN enhances the speed and precision of text extraction, improving the overall efficiency and accuracy of the network. Experimental results demonstrate its effectiveness in robust image analysis. This framework meets the growing need for advanced image processing by offering efficient solutions for object detection and text extraction in various applications. The process includes preprocessing to reduce noise and improve visual clarity, followed by object detection using CNNs to categorize and locate objects with bounding boxes. Post-processing techniques further refine these detections by filtering based on confidence scores, thereby increasing accuracy and eliminating redundancies.

Text extraction from images involves identifying and isolating text within visual content to convert it into a computer-readable format. The process typically begins with an image input, where advanced techniques of object detection are employed to locate regions containing text. Utilizing algorithms like YOLO (You Only Look Once), the system detects and classifies these text-containing objects in real-time, ensuring accurate identification of various fonts and orientations. Once the text regions are isolated using a Novel YOLO v3 module, Optical Character Recognition (OCR) techniques may be applied to analyze the pixel data and convert the detected text into editable and searchable text formats. This methodology not only enhances data accessibility but also streamlines processes in various applications, such as document digitization in health care industry, automated data entry, and real-time translation services in similar industry. By integrating text detection with Scut Foru trained model with the existing YOLOv3 module where feature classification is done using a Hybrid Activation Function, the text extraction system achieves accuracy and efficiency in extracting meaningful information from complex visual environments.

**Areas to be explored**:

The author expresses that, in the realm of text extraction from images, there exists a notable research gap which hinders the advancement of robust and efficient systems. While significant progress has been made in traditional Optical Character Recognition (OCR) techniques, challenges persist in accurately detecting and extracting text from complex backgrounds, varying fonts, and multiple orientations. Current methodologies often struggle with noisy images or those containing overlapping text, which diminishes extraction quality. Moreover, most existing approaches are heavily reliant on pre-processed images, limiting their applicability in real-world scenarios where conditions are unpredictable. The integration of advanced object detection algorithms, such as YOLO, has shown promise in improving localization, yet further research is needed to enhance their effectiveness in diverse contexts. Thus, addressing these limitations through innovative techniques that combine object detection with adaptive OCR methodologies is essential for achieving higher accuracy and broader applicability in text extraction tasks.

**Uses of YOLO and CNN in healthcare**:

Both YOLO (You Only Look Once) and CNNs (Convolutional Neural Networks) play significant roles in healthcare applications, particularly in medical imaging and diagnostics. The following are their uses in healthcare:

**YOLO**

1. Real-Time Object Detection: YOLO is used for detecting and localizing objects in medical images quickly. This can be particularly useful in identifying tumors, lesions, or abnormalities in scans like X-rays, MRIs, and CTs.
2. Automated Diagnosis: By integrating YOLO with diagnostic tools, healthcare providers can automate the identification of conditions, reducing the workload on radiologists and improving diagnostic speed.
3. Surgical Assistance: YOLO can assist in real-time during surgeries by identifying critical structures or potential areas of concern, aiding surgeons in making informed decisions.

**CNN**

1. Image Classification: CNNs excel in classifying medical images, such as differentiating between healthy and diseased tissues. This is crucial in fields like oncology for cancer detection.
2. Segmentation: CNNs are used to segment images into meaningful parts, helping to delineate organs or tumours in medical scans. This enhances the accuracy of treatments and planning for surgeries.
3. Predictive Analytics: Beyond image analysis, CNNs can be applied to patient data for predicting outcomes or potential complications based on historical data patterns.
4. Telemedicine: CNNs enable the analysis of images in telemedicine applications, allowing remote diagnosis and monitoring of patients by specialists.

In summary, both YOLO and CNNs provide powerful tools for improving diagnostic accuracy, enhancing treatment planning, and streamlining workflows in healthcare settings.

## 3 Results

The HwaMei-500 corpus was implemented on a Python platform using a Neural Network model. It was trained with YOLOv3 and CNN architectures, and their performance was compared to evaluate efficiency. The development included testing with selected hyperparameters, and metrics such as mAP (Mean Average Precision) were calculated.

For evaluating text extraction with our proposed method, metrics such as Character Error Rate (CER) and Word Error Rate (WER) were used.

1. **The Character Error Rate (CER)** is calculated using the Levenshtein distance, which measures the minimum number of character-level edits required to convert the ground truth text into the OCR output.

   The expression for CER as given in Eq. (1)

$$CER = \frac{S + D + I}{N} \tag{1}$$

where

- $S$ is the number of substitutions;
- $D$ is the number of deletions;
- $I$ is the number of insertions;
- $N$ is the number of characters in the ground truth texts.

2. **Word Error Rate (WER)** may be more relevant when dealing with transcriptions of paragraphs and sentences where the words convey meaningful content. the expression for WER as given in Eq. (2):

$$WER = \frac{S_w + D_W + I_W}{N_w} \tag{2}$$

## 4 Discussion

In today's digital age, image processing and computer vision have seen significant advancements. Object detection and scene recognition stand out as primary research areas within computer vision. While there has been substantial research in these domains, implementing real-time text extraction and object detection presents its own set of challenges. The presence of various objects in dynamic backgrounds can complicate detection and text extraction processes.

In specific scenarios, object detection mechanisms identify particular objects within a scene. Recently, techniques like Neural Networks (NNs) and You Only Look Once (YOLOv3) have been employed to enhance classification accuracy. These techniques are essential for training classifiers to recognize objects in a wide variety of images. Text extraction is vital for retrieving text from images, playing a key role in document image analysis. Automatic text extraction focuses on identifying and extracting areas that contain only text, without performing character recognition.

The text extraction process involves four main stages:

- Detection
- Localization
- Segmentation
- Enhancement

Detecting text within an image can be difficult for the human eye. To locate text, images are scanned to detect the presence of textual elements. Automating text extraction has led to the development of applications such as text classification, attribute extraction, and product matching.

Text extraction involves several critical steps: detection, localization, augmentation, and recognition. However, challenges arise from variations in text shape, size, style, orientation, alignment, and background contrast.

Document analysis and image indexing are increasingly important in the field of text extraction. Pre-processing includes basic image operations where both input and output are intensity images. The quality of an image significantly affects the accuracy of object recognition or classification. Generally, high-quality images result in better recognition rates compared to noisy or low-quality images. Extracting features from noisy images can be difficult, often leading to lower recognition rates.

To mitigate issues from poor-quality images, pre-processing is performed before feature extraction. The performance of AI or machine learning algorithms can improve accuracy or reduce the need for extensive training data. Intensity images, which represent the raw sensor data, are often displayed as matrices of brightness values. Although geometric transformations such as rotation, scaling, and translation are included in pre-processing because they use similar techniques, the main objective is to refine the image data by reducing distortions and enhancing features critical for subsequent analysis. Over time, we have tackled numerous challenging projects involving object detection. Nowadays, training and deploying a model can be accomplished in minutes. We can easily upload our data and with just a few clicks, train and deploy our model as an API. Our study delves into the effectiveness of object detection and text extraction from images.

We may apply Neural Networks to train the model using a large dataset of images and text. Performance evaluation of the model reveals its capability to accurately detect text when presented with new images.

The mean Average Precision (mAP) was used to record the performance metrics for evaluation of the proposed detection method. Precision is defined as the percentage of a predicted region that belongs to the ground truth. The expression for Precision is given in Eq. (3):

$$Precision = \frac{TP}{TP + FP} \tag{3}$$

The Precision calculation is based on the following metrics:

- **True Positive (TP)**: An instance for which both actual and predicted values are positive.
- **True Negative (TN)**: An instance for which both actual and predicted values are negative.
- **False Positive (FP)**: An instance for which the actual value is negative and the predicted value is positive.

- **False Negative (FN)**: An instance for which the actual value is positive and the predicted value is negative.

Average Precision (AP) is calculates the average value of precision over various level of recall. If the values of AP were high, it denotes the performance was better. The formula for the AP was given in Eq. (4):

$$AP = \sum_{n} (R_n - R_{n-1}) P_n \tag{4}$$

where $P_n$ is the precision and $R_n$ is the recall at $n$th threshold. Recall is determined as the percentage of the ground truth that is available in the predicted region. The formula for the AP was given in Eq. (5):

$$Recall = \frac{TP}{TP + FN} \tag{5}$$

mean Average Precision (mAP) was widely used metric for performance of object detection. It is the mean of average precision calculated over all classes. The expression for mean Average Precision (mAP) is given in Eq. (6):

$$mAP = \frac{1}{N} \sum_{I=1}^{N} AP_i \tag{6}$$

where $AP_i$ denotes the average precision and N denotes the total number of classes.

The performance made during execution of the proposed detection method was examined under detection precision (mAP) using the COCO data set. Table 1 list out the (mAP) values of the proposed method and existing methods. The detection precision (mAP) of the newly proposed and existing method was graphically shown in Fig. 5.

**YOLO outperform CNN**:

**Table 1** mAP of proposed and existing method on COCO data set

| Method | References | mAP (%) |
|---|---|---|
| Weighted Boxes Function | Solovyev et al. (2021) | 56.4 |
| YoloV3 | Zhao and Li (2020) | 53.2 |
| CBNET | Zhao and Li (2020) | 53.3 |
| POTO-ResNext-101-DCN | Wang et al. (2021) | 47.6 |
| DETR | Carion et al. (2020) | 44.9 |
| PG-PS-FR-CNN | Cheng et al. (2020) | 20.7 |
| Proposed YOLO V3 | – | 58.7 |

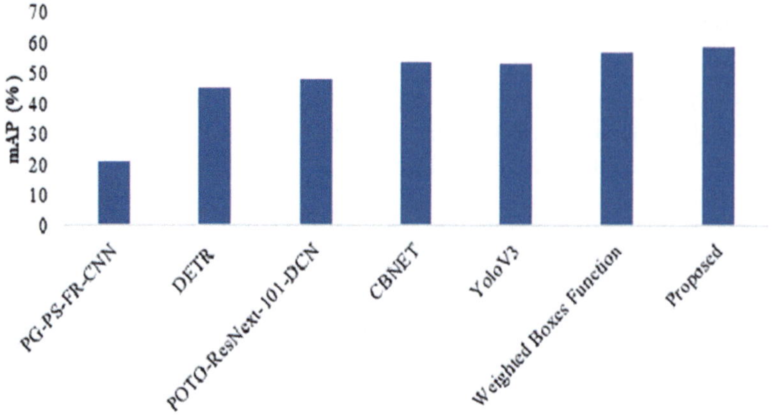


**Fig. 5** Comparison of mAP for proposed and existing methods on COCO data set

YOLO (You Only Look Once) has emerged as a leading technique in object detection, often outperforming traditional Convolutional Neural Networks (CNNs) in both speed and efficiency. Unlike CNNs, which typically require multiple stages for detection and classification, YOLO processes images in a single pass, enabling real- time object detection without compromising accuracy.

This efficiency is particularly advantageous in applications such as autonomous driving and video surveillance, where rapid decision-making is crucial. YOLO's architecture divides the input image into a grid, predicting bounding boxes and class probabilities simultaneously, which streamlines the detection process. As a result, YOLO demonstrates superior performance in scenarios demanding immediate feedback and high processing speed, making it a preferred choice over conventional CNNs for tasks requiring swift object localization and identification. This capability not only enhances operational efficiency but also expands the scope of practical applications across various fields.

## 5 Conclusion

The work can be said as a contribution to the society. Network language models are useful if pre-trained. The pre-processing of the network model for extracting information from the clinical reports as texts have enabled to view the results which are not able to be viewed and understand easily with human eye. The human intervention has been kept in loop to minimize the risk considering this as a under progressed work. Future work could involve assessing the performance of the proposed method using additional metrics such as execution time, sensitivity, specificity, and error rate.

## References

Adarsh, P., Rathi, P. & Kumar, M. (2020), Yolo v3-tiny: Object detection and recognition using one stage improved model, in '2020 6th international conference on advanced computing and communication systems (ICACCS)', IEEE, pp. 687–694.

Albright, D., Lanfranchi, A., Fredriksen, A., Styler IV, W. F., Warner, C., Hwang, J. D., Choi, J. D., Dligach, D., Nielsen, R. D., Martin, J. et al. (2013), 'Towards comprehensive syntactic and semantic annotations of the clinical narrative', Journal of the American Medical Informatics Association 20(5), 922–930.

Campillos, L., Deléger, L., Grouin, C., Hamon, T., Ligozat, A.-L. & Névéol, A. (2018), 'A french clinical corpus with comprehensive semantic annotations: development of the medical entity and relation limsi annotated text corpus (merlot)', Language Resources and Evaluation 52, 571–601.

Carion, N., Massa, F., Synnaeve, G., Usunier, N., Kirillov, A. & Zagoruyko, S. (2020), End-to-end object detection with transformers, in 'European conference on computer vision', Springer, pp. 213–229.

Casillas, A., Pérez, A., Oronoz, M., Gojenola, K. & Santiso, S. (2016), 'Learning to extract adverse drug reaction events from electronic health records in spanish', Expert Systems with Applications 61, 235–245.

Cheng, G., Yang, J., Gao, D., Guo, L. & Han, J. (2020), 'High-quality proposals for weakly supervised object detection', IEEE Transactions on Image Processing 29, 5794–5804.

Devlin, J. (2018), 'Bert: Pre-training of deep bidirectional transformers for language understanding', arXiv preprint arXiv:1810.04805.

Gu, J. & Liu, C. (2012), A new feature local binary patterns (flbp) method, in 'Proceedings of the International Conference on Image Processing, Computer Vision, and Pattern Recognition (IPCV)', The Steering Committee of The World Congress in Computer Science, Computer ..., p. 1.

He, B., Dong, B., Guan, Y., Yang, J., Jiang, Z., Yu, Q., Cheng, J. & Qu, C. (2017), 'Building a comprehensive syntactic and semantic corpus of Chinese clinical texts', Journal of biomedical informatics 69, 203–217.

Kumar, C.J, Das, P.R. and Hazarika, A (2022), Autism spectrum disorder diagnosis and machine learning: a review, International Journal of Medical Engineering and Informatics, Inderscience, vol. 14(6).

LeCun, Y., Bengio, Y. & Hinton, G. (2015), 'Deep learning', nature 521(7553), 436–444.

Ravì, D., Wong, C., Deligianni, F., Berthelot, M., Andreu-Perez, J., Lo, B. & Yang, G.-Z. (2016), 'Deep learning for health informatics', IEEE journal of biomedical and health informatics 21(1), 4–21.

Solovyev, R., Wang, W. & Gabruseva, T. (2021), 'Weighted boxes fusion: Ensembling boxes from different object detection models', Image and Vision Computing 107, 104117.

Sung, S.-F., Lin, C.-Y. & Hu, Y.-H. (2020), 'Emr-based phenotyping of ischemic stroke using supervised machine learning and text mining techniques', IEEE journal of biomedical and health informatics 24(10), 2922–2931.

Wang, K., Liu, M. & Ye, Z. (2021), 'An advanced yolov3 method for small-scale road object detection', Applied Soft Computing 112, 107846.

Zhao, L. & Li, S. (2020), 'Object detection algorithm based on improved yolov3', Electronics 9(3), 537.

# A Study on Performance of Ensemble Based Classifiers on Healthcare Data

**Irani Hazarika, Debashis Saikia, and Anjana Kakoti Mahanta**

**Abstract** This chapter focuses on a study of the performance of a set of ensemble-based classifiers on healthcare data, for example accuracy of classifying a patient as Covid-19 positive or negative based on diverse medical parameters. In healthcare data it is seen that data come for classification may be imbalanced or the considered parameters are heterogeneous or there may be lots of missing values or may have large numbers of attributes. To deals with these issues in this work ensemble based classifier has been considered. Thus, four numbers of ensemble based classifiers (EBC-1, EBC-2, EBC-3 and EBC-4) have been designed. Here, EBC-1 is the basic ensemble based classifier which can works on heterogeneous data. Based on EBC-1, other classifiers have been designed to deals with other issues. Results obtained by all the ensemble-based classifiers on a COVID-19 dataset are compared in terms of classification accuracy and discussed the efficiency of the classifiers on the data.

**Keywords** Classification · Ensemble based learning · Imbalance data · Healthcare data

## 1 Introduction

In healthcare, patient outcomes are primarily dependent on accurate predictions and decisions. However, high dimensionality, class imbalance, noisy features, and complex relationships are the major challenges associated with healthcare data (As shown in Table 1). For improving overall performance, robustness and generalization ability, ensemble-based classifiers (Rokac 2010; Reddy et al. 2020; Lee et al. 2020) which combine predictions of multiple base classifiers are considered as a powerful tool. Addressing healthcare data too, ensemble-based classifiers by

I. Hazarika (✉) · A. K. Mahanta
Department of Computer Science, Gauhati University, Guwahati, Assam, India
e-mail: queensarathi@gmail.com

D. Saikia
Department of Instrumentation and USIC, Gauhati University, Guwahati, Assam, India



T. P. Singh et al. (eds.), *Revolutionizing Healthcare: Impact of Artificial Intelligence on Diagnosis, Treatment, and Patient Care*, Studies in Computational Intelligence 1182,
https://doi.org/10.1007/978-3-031-80813-5_4

combining multiple single classifiers are deemed for improvement of the predictive accuracy, robustness, and generalization ability. In literature there are several works on classification that use ensemble-based classifiers to classify heterogeneous data, imbalance data etc. (Parikh and Polikar 2007; Liu et al. 2009; Hazarika et al. 2016) obtained from various domain (Bhadra and Kumar 2023).

In the healthcare domain, a few applications of ensemble-based classifier are—to study the infections acquired in intensive care units (Yadav et al. 2020), detecting the Covid-19 disease (García et al. 2016) or other medical disease diagnosis (Singh et al. 2021) etc. In such applications also data may be imbalanced. In Yadav et al. (2020), the authors introduce a new way of addressing imbalanced data. In García et al. (2016), the authors proposed a new ensemble-based classifier to detect COVID-19 disease for infected patients, which forecasts the spread of epidemic transmission. It is suitable for applications that need precise classifications, real-time responses, and short reaction times. In Singh et al. (2021), the authors study on the performances of ensemble-based classifier and single classifier in Medical Disease Diagnosis.

Again, deep learning can be incorporated with ensemble learning techniques to improve the classification output. In Nguyen et al. (2021), the authors have proposed three such approaches—deep-stacked generalization ensemble learning, gradient deep learning boosting, and Deep aggregation learning (Bhadra et al. 2024). These approaches are collectively known as deep ensemble learning, which works on statistical data, image-based data and sequential data. All the approaches perform well with higher efficiency by replacing the core learning units of the corresponding ensemble technique with suitable DL models. In the case of healthcare data, it is seen that data come for classification may be imbalanced, attributes may be heterogeneous with lots of missing values or maybe large numbers of attributes present. By considering these issues, in this work a set of ensemble-based classifier i.e. EBC-1, EBC-2, EBC-3, EBC-4 have been designed to works on healthcare data. Here, EBC-1 is the basic ensemble based classifier which can works on heterogeneous data. To take care of imbalance data, EBC-2 has been designed using EBC-1. Again, using EBC-1, EBC-3 has been developed to handles attributes with many missing values, low numbers of attributes etc. EBC-4 has been developed using EBC-2 and EBC-3 to handles imbalance data, data with many missing values, low numbers of attributes

**Table 1** A sample dataset of different patients tested for Covid 19

| Covid 19 | A1 | A2 | … | An | B1 | B2 | … | Bm | C1 | C2 | … | Cp | D | E | F |
|---|---|---|---|---|---|---|---|---|---|---|---|---|---|---|---|
| Positive + ICU | 0.2 | 0.3 | … | 0.4 | 60 | 460 | … | 900 | ? | ? | … | ? | Active | ? | 10.7 |
| Positive + Semi-ICU | ? | ? | … | ? | 1200 | 520 | … | 80 | 0 | 1 | … | 1 | Not active | ? | 10.7 |
| Negative + ICU | 0.5 | 0.1 | … | 0.8 | 100 | 30 | … | 60 | ? | ? | … | ? | Active | ? | ? |
| Negative | 0.6 | ? | … | 0.2 | 300 | 700 | … | 130 | 0 | 1 | … | 1 | ? | ? | 10.7 |

etc. The efficiency of the classifiers has been compared and discussed in terms of classification accuracy.

In Sect. 2, preliminaries descriptions related to the work have been given. In Sect. 3, workings of the set of ensemble-based classifiers on healthcare data have been discussed. In Sect. 4, experimental results have been discussed and in Sect. 5, conclusion is given.

## 2 Preliminaries

### *2.1 Introduction to Ensemble Based Classifier*

Ensemble learning represents a powerful methodology in machine learning, wherein the predictions of multiple individual models are combined to yield a final prediction. The core principle behind ensemble learning lies in its ability to harness diversity among models, thereby often surpassing the predictive performance of any single model. By integrating varied perspectives of the data and mitigating the risk of overfitting, ensemble learning stands as a pivotal technique in the machine learning domain. Several prominent ensemble learning methods exist, each presenting its unique approach to amalgamating base classifiers. Notable among these are:

**Bagging (Bootstrap Aggregating)**. Bagging involves training multiple base classifiers independently on different subsets of the training data, typically sampled with replacement. The final prediction is obtained by averaging or voting among the predictions of these classifiers. Examples of bagging-based ensemble classifiers include Random Forest.

**Boosting**. Boosting algorithms, such as AdaBoost (Adaptive Boosting) and Gradient Boosting, sequentially train weak learners where each subsequent model focuses on the examples misclassified by its predecessors. Boosting aims to improve prediction accuracy by giving more weight to difficult-to-classify instances. Boosting-based ensemble classifiers can adaptively combine the predictions of weak learners to produce a strong overall classifier.

**Stacking**. Stacking involves training multiple diverse base classifiers and then using a meta-classifier to learn how to combine their predictions effectively. Unlike bagging and boosting, stacking can incorporate various types of base classifiers, such as decision trees, support vector machines, and neural networks. The meta-classifier learns to weigh the predictions of the base classifiers based on their performance on a validation set, improving overall prediction accuracy and robustness.

**Voting Classifiers**. Voting classifiers combine the predictions of multiple base classifiers by either majority voting (hard voting) or averaging the class probabilities (soft voting). In hard voting, the final prediction is the majority class predicted by the base classifiers, while in soft voting, the final prediction is the class with the highest average probability across all base classifiers. Voting classifiers are straightforward to implement and can be effective in improving prediction accuracy.

Ensemble-based classifiers find widespread utility across diverse machine learning applications, encompassing classification, regression, and anomaly detection tasks. Particularly advantageous when individual classifiers exhibit distinct strengths and weaknesses, ensemble learning effectively exploits the complementary nature of these classifiers to achieve superior performance. Moreover, ensemble-based classifiers exhibit resilience to overfitting and noisy data, rendering them well-suited for real-world applications where data quality and reliability are paramount.

## *2.2 Homogeneous Versus Heterogeneous Data*

Based on the values of the attributes present in the data, data can be divided into two categories as described below—

**Homogeneous Data**. Homogeneous data refers to a dataset comprising elements having similarities or belonging to the same type. Within a homogeneous dataset, all attributes usually fall within the same type.

**Heterogeneous Data**. Heterogeneous data refers to a dataset composed of elements that exhibit dissimilarities or belong to different types. Within a heterogeneous dataset, the attributes describing the data points can vary in type and structure.

## *2.3 Imbalance Data*

Imbalanced data signifies a scenario where the classes or categories within a dataset are not evenly distributed. In binary classification problems, this typically means that one class (the minority class) is significantly less frequent than the other class (the majority class). In multi-class classification problems, this means that some classes (the minority classes) have significantly less numbers of data than the other classes (the majority classes). In case of imbalanced data, machine learning models might show biases towards the majority class, leading to difficulties in accurately categorizing instances from the minority class (Bhadra et al. 2024). Below are some prevalent techniques for addressing imbalanced data:

**Re-sampling Techniques**. Re-sampling is a technique which deals with highly imbalance dataset as given below—

*Over-sampling*. The number of instances in the minority class can be increased by generating synthetic samples (e.g., using SMOTE—Synthetic Minority

Over-sampling Technique) or by randomly duplicating existing minority class samples.
*Under-sampling.* The number of instances in the majority class can be decreased by randomly removing samples which leads to a more balanced dataset.
*Combining Over-sampling and Under-sampling.* Utilize a combination of over-sampling and under-sampling techniques to balance the class distribution while minimizing the potential for overfitting or information loss.

**Algorithmic Approaches**. There are two ways where adjustments are done within the learning algorithm for minimizing misclassification problem in imbalanced data—

*Adjustments of class weights.* By modifying the class weights within the learning algorithm to impose greater penalties for misclassifications of the minority class.
*Selection of algorithm.* By choosing machine learning algorithms that are inherently robust to imbalanced data, such as decision trees, random forests, gradient boosting machines and ensemble methods.

**Ensemble Methods**. Ensemble learning techniques, such as bagging and boosting, can effectively handle imbalanced data by combining the predictions of multiple models trained on different subsets of the data or by focusing on the misclassified instances of the minority class during training (Sharma et al. 2024).
**Data Pre-processing**. Another way to handle imbalance data is data pre-processing. In data pre-processing, feature engineering is performed to extract more informative features or reduce the dimensionality of the dataset, which can help the model better discriminate between classes (Kumar et al. 2022). Normalization or scaling of the features is also done to ensure that the algorithm is not biased towards features with larger magnitudes.

## 3 Classification of Healthcare Data Using Ensemble Based Classifiers

It is seen that the healthcare data may contain different types of attributes which denote different laboratory tests. Thus, the attribute values can be of type numeric, categorical etc. Again, if a patient has not taken a particular test, then the value of those attributes will be NULL. Thus, sometimes it is seen that healthcare data contains different types of attributes as well as lots of missing values for the attributes (Bhadra and Kumar 2023). Also, sometimes the data may be imbalanced, due to the facts like a disease present in the dataset is from rare category etc. By taking care of these observations, the aim of this work is to classify healthcare data, especially Covid-19 data using a set of ensemble-based classifiers EBC-1, EBC-2, EBC-3, EBC-4 as shown in Fig. 1. These ensemble-based classifiers consider the factors like—Data

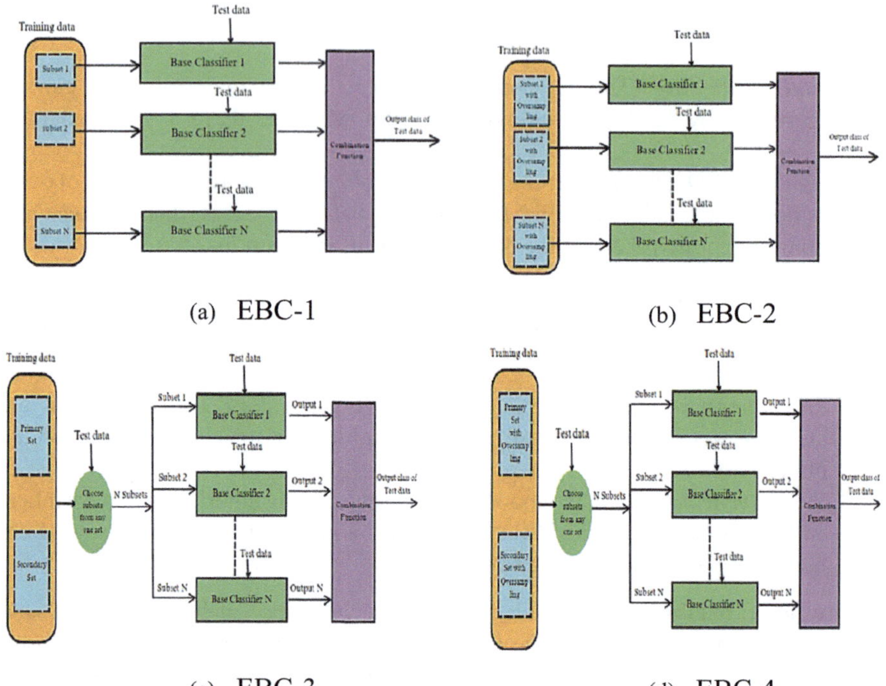


(a) EBC-1 (b) EBC-2 (c) EBC-3 (d) EBC-4

**Fig. 1** Frameworks of ensemble based classifiers **a** EBC-1, **b** EBC-2, **c** EBC-3 and **d** EBC-4

pre-processing, dividing the data into homogeneous subsets, oversampling to handle class imbalance problem etc. as mentioned below.

## *3.1 Apply Data Pre-processing on the Training Data*

In the first phase of data pre-processing, the input training data will be pre-processed using the Step A, Step B and Step C as mentioned below. The pre-processed data will contain N numbers of homogeneous subsets, which may be further divided into two sets as primary set and secondary set in the second phase as described below—

**Phase I: Pre-processing Phase**

*Step A. Remove unwanted attributes from the training data*

In this step unwanted attributes are removed i.e. attributes having same values for all data, NULL values for all data, attributes whose values denotes serial number etc. are removed. For example, from Table 1, attributes E and F will be removed as E has all values NULL and F has all values same.

*Step B. Dividing the training data into homogeneous subsets*

In this step the data obtained from Step A, are divided into some homogeneous subsets. Types of attributes and dependencies of the attributes are considered while making these subsets. Thus, in each homogeneous subset, all attributes' values are of same type and they are dependent. Again, any two homogeneous subsets may have same types of attributes, but the attributes between the subsets are not dependent. Two attributes A1, A2 of same types are called dependent, if, for any data D1 present in the dataset—

- If D1 does not have missing value for attributes {A1} then it does not have a missing value for attribute {A2} also.
- If D1 has missing value for attributes {A1} then it has a missing value for attribute {A2} also.

If Θ% data in the datasets satisfies this dependency for attributes {A1} and {A2} then attributes {A1, A2} comes under the same homogeneous subsets. Otherwise, they will be put as attributes of two different homogeneous subsets. The value of Θ could be as high as 100%. For example—After applying Step A and Step B, the data in Table 1 will be divided into the homogeneous subsets Subset 1, Subset 2, Subset 3 and Subset 4 in its attributes spaces as shown in Table 2, where Θ is considered as 75%.

**Table 2** Homogeneous subsets obtained from the training data after applying Step A and Step B of Phase I of data pre-processing

| **Subset 1** | | | | | **Subset 2** | | | | |
|---|---|---|---|---|---|---|---|---|---|
| Covid 19 | A1 | A2 | … | An | Covid 19 | B1 | B2 | … | Bm |
| Positive + ICU | 0.2 | 0.3 | … | 0.4 | Positive + ICU | 60 | 460 | … | 900 |
| Positive + Semi-ICU | ? | ? | … | ? | Positive + Semi-ICU | 1200 | 520 | … | 80 |
| Negative + ICU | 0.5 | 0.1 | … | 0.8 | Negative + ICU | 100 | 30 | … | 60 |
| Negative | 0.6 | ? | … | 0.2 | Negative | 300 | 700 | … | 130 |
| **Subset 3** | | | | | **Subset 4** | | | | |
| Covid 19 | C1 | C2 | … | Cp | Covid 19 | D | | | |
| Positive + ICU | ? | ? | … | ? | Positive + ICU | Active | | | |
| Positive + Semi-ICU | 0 | 1 | … | 1 | Positive + Semi-ICU | Not Active | | | |
| Negative + ICU | ? | ? | … | ? | Negative + ICU | Active | | | |
| Negative | 0 | 1 | … | 1 | Negative | ? | | | |

*Step C. Handling attributes having missing values*

If a homogeneous subset contains any data having NULL attribute value, then remove the data from the subset. For example, after removing the missing values, the subsets Subset 1, Subset 2, Subset 3 and Subset 4 are shown in Table 3. After applying Step C, the size of the obtained subsets may not be same, because all the subsets may not contain the same data. Suppose the data d1 has NULL values for all the attributes present in the ith subset and all attributes' values are present for the jth subset. Then d1 will not be present in the ith subset, but it will present in jth subset after applying Step C.

**Phase II: Dividing the homogeneous subsets of the training data into two sets: Primary Set and Secondary Set**

In the case of high dimensional data there are many attributes present to describe a data. In such data, it is seen that the classifier must be processed all such large numbers of attributes to classify a test data, which may affect its execution time. To reduce the execution time, instead of considering all the attributes for classification, in this phase, the homogeneous subsets obtained from Phase 1 are again divided into two sets—Primary Set and Secondary Set. The primary set contains the subsets whose number of attributes is more than a minimum threshold, Attr_threshold and number of data is also more than $\partial\%$ of total data present in the dataset. Otherwise, the subset is considered as a secondary subset and added into the secondary set. From the above subsets {Subset 2} will be added into the primary set and {Subset 1,

**Table 3** Homogeneous subsets obtained from the training data after applying Step C of Phase I of data pre-processing

| **Subset 1** | | | | | **Subset 2** | | | | |
|---|---|---|---|---|---|---|---|---|---|
| Covid 19 | A1 | A2 | … | An | Covid 19 | B1 | B2 | … | Bm |
| Positive + ICU | 0.2 | 0.3 | … | 0.4 | Positive + ICU | 60 | 460 | … | 900 |
| Negative + ICU | 0.5 | 0.1 | … | 0.8 | Positive + Semi-ICU | 1200 | 520 | … | 80 |
| | | | | | Negative + ICU | 100 | 30 | … | 60 |
| | | | | | Negative | 300 | 700 | … | 130 |
| **Subset 3** | | | | | **Subset 4** | | | | |
| Covid 19 | C1 | C2 | … | | Covid 19 | D | | | |
| Positive + Semi-ICU | 0 | 1 | … | | Positive + ICU | Active | | | |
| Negative | 0 | 1 | … | | Positive + Semi-ICU | Not Active | | | |
| | | | | | Negative + ICU | Active | | | |

Subset 2, Subset 4} will be added into the secondary set, if value of Attr_threshold is considered as 2 and ∂ is considered as 60%.

### 3.2 Handling Class Imbalance Problem: Applying Oversampling Method on the Training Subsets

The pre-processed subsets obtained after data pre-processing may have class imbalance problems. If any subset is found to have class imbalance problem, then oversampling method may be applied on the subset to overcome it.

### 3.3 Classify the Test Data T Using Ensemble Based Classifiers

To classify a test data T, a set of four numbers of ensemble-based classifiers EBC-1, EBC-2, EBC-3, EBC-4 (Fig. 1) have been applied on N numbers of subsets picked from the pre-processed training subsets obtained from Phase I or Phase II of data pre-processing. There are two ways to select these N subsets as described below—

**Choose N subsets from all the training subsets obtained from Phase I**. When test data T is coming for classification, first check the values of the attributes of T, which are presented in each training subset obtained from Phase I of data pre-processing. If any attribute present in a training subset has NULL value for the test data T, then skip that subset from classify task. The rest of the training subsets will be act as the training set for classification task.

**Choose N subsets from any one set i.e. either from Primary set or Secondary set obtained from Phase II**. When test data T is coming for classification, first check the values of the attributes of T, which are presented in each training subset of primary set, obtained from Phase II of data pre-processing. If there is no subset in the primary set such that each attribute presents in it has a value in test data T, then pick the secondary set. Otherwise, keep the primary set. After choosing any one set i.e. primary set or secondary set, if attributes present in any subset of the selected set has NULL value for test data T, then skip that subset from classify task. The rest of the training subsets will be act as the training set for classification task. On this training set oversampling may be applied to take care of imbalance problems.

After obtaining the training set, the base classifiers for the ensemble-based classifiers are selected. All the models have used k-nearest neighbor (k-NN) as base classifiers and majority voting as combination function as described below—

**Set of k-NN with majority voting.** If N numbers of subsets are chosen as training set to classify the test data T, then a set of N numbers of base classifiers (i.e. N numbers of k-NN) are used for classification. Each base classifier is applied to an individual subset. Thus, k-NN is applied on each of the chosen subsets to classify the test data T. After that N numbers of outputs obtained from all the k-NN classifiers for the test data T are combined using majority voting method to get the final class label of T.

All the ensemble-based classifiers EBC-1, EBC-2, EBC-3, EBC-4 have used the above elements as described below—

**Ensemble Based Classifier-1 (EBC-1)**. In this model, the N subsets used as training set are selected from all the training subsets obtained from Phase I. Oversampling has not been used on the training set to handle class imbalance problem and a set of k-NN with majority voting is used for ensemble-based classification.

**Ensemble Based Classifier-2 (EBC-2)**. In this model, the N subsets used as training set are selected from all the training subsets obtained from Phase I. Oversampling has been done on the training set to handle class imbalance problem and a set of k-NN with majority voting is used for ensemble-based classification.

**Ensemble Based Classifier-3 (EBC-3)**. In this model the N subsets used as training set are selected either from primary set or secondary set obtained from Phase II. Oversampling has not been used on the training set to handle class imbalance problem and a set of k-NN with majority voting is used for ensemble-based classification.

**Ensemble Based Classifier-4 (EBC-4)**. In this model the N subsets used as training set are selected either from primary set or secondary set obtained from Phase II. Oversampling has been used on the training set to handle class imbalance problems and a set of k-NN with majority voting is used for ensemble-based classification.

## 4 Results and Discussion

For experimental analysis, the ensemble-based classifiers EBC-1, EBC-2, EBC-3, EBC-4 have been applied on the Covid-19 dataset obtained from Kaggle. The dataset contains 111 numbers of features and 5644 numbers of data with many missing values. Attributes, i.e. SARS-Cov-2 exam result, Patient admitted to regular ward, Patient admitted to semi-intensive unit and Patient admitted to intensive care unit are together considered as the decision attributes. Here, attribute SARS-Cov-2 exam result has values positive or negative and the rest of the attributes have values either 0 or 1. From these four attributes 8 classes {Negative000, Negative001, Negative010, Negative100, Positive000, Positive001, Positive010, Positive100} can be derived. The Negative000 denotes patient with only covid negative, Negative001 denotes patient with covid negative and admitted to regular ward, Negative010 denotes patient with covid negative and admitted to semi-intensive unit, Negative100 denotes patient with covid negative and admitted to intensive care unit. Similarly, Positive000 denotes patient with only covid positive, Positive001 denotes patient with covid positive and admitted to regular ward, Positive010 denotes patient with covid positive and admitted to semi-intensive unit, Positive100 denotes patient with covid positive and admitted to intensive care unit. The data distribution of each class is shown in Table 2.

In this experiment, for the purpose of this classification, 90% data are used as training data and 10% are used as test data from the Covid-19 dataset. In these 90% training data, Phase I and Phase II of data pre-processing have been applied. In the Step A of Phase I of data pre-processing, the following unwanted attributes are removed-

- Attributes denote serial numbers i.e. Patient ID, Relationship (Patient/Normal) (Numeric), International normalized ratio (INR) (Numeric) denotes serial number.
- Attributes having all values NULL i.e. Mycoplasma pneumonia, Fio2 (venous blood gas analysis), Urine-Nitrite, Urine-Sugar, Partial thromboplastin time (PTT), Prothrombin time (PT) Activity, Vitamin B12, D-Dimer, Albumin.
- Attributes having same values for all data i.e. Myeloblasts, Urine-Esterase, Urine-Bile pigments, Urine-Ketone Bodies, Urine-Urobilinogen, Urine-Protein.

In Step B of Phase I of data pre-processing, training data are divided into some homogeneous subsets (shown in Table 5) as discussed in Sect. 3.1. In these homogeneous subsets attributes are of types Numeric and Categorical (Table 4).

In Step C of Phase I of data pre-processing, data having NULL values are removed from the homogeneous subsets. In Figs. 2 and 3, the number of attributes and number of data present in each homogeneous subset obtained from pre-processed training set have been shown.

In Phase II of data pre-processing, the homogeneous subsets are again divided into two sets-Primary Set and Secondary Set. All the subsets having more than two attributes (including class attribute) and data more than 1% of total training data (Fig. 4) are kept in the Primary Set and rest are kept in the Secondary Set. Thus, the subsets shown in Table 5 are divided as follows—

- **Primary Set**: {2, 3, 4, 10, 12, 16, 25, 27}
- **Secondary Set**:{1, 5, 6, 7, 8, 9, 11, 13, 14, 15, 17, 18, 19, 20, 21, 22, 23, 24, 26}

**Table 4** Data distribution of the classes of Covid-19 dataset

| Classes present | Number of data present |
|---|---|
| Negative000 | 4968 |
| Negative001 | 33 |
| Negative010 | 42 |
| Negative100 | 43 |
| Positive000 | 506 |
| Positive001 | 8 |
| Positive010 | 8 |
| Positive100 | 36 |

**Table 5** Subsets obtained by dividing the attributes of Covid-19 dataset

| Subsets | Attributes | Types |
|---|---|---|
| 1 | Patient age quintile | Numeric |
| 2 | Hemoglobin, Platelets, Mean platelet volume, Red blood Cells, Lymphocytes, Mean corpuscular hemoglobin concentration (MCHC), Leukocytes, Basophils, Mean corpuscular hemoglobin (MCH), Eosinophils, Mean corpuscular volume (MCV), Monocytes, Red blood cell distribution width (RDW) | Numeric |
| 3 | Serum Glucose | Numeric |
| 4 | Neutrophils | Numeric |
| 5 | Urea | Numeric |
| 6 | Proteina C reativa mg/dL | Numeric |
| 7 | Creatinine | Numeric |
| 8 | Potassium | Numeric |
| 9 | Sodium | Numeric |
| 10 | Alanine transaminase, Aspartate transaminase | Numeric |
| 11 | Gamma-glutamyltransferase | Numeric |
| 12 | Total Bilirubin, Direct Bilirubin, Indirect Bilirubin | Numeric |
| 13 | Alkaline phosphatase | Numeric |
| 14 | Ionized calcium | Numeric |
| 15 | Magnesium | Numeric |
| 16 | $pCO_2$ (venous blood gas analysis), Hb saturation (venous blood gas analysis, Base excess (venous blood gas analysis), $pO_2$ (venous blood gas analysis), Total $CO_2$ (venous blood gas analysis), pH (venous blood gas analysis), $HCO_3$ (venous blood gas analysis) | Numeric |
| 17 | Rods #, Segmented, Promyelocytes, Metamyelocytes, Myelocytes | Numeric |
| 18 | Urine-Aspect, Urine-Hemoglobin, Urine-Crystals, Urine-Color | Categorical |
| 19 | Urine-pH, Urine-Density, Urine-Leukocytes, Urine-Red blood cells | Numeric |
| 20 | Creatine phosphokinase (CPK) | Numeric |
| 21 | Lactic Dehydrogenase | Numeric |
| 22 | Hb saturation (arterial blood gases), $pCO_2$ (arterial blood gas analysis) Base excess (arterial blood gas analysis), pH (arterial blood gas analysis), Total $CO_2$ (arterial blood gas analysis), $HCO_3$ (arterial blood gas analysis), $pO_2$ (arterial blood gas analysis), $ctO_2$ (arterial blood gas analysis) | Numeric |
| 23 | $ctO_2$ (arterial blood gas analysis) | Numeric |
| 24 | Phosphor | Numeric |
| 25 | Respiratory Syncytial Virus, Influenza A, Influenza B, Parainfluenza 1, CoronavirusNL63, Rhinovirus/Enterovirus, Coronavirus HKU1, Parainfluenza 3, Chlamydophila pneumonia, Adenovirus, Parainfluenza 4, Coronavirus229E, CoronavirusOC43, Inf A H1N1 2009, Bordetella pertussis, Metapneumovirus, Parainfluenza 2 | Categorical |

(continued)

**Table 5** (continued)

| Subsets | Attributes | Types |
|---|---|---|
| 26 | Strepto A | Categorical |
| 27 | Influenza B rapid test, Influenza A rapid test | Categorical |

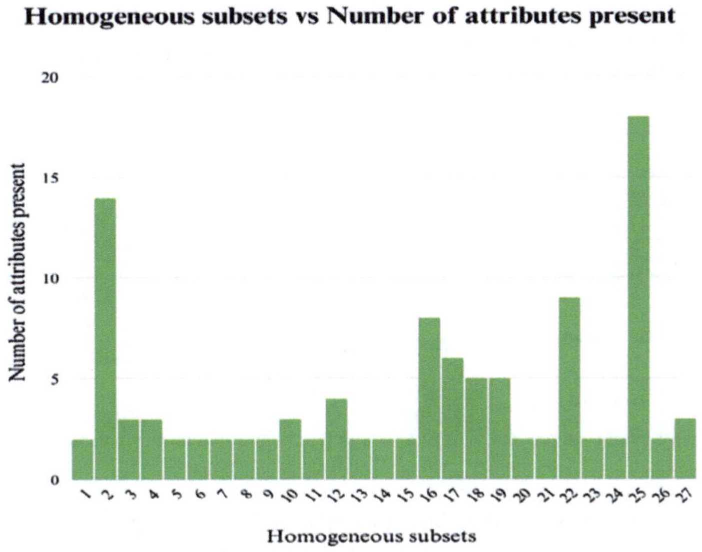


**Fig. 2** Number of attributes present in the homogeneous subsets

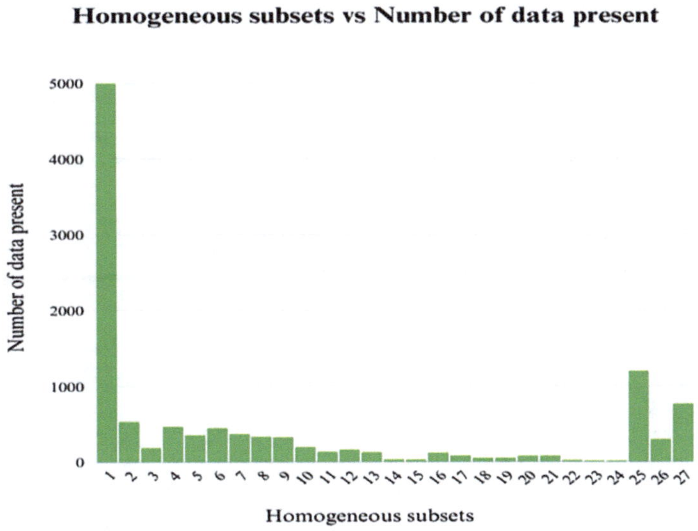


**Fig. 3** Number of data present in the homogeneous subsets

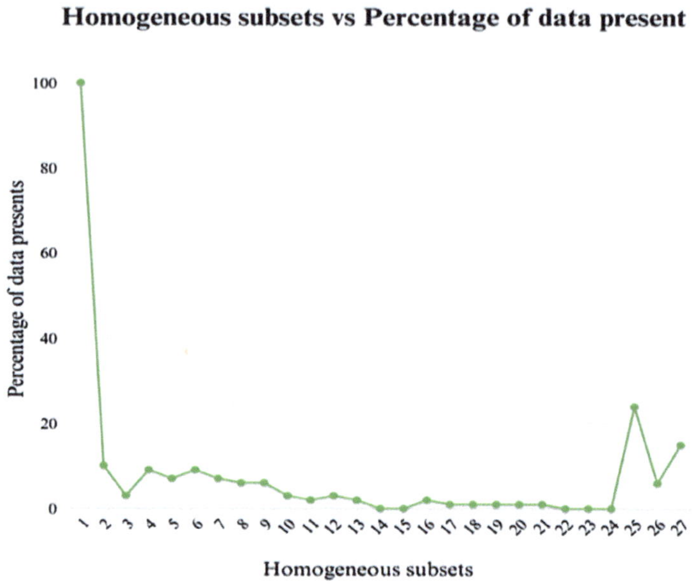


**Fig. 4** Percentage of data present w.r.t total training data in the homogeneous subsets

The data distributions of the classes in each of the subsets present in the Primary Set and Secondary Set are shown in Fig. 5 and it is seen that the data present in the subsets are highly imbalance.

In Tables 6, 7, 8 and 9, accuracy values obtained by the ensemble-based classifiers EBC-1, EBC-2, EBC-3 and EBC-4 on each class of Covid-19 dataset have been

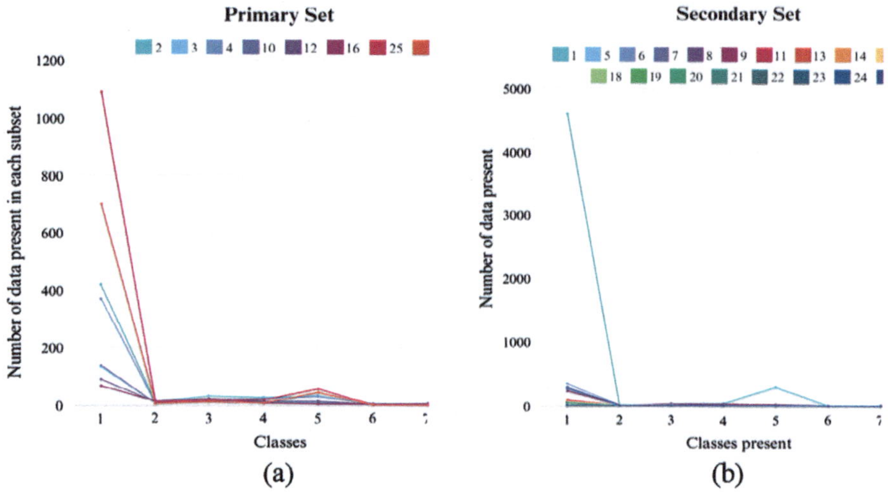


**Fig. 5** Data distributions of the classes in each of the subset of primary set and secondary set

shown. As mentioned earlier, k-NN has been used as base classifiers, so the accuracy values for different values of k have been shown.

From Table 6, it is seen that, the ensemble-based classifier EBC-1 gives highest overall accuracy 1 when the value of k is 1. It is also observed that the overall accuracy values given by the classifier EBC-1 decreases as the value of k increases. This classifier EBC-1 gives accuracy values 1 for all possible values of k for the classes—Class 1, Class 3, Class 4, Class 7.

Again, from Table 7, it is seen that the classifier EBC-2 gives overall accuracy values 1 for all possible values of k, thus for all the classes also.

From Table 8, it is seen that the classifier EBC-3 gives the highest overall accuracy 0.96, when the value of k is 1. As the classifier EBC-1, overall accuracy values of

**Table 6** Accuracy obtained from ensemble based classifier-1 (EBC-1)

| K | Accuracy of Classes | | | | | | | | Overall accuracy |
|---|---|---|---|---|---|---|---|---|---|
| | 1 | 2 | 3 | 4 | 5 | 6 | 7 | 8 | |
| 1 | 1 | 1 | 1 | 1 | 1 | 1 | 1 | 1 | 1 |
| 2 | 1 | 1 | 1 | 1 | 0.995 | 0.75 | 1 | 0.954 | 0.962 |
| 3 | 1 | 1 | 1 | 1 | 1 | 0.75 | 1 | 1 | 0.968 |
| 4 | 1 | 1 | 1 | 1 | 0.995 | 0.75 | 1 | 0.954 | 0.962 |
| 5 | 1 | 1 | 1 | 1 | 1 | 0.75 | 1 | 0.954 | 0.963 |
| 6 | 1 | 0.875 | 1 | 1 | 1 | 0.25 | 1 | 0.954 | 0.884 |
| 7 | 1 | 0.8125 | 1 | 1 | 1 | 0 | 1 | 0.954 | 0.845 |
| 8 | 1 | 0.75 | 1 | 1 | 1 | 0 | 1 | 0.954 | 0.838 |
| 9 | 1 | 0.625 | 1 | 1 | 1 | 0 | 1 | 0.954 | 0.822 |
| 10 | 1 | 0.5 | 1 | 1 | 1 | 0 | 1 | 0.954 | 0.806 |

**Table 7** Accuracy obtained from ensemble based classifier-2 (EBC-2)

| K | Accuracy of classes | | | | | | | | Overall accuracy |
|---|---|---|---|---|---|---|---|---|---|
| | 1 | 2 | 3 | 4 | 5 | 6 | 7 | 8 | |
| 1 | 1 | 0.9375 | 1 | 1 | 1 | 0.75 | 1 | 1 | 0.96 |
| 2 | 1 | 0.6875 | 0.9 | 1 | 0.995 | 0 | 1 | 1 | 0.822 |
| 3 | 1 | 0.812 | 1 | 1 | 1 | 0.25 | 1 | 1 | 0.88 |
| 4 | 1 | 0.375 | 1 | 1 | 0.995 | 0.25 | 1 | 0.954 | 0.821 |
| 5 | 1 | 0.375 | 0.9 | 1 | 1 | 0 | 1 | 0.954 | 0.778 |
| 6 | 1 | 0.125 | 0.9 | 1 | 1 | 0 | 0.6 | 0.954 | 0.697 |
| 7 | 1 | 0.1875 | 0.9 | 1 | 1 | 0 | 0.4 | 0.954 | 0.68 |
| 8 | 1 | 0.125 | 0.9 | 1 | 1 | 0 | 0.2 | 0.954 | 0.647 |
| 9 | 1 | 0.125 | 0.9 | 1 | 0.995 | 0 | 0 | 0.954 | 0.621 |
| 10 | 1 | 0.0625 | 0.9 | 1 | 1 | 0 | 0 | 0.954 | 0.614 |

**Table 8** Accuracy obtained from ensemble based classifier-3 (EBC-3)

| K | Accuracy of classes | | | | | | | | Overall accuracy |
|---|---|---|---|---|---|---|---|---|---|
| | 1 | 2 | 3 | 4 | 5 | 6 | 7 | 8 | |
| 1 | 1 | 1 | 1 | 1 | 1 | 1 | 1 | 1 | 1 |
| 2 | 1 | 1 | 1 | 1 | 1 | 1 | 1 | 1 | 1 |
| 3 | 1 | 1 | 1 | 1 | 1 | 1 | 1 | 1 | 1 |
| 4 | 1 | 1 | 1 | 1 | 1 | 1 | 1 | 1 | 1 |
| 5 | 1 | 1 | 1 | 1 | 1 | 1 | 1 | 1 | 1 |
| 6 | 1 | 1 | 1 | 1 | 1 | 1 | 1 | 1 | 1 |
| 7 | 1 | 1 | 1 | 1 | 1 | 1 | 1 | 1 | 1 |
| 8 | 1 | 1 | 1 | 1 | 1 | 1 | 1 | 1 | 1 |
| 9 | 1 | 1 | 1 | 1 | 1 | 1 | 1 | 1 | 1 |
| 10 | 1 | 1 | 1 | 1 | 1 | 1 | 1 | 1 | 1 |

**Table 9** Accuracy obtained from ensemble based classifier-4 (EBC-4)

| K | Accuracy of classes | | | | | | | | Overall accuracy |
|---|---|---|---|---|---|---|---|---|---|
| | 1 | 2 | 3 | 4 | 5 | 6 | 7 | 8 | |
| 1 | 1 | 1 | 1 | 1 | 1 | 0.75 | 1 | 1 | 0.968 |
| 2 | 1 | 1 | 1 | 1 | 1 | 0.75 | 1 | 1 | 0.968 |
| 3 | 1 | 1 | 1 | 1 | 1 | 0.75 | 1 | 1 | 0.968 |
| 4 | 1 | 1 | 1 | 1 | 1 | 1 | 1 | 1 | 1 |
| 5 | 1 | 1 | 1 | 1 | 1 | 1 | 1 | 1 | 1 |
| 6 | 1 | 1 | 1 | 1 | 1 | 1 | 1 | 1 | 1 |
| 7 | 1 | 1 | 1 | 1 | 1 | 1 | 1 | 1 | 1 |
| 8 | 1 | 1 | 1 | 1 | 1 | 1 | 1 | 1 | 1 |
| 9 | 1 | 1 | 1 | 1 | 1 | 1 | 1 | 1 | 1 |
| 10 | 1 | 1 | 1 | 1 | 1 | 1 | 1 | 1 | 1 |

EBC-3 decreases when the value of k increases. For Class 1 and Class 4, it always produces accuracy values 1.

From Table 9, it is seen that the classifier EBC-4 gives highest overall accuracy 1, when the value of k is more than equal 4. Unlike the classifiers EBC-1 and EBC-3, here accuracy values increase when the value of k increases. It will produce accuracy values 1 for all classes except Class 6 for all values of k.

From the above results it can be concluded that, as the training data present in the subsets are highly imbalance, so the ensemble-based methods EBC-2 and EBC-4, those incorporate class imbalance issues are, able to produce accuracy values 1 in most of the cases. Classifier EBC-1, which does not incorporate class imbalance issue, produces accuracy value 1 when value of k is 1 only. For other values of k its performance decreases. Again, it is seen that EBC-4 uses training subsets as two sets—Primary Set and Secondary Set and its accuracy reaches 1, when value of k is more than equal 4 only. Thus, this classifier may be used to classify imbalance data with large numbers of attributes by setting high value for k. Because if the number of attributes is high then in classification EBC-4 can use a smaller set of attributes by dividing the attributes as primary and secondary sets. In contrast to this method, EBC-2 does not divide the subsets as Primary Set or Secondary Set and its accuracy values are always 1. Thus, this classifier may be used to classify imbalance data with low number of attributes.

## 5 Conclusion

This work focuses on the study of four numbers of ensemble-based classifiers EBC-1, EBC-2, EBC-3, EBC-4 to classify healthcare data. These classifiers are considered to deal with factors like heterogeneous data, class imbalance problems, missing values, large number of attributes etc. For experimental purpose, these classifiers are applied on a Covid-19 dataset obtained from Kaggle. It is seen that all the classifiers except EBC-3 produce accuracy values 1 for different values of k, where k is the input for the base classifiers k-NN. Finally, it can be concluded that classifier EBC-2 may be suitable to classify imbalance data with low number of attributes by setting any values for k because it does not partition the attributes of the training data into primary and secondary sets. Again, classifier EBC-4 may be suitable to classify imbalance data with larger number of attributes by setting high value for k because it partitions the attributes of the training data into primary and secondary sets. One application of this work could be in healthcare sector. Suppose doctors have information of different lab test data for patients suffering from different diseases. Now, for a new patient if the parameter values of different lab test data are known, then using the suitable ensemble based method, doctor can predict the disease of the patient. One limitation of this method is that to deals with missing values, it removes the data having missing values instead of computing the missing values. In future works this limitation will be taken care of.

## References

Bhadra, S.; Kumar, C.J.; Bhattacharyya, D.K. (2024). Multiview EEG signal analysis for diagnosis of schizophrenia: an optimized deep learning approach. Multimedia Tools and Applications. https://doi.org/10.1007/s11042-024-20205-y.

Bhadra, S., & Kumar, C. J. (2023). Enhancing the efficacy of depression detection system using optimal feature selection from EHR. Computer Methods in Biomechanics and Biomedical Engineering, 27(2), 222–236. https://doi.org/10.1080/10255842.2023.2181660.

García, M.N.M., Herráez, J.C.B., Barba, M.S., & Hernández, F.S. (2016). Random Forest Based Ensemble Classifiers for Predicting Healthcare-Associated Infections in Intensive Care Units.

In: Omatu, S., et al. Distributed Computing and Artificial Intelligence, 13th International Conference. Advances in Intelligent Systems and Computing, (vol 474). SPRINGER.
Hazarika, I., Chodhury, R., & Mahanta, A. K. (2016). Design of a weighted meta classifier for imbalance data having heterogeneous features. In Proceedings of International Conference on Computing Communication Control and automation (pp. 1–6). IEEE.
Kumar, C.J.; Das, P.R.; Hazarika, A. (2022). Autism spectrum disorder diagnosis and machine learning: a review. International Journal of Medical Engineering and Informatics, Inderscience, 14(6).
Lee, J., Wang, W., Harrou, F., & Sun, Y. (2020). Wind power prediction using ensemble learning-based models. IEEE Access, 8, 61517–61527.
Liu, X. Y., Wu, J., & Zhou, Z. H. (2009). Exploratory Undersampling for Class- Imbalance Learning. IEEE Transactions on Systems, Man, and Cybernetics - Part B: Cybernetics, 39(2), 539–550.
Nguyen, D.K., Lan, C.H., & Chan, C.L. (2021). Deep Ensemble Learning Approaches in Healthcare to Enhance the Prediction and Diagnosing Performance: The Workflows, Deployments, and Surveys on the Statistical, Image-Based, and Sequential Datasets. International Journal of Environmental Research and Public Health, 18(20):10811.
Parikh, D., & Polikar, R. (2007). An Ensemble-Based Incremental Learning approach to data fusion. IEEE Transactions on Systems, Man, and Cybernetics - Part B: Cybernetics, 37(2), 437–450.
Reddy, G.T., Battacharya, S., Ramakrishnan, S. S, Chowdhary, C. L., Hakak, S., Kaluri, R., Reddy, & M. P. K. (2020). An Ensemble-based Machine Learning model for Diabetic Retinopathy Classification. In Proceedings of International Conference on Emerging Trends in Information Technology and Engineering, (pp. 1–6). IEEE.
Rokac, L. (2010). Ensemble-based classifiers. Artificial Intelligence Review, 33, 1–39.
Sharma, M.; Kumar, C.J.; Bhattacharyya, D.K. (2024). Machine/deep learning techniques for disease and nutrient deficiency disorder diagnosis in rice crops: A systematic review. Biosystems Engineering, 244, 77–92.
Singh, P.D., Kaur, R., Singh, K.D., & Dhiman, G. (2021). A Novel Ensemble-based Classifier for Detecting the COVID-19 Disease for Infected Patients. Information Systems Frontiers, 23, 1385–1401.
Yadav, S.S., Kadam, V.J., & Jadhav, S.M. (2020). Comparative Analysis of Ensemble Classifier and Single Base Classifier in Medical Disease Diagnosis. In: Bansal, J., Gupta, M., Sharma, H., Agarwal, B. (eds) Communication and Intelligent Systems. ICCIS 2019. Lecture Notes in Networks and Systems, (vol. 120). SPRINGER.

# Pre-diagnosis of Cardiovascular Diseases Through Machine Learning and Deep Learning Techniques Using Clinical Parameters

**Subrat Chetia and Chandan Jyoti Kumar**

**Abstract** Cardiovascular Disease (CVD) remains a primary cause of the death in the world, which underlines the importance of early detection and intervention of the illness. By using a dataset comprising various clinical parameters, including the demographic information, previous medical reports, results of diagnostic tests etc., Machine learning (ML) and Deep learning (DL) models can be trained and evaluated to accurately classify individuals into CVD and non-CVD groups. Available heart health data can be analyzed using different types of classifiers including the support vector machines, random forest classifiers, logistic regression, convolutional neural networks. These ML and DL approaches enhance CVD detection and thereby offer valuable insights for improving diagnostic accuracy and patient outcomes in clinical practice. These techniques can improve heart disease diagnosis, but they must be optimized and interpretability issues must be addressed for practical deployment in healthcare settings.

**Keywords** Cardiovascular disease (CVD) · Machine learning (ML) · Deep learning (DL) · Convolutional neural networks · Echocardiogram (ECG) · Phonocardiogram (PCG) · MRI · CT scan

## 1 Introduction

Cardiovascular disease (CVD) is universally recognized as one of the most substantial worldwide health concerns. CVD comprises of diseases like peripheral artery disease, stroke, coronary artery disease and heart failure among others. This underlines the need to diagnose such disorders early so as to avoid negative consequences that

S. Chetia (✉)
Pandit Deendayal Upadhyaya Adarsha Mahavidyalaya, Dalgon, Assam 784116, India
e-mail: subrat.chetia@gmail.com

C. J. Kumar
Department of Computer Science and IT, Cotton University, Guwahati, Assam 781001, India
e-mail: chandan.kumar@cottonuniversity.ac.in

 
T. P. Singh et al. (eds.), *Revolutionizing Healthcare: Impact of Artificial Intelligence on Diagnosis, Treatment, and Patient Care*, Studies in Computational Intelligence 1182,
https://doi.org/10.1007/978-3-031-80813-5_5

may result from these disorders. Traditional diagnostic methods often rely on non-invasive and invasive procedures, which may pose practical limitations or inherent risks for routine screening. Non-invasive techniques for diagnosing heart conditions include magnetic resonance imaging (MRI) and computed tomography (CT), both commonly used yet carrying limitations. These methods often require patients to maintain specific positions, expose them to ionizing radiation, demand significant financial investment, and may fail to detect cardiac diseases in their early stages (Hamilton-Craig et al. 2012). Conventional methods of invasive cardiac disease diagnosis involve a range of procedures including physical examinations, laboratory tests, physician assessments, and similar techniques. Among these, angiography and catheterization stand out as widely used invasive procedures. Although these approaches are beneficial, they have limitations such as invasiveness (which may in turn result in infection or bleeding), complexity, high cost, and limited scope for visualization etc. which collectively reduce the accuracy of real-time scenarios. Illustration of the common methods used in diagnosing cardiovascular diseases is provided in Fig. 1.

Over the past few years, there has been an increasing trend in applying both ML and DL techniques in analyzing clinical data, making it the fusion between technology and healthcare. The prospective advancement of ML and DL techniques is set to redefine diagnosis and treatment of cardiovascular diseases a new paradigm (Gala et al. 2024). These ML and DL algorithms are sophisticated and more informative ways of diagnosing diseases as compared to the traditional healthcare diagnosis models. They perform exceptionally when it comes to processing large and complex datasets, such as medical images and electrocardiograms (ECGs). Depending on the modality, ML algorithms use coronary angiography acquisition and cardiac Magnetic Resonance Imaging (MRI) data to analyze parameters related to ischemic heart disease, identify features of plaque and assess myocardial viability. Deep learning models are employed to discover Arrhythmias by recognising small abnormalities in ECG waveforms that are suggestive of diseases such as atrial fibrillation or ventricular tachycardia. These capabilities significantly improve the accuracy of pathological diagnosis and classification. Successful integration of computational and clinical

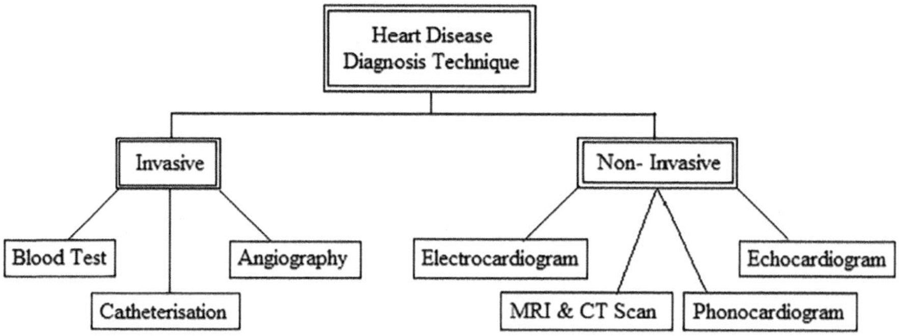


**Fig. 1** Cardiovascular disease (CVD) diagnosis techniques

expertise has great potential for innovation in CVD treatment, giving physicians predictive power capable of radically improving patient outcomes as well through better decision support.

## 2 Classification of Cardiovascular Disease Using ML and DL

Classification of cardiovascular diseases using machine learning (ML) and deep learning (DL) techniques are different from manual based methods like clinical examination findings, test results etc., because ML and DL algorithms can identify the feature contributions which come as a black box to human interpretation. The common approaches to diagnosing cardiac diseases using ML and DL, alongside clinical or biological data, can be categorized into four groups (Table 1).

- Group A focuses on biomedical signals including ECG and PPG as inputs (Alshayeji et al. 2021). These signals are chosen as inputs due to their ease of acquisition and the availability of compact sensors with robust processing capabilities. Following signal acquisition, preprocessing is employed to eliminate noise and achieve normalization (Charleonnan et al. 2016). The process of extracting crucial attributes from the preprocessed signal may involve utilizing data from the time domain, the frequency domain, or a hybrid of the two (Ashiquzzaman et al. 2017; Asri et al. 2016). Optimization is used to ensure that every completed feature is pertinent and not redundant. These refined attributes are subsequently inputted into machine learning algorithms in order to generate a classification for heart disease. The investigation is constrained by its reliance on biological signals which necessitates supplementary information to ensure a precise diagnosis.

**Table 1** Various techniques for classification of CVD using ML and DL

| Group | Input parameters | Processing | Feature extraction | Feature selection | Results |
|---|---|---|---|---|---|
| A | Vital Signal (ECG or PPG) | Filter and Normalization | Time domain, Frequency domain | Feature optimization algorithms | Classification of CVD using ML |
| B | Medical Image (ECHO, MRI, CT Scan) | Image Segmentation | Statistical and texture feature | Feature optimization algorithms | Classification of CVD using ML/DL |
| C | Clinical results from diagnostic lab tests | Normalize and check for unusual or missing values | Based on how relevant the data is and no redundancy | | Classification of CVD using ML/DL |
| D | Raw Vital Signal (ECG, PPG, PCG etc.) | Removal of Noise and other artifacts and Normalization | Neural Network | | Classification of CVD using ML/DL |

- Group B commences its investigation by utilizing medical photographs as primary sources of data. This technique adheres to the same sequence of operations as Group A, which includes image flattening and noise reduction as preprocessing steps. It then selects and extracts features, including statistical features, textures, and the cropped or chosen region of the image. ML/DL algorithms are subsequently applied to these characteristics in order to categorize various cardiac diseases and disorders (Chaki et al. 2020; Fernández-Edreira et al. 2021; Esmaeilzadeh et al. 2018; Fitriyani et al. 2019). The availability of large datasets continues to be the greatest obstacle for investigators, as training these models requires a substantial number of high-quality images.
- Scientists are inclined to conduct investigation in Group C due to the abundance of datasets containing clinical parameter information (Eshtay et al. 2018; Desai et al. 2018; Acharya et al. 2017). The system is furnished with a predetermined list of clinical parameters, the quantity and variety of which are reliant upon the amount of data that was chosen. After the elimination of outliers and the completion of absent data, the exclusive features are optimized through the utilization of choice techniques. Precise measurement of clinical indicators within the context of patients with cardiac disease is an advantage for carrying out such investigation.
- In Group D, biomedical signals serve as the primary input. These signals subsequently gone through systematic processing involving filtration, normalization, cleansing, and smoothing. Convolutional neural networks (CNN) use this processed information to yield classification outcomes (Dezaki et al. 2018; Alizadehsani et al. 2018; An et al. 2019). Studies indicates the remarkable effectiveness of 1D CNN in analyzing biomedical data (e.g., ECG, PPG, and PCG), alike to 2D CNNs' feature extraction and classification capabilities for image inputs, facilitating the categorization of cardiac conditions like arrhythmia (Ahmed et al. 2023). Biomedical data, comprising sequential and time series information from sources like electrocardiograms (ECGs), X-rays, and MRIs, highlight the practical utility of 1D/2D CNNs i.e., proficiency in extracting local patterns.

# 3 ML and DL Algorithms for Cardiovascular Disease Detection

## *3.1 Machine Learning Algorithms*

Machine learning algorithms involve training a model on a dataset of labeled examples (in this case, clinical parameters and corresponding CVD diagnoses) to learn patterns and generate predictions on unfamiliar data. Here are some common machine learning algorithms that could be applied to this task:

(a) **Logistic Regression**

- Logistic regression is a simple but very efficient algorithm mainly used for binary classification tasks like CVD detection.
- It models the probability of a result that can only have two possible outcomes (e.g., having CVD or not) based on one or more independent variables (clinical parameters).
- Clinical parameters can be features such as blood pressure, cholesterol levels, age, gender, etc.
- The model learns the association between these features and the possibility of having CVD.

(b) **Decision Trees and Random Forests**

- Decision trees and random forests are non-linear models that can handle both classification and regression tasks.
- They are capable of capturing complex relationships between features and outcomes.
- In Decision Trees the data is divided into subsets depending on the value of the input features.
- Random forests use several decision trees to get better results and to avoid problems of overfitting.

(c) **Support Vector Machines (SVM)**

- SVM is one of the ideal algorithms for the classification tasks.
- It operates by seeking the hyperplane that most of optimally separates the classes in the features space.
- SVM can handle high-dimensional data and is effective even when the number of features exceeds the number of samples.

## 3.2 *Deep Learning (DL) Algorithms*

In recent years, DL algorithms, particularly neural networks, are gaining more popularity of recent due to their capability to train models with layered data representations. The subsequent sections describe the methods and deployment of DL algorithms in diagnosing CVD.

(a) **Convolutional Neural Networks (CNNs)**

- CNNs are usually used for image classification, but we can also use them for tabular data by channel-wise separation of features.
- For CVD detection it can be useful to represent clinical parameters as input channels and the CNN can extract relevant features.
- CNNs can also learn features from raw data hence decreasing the chances of having to manually design the features.

(b) **Long Short-Term Memory (LSTM) and Recurrent Neural Networks (RNNs)**

**Table 2** Comparison of ML versus DL for prediction of CVD based on clinical parameters

| Feature | Machine learning | Deep learning |
|---|---|---|
| Model complexity | Less complex, easier to interpret | More complex, black-box nature |
| Feature engineering | Requires manual feature selection and engineering | Can learn features automatically |
| Data requirements | Generally, works well with smaller datasets | Often requires larger datasets for optimal performance |
| Accuracy | Can achieve good accuracy, but may be limited by feature selection | Potentially higher accuracy, especially with complex relationships |
| Interpretability | Easier to understand how the model arrives at a decision | Difficult to understand the inner workings of the model |
| Common techniques | Logistic Regression, Decision Trees, Support Vector Machines | Convolutional Neural Networks (CNNs), Recurrent Neural Networks (RNNs) |

- RNNs and LSTM networks are fundamentally designed for sequential data and this is why time series and sequences of clinical measurements might be easier to handle in such networks.
- They can capture temporal dependencies between clinical parameters over time.
- These networks can be applied to data where the order of measurements is important, such as electrocardiogram (ECG) data.

(c) **Hybrid Architectures**

- The use of additional architectures or connecting neural networks, deep reinforcement learning or embedding old-school machine learning algorithms in deep learning can also work.
- For instance, the integration of CNNs for feature extraction and a typical conventional classifier-ML such as logistic regression for the final prediction (Table 2).

## 4 Role of ML and DL Techniques in CVD Prevention

Both ML and DL have become a highly potent tool in the CVD prevention in recent decades. It must be said that application of such techniques has brought about more selective and effective treatment methods specifically designed at curbing cases of CVDs and dealing with possible complications.

Risk assessment or prediction, is one of the most widely used applications of ML and DL in cardiovascular disease (CVD) prevention. The use of predictive analytics involves the use of sophisticated models that analyse the patient records to come up with the patient cohort that is most likely to develop cardiovascular disease (CVD). Despite the fact that medical imaging is an efficient method of using in diagnosing CVD, first signs of the condition may remain undetected by human observers. Another way that the development of machine learning proved helpful is its capacity to analyse medical pictures—for instance, echocardiograms—that may contain obscure signs of CVD to which observers would possibly fail to take notice. Details about the risk factors which include age, gender, blood pressure, cholesterol levels and history of cardiovascular disease in the family; and history of diet and exercise are also taken into account. This makes it possible for healthcare practitioners to quickly and accurately respond to the development of risk factors in order to counteract the continuing development of cardiovascular disease (CVD).

Many studies show that the application of machine learning is also entering the field of personalised medicine for patients with cardiovascular risks. Because of variations in the type and nature of CVD, the causative treatments that are beneficial for one individual may not be beneficial for the other. Even more, when patients' records are closely analysed, machine learning algorithms are capable of differentiating which persons are most likely to benefit from certain therapies (Bhadra et al. 2024). This feature allows the treatment provider to give individual care solutions to meet the exact needs of the patient to enhance the success rates of the delivery of care and individuality.

Tele monitoring and telemedicine are also other areas that have been heavily enhanced by machine learning (ML) and deep learning (DL) (Das et al. 2024). The patient is now able to be supervised and given medical advice from the comforts of his or her residence. As sensor technology progress further, it is gradually becoming more feasible to get real time ECG or phonocardiogram data. These technologies, together with the use of, machine learning (ML) as well as deep learning (DL) help in giving out the best strategies for the management of patients' data in real time. For example, in real-time manner, data from an ECG of a patient can be evaluated and the existence of arrhythmia can be established, or further treatment can be recommended. At the moment, various companies are using ML to fight CVD or cardiovascular diseases, for short. Among these entities are

- **Viz.ai**—A business called Viz.ai employs ML to assist doctors in identifying and treating stroke-risk individuals. With their software, they analyses medical imaging data in order to identify warning signs of early stroke and hence enabling better patient outcomes and quicker intervention.
- **AliveCor**—AliveCor is a startup firm that designs new software and medical device which is principally used for the observation of heart pulse. As for their products, they use machine learning (ML) algorithms to analyse the data of electrocardiogram (ECG) and make health related recommendations to doctors.
- **Cardiogram**—Cardiogram is another firm that employs ML to detect preliminary indications of cardiovascular disease (CVD) based on information acquired from wearables, including smartwatches. Its program has employed data on heart rate, heart rate variability, and other parameters to come up with a list of people who've high risk of developing CVD.

- **Bay Labs**—Bay Labs uses artificial neural networks to analyse of echocardiography and detect precursors of heart diseases. They have designed their software to ensure that physicians obtain timely and recent data about their patients' condition so as to administer appropriate and timely care.
- **Zebra Medical Vision**—Zebra Medical Vision is a company that employs the use of artificial neuron network to process medical imaging data for identifying the early signs of cardiovascular diseases (CVD). Their program has the capacity to alert symptoms characteristic of heart failure, coronary artery disease and many other diseases.

In recent years the role of ML and DL has been used to fight cardiovascular illnesses (CVDs) by several firms and this paradigm will only expand as these ML and DL field advances.

## 5 Benefits of Machine Learning and Deep L in the Diagnosis of CVD

There are many advantages of utilizing machine learning as well as deep learning in the diagnosis of cardiovascular diseases. Here are some major benefits which have been considered for this purpose.

- **Automatic Feature Extraction**: The aspect of ML and DL models are that they can learn features from the raw data and they do not require hand crafted features.
- **Improved Accuracy**: Deep learning models can attain higher precision compared to traditional huma diagnosis, especially when dealing with large and complex datasets.
- **Integration with Medical Imaging**: CNNs can analyze medical images to identify subtle abnormalities indicative of cardiovascular diseases.

## 6 Challenges of ML and DL

Although, ML and DL holds significant advantages in cardiovascular disease prevention, there are still challenges that must be addressed. Few major concerns are—

- **Data Requirements**: The general application of the ML and DL models requires large datasets for ideal training results to be obtained. But in healthcare, privacy concerns and data availability limitations may confine the accessibility of such a range of data. Federated Learning (FL) approach may be considered as a viable solution to this problem is where the models are trained collectively on decentralized data but the original data remains local and private.

- **Interpretability**: One more concern related to deep learning models is that they might be considered to be 'black boxes' as the theoretical foundation of the forecasts becomes rather difficult to provide. This may lead to low acceptability of results in medicinal practices and low trust in all facilities that employ ML and DL methods in interpretation. In order to overcome this issue, different ways are investigated to improve model interpretability. SHAP (SHapley Additive exPlanations) or LIME (Local Interpretable Model-agnostic Explanations) can be used to get additional insights about the model predictions by focusing on the relevance of each feature.
- **Overfitting**: Another serious problem of deep learning models is overlearning which occurs when the model learns the training data excessively and performs poorly when the model is exposed to new data. To minimize such risk transfer learning may be an effective approach which involves training the model on a large dataset in advance and then fine-tuning that model on a smaller dataset that is specific to the task (Chun et al. 2021). Additionally, there are techniques such as dropout, regularization, and early stopping, which help improve the model's resistance to over learning the data set. Researchers have also pointed out that finding a balance of hyperparameters like learning rates also help in minimizing the overfitting problem without compromising the performance of the model.

## 7 The Future of ML and DL in Detection of Cardiovascular Disease

Machine learning (ML) combined with deep learning (DL) with the use of electronic health records (EHRs) as well the wearable health devices seem a promising approach to more accurate personalized risk prediction and timely diagnostics of the cardiovascular diseases. Here are some promising possibilities:

- **Development of Clinical Decision Support Systems**: ML models can be integrated into clinical workflows to assist physicians in making informed decisions about cardiovascular disease diagnosis and treatment plans.
- **Remote Patient Monitoring**: Wearable technologies also have the ability of monitoring real-time health information that could be used to frequently predict cardiovascular events.
- **Precision Medicine**: To prevent cardiovascular diseases, there is a possibility of using the patient data and genetic data to develop preventive strategies and medical management of such diseases using ML models.

## 8 Summary

Cardiovascular disease or CVD is one of the most prevalent diseases which include conditions such as stroke, heart failure and others. Integration of computational methods and clinical approaches opens new opportunities to enhance the diagnostic and treatment of CVD as well as the increase of predictive capabilities for the better outcomes of the patient. Generally, the efficiency of ML and DL models is dependent on the size of the data set required for building the model; which is a challenge in healthcare since data sharing and collection is highly regulated due to privacy concerns. The methods of identifying CVD using ML and DL can be divided into several categories based on the data sources. Some approaches are based on the use of biological signals while the others are based on medical images. By incorporating ML models into clinical workflows, health care providers will be able to extend their capabilities to diagnose and treat CVD cases in a more efficient manner and in ways that benefits the patient in near future.

## References

Acharya, U.R.; Oh, S.L.; Hagiwara, Y.; Tan, J.H.; Adam, M.; Gertych, A.; San Tan, R. (2017). A deep convolutional neural network model to classify heartbeats. *Computers in Biology and Medicine*, 89, 389–396.

Ahmed, A.A.; Ali, W.; Abdullah, T.A.; Malebary, S.J. (2023). Classifying cardiac arrhythmia from ECG signal using 1D CNN deep learning model. *Mathematics*, 11, 562.

Alizadehsani, R.; Hosseini, M.J.; Khosravi, A.; Khozeimeh, F.; Roshanzamir, M.; Sarrafzadegan, N.; Nahavandi, S. (2018). Non-invasive detection of coronary artery disease in high-risk patients based on the stenosis prediction of separate coronary arteries. *Computer Methods and Programs in Biomedicine*, 162, 119–127.

Alshayeji, M.H.; Ellethy, H.; Abed, S.; Gupta, R. (2021). Computer-aided detection of breast cancer on the Wisconsin dataset: An artificial neural networks approach. *Biomedical Signal Processing and Control*, 71, 103141.

An, Y.; Huang, N.; Chen, X.; Wu, F.; Wang, J. (2019). High-Risk Prediction of Cardiovascular Diseases via Attention-Based Deep Neural Networks. *IEEE/ACM Transactions on Computational Biology and Bioinformatics*, 18, 1093–1105.

Ashiquzzaman, A.; Tushar, A.K.; Islam, R.; Shon, D.; Im, K.; Park, J.-H.; Lim, D.-S.; Kim, J. (2017). Reduction of Overfitting in Diabetes Prediction Using Deep Learning Neural Network. In *IT Convergence and Security*; Springer: Berlin/Heidelberg, Germany, pp. 35–43.

Asri, H.; Mousannif, H.; Al Moatassime, H.; Noel, T. (2016). Using Machine Learning Algorithms for Breast Cancer Risk Prediction and Diagnosis. *Procedia Computer Science*, 83, 1064–1069.

Bhadra,S and Kumar,C.J. and Bhattacharyya,D.K.( 2024), *Multiview EEG Signal Analysis for Diagnosis of Schizophrenia: An Optimized Deep Learning Approach.*, Multimedia Tools and Applications, Springer, https://doi.org/10.1007/s11042-024-20205-y.

Chaki, J.; Ganesh, S.T.; Cidham, S.; Theertan, S.A. (2020). Machine learning and artificial intelligence based Diabetes Mellitus detection and self-management: A systematic review. *Journal of King Saud University - Computer and Information Sciences*, 34, 3204–3225.

Charleonnan, A.; Fufaung, T.; Niyomwong, T.; Chokchueypattanakit, W.; Suwannawach, S.; Ninchawee, N. (2016). Predictive analytics for chronic kidney disease using machine learning techniques. In *Proceedings of the 2016 Management and Innovation Technology International Conference (MITicon)*, Bang-San, Thailand; pp. MIT-80–MIT-83.

Jaehee Chun, Justin C. Park, Sven Olberg et al. (2021). Intentional deep overfit learning (IDOL): A novel deep learning strategy for adaptive radiation therapy in breast cancer. *Medical Physics*, 48.9, 5161–5171.

Das, P. R., Talukdar, R. R., & Kumar, C. J. (2024). Exploring the interplay of abstinence self-efficacy, locus of control, and perceived social support in substance use disorder recovery. Current Medical Research and Opinion, 40(9), 1625–1635. https://doi.org/10.1080/03007995.2024.2390046.

Desai, S.D.; Giraddi, S.; Narayankar, P.; Pudakalakatti, N.R.; Sulegaon, S. (2018). Back-Propagation Neural Network Versus Logistic Regression in Heart Disease Classification. In *Advanced Computing and Communication Technologies*; Springer: Berlin/Heidelberg, Germany, pp. 133–144.

Dezaki, F.T.; Liao, Z.; Luong, C.; Girgis, H.; Dhungel, N.; Abdi, A.H.; Behnami, D.; Gin, K.; Rohling, R.; Abolmaesumi, P.; et al. (2018). Cardiac Phase Detection in Echocardiograms With Densely Gated Recurrent Neural Networks and Global Extrema Loss. *IEEE Transactions on Medical Imaging*, 38, 1821–1832.

Eshtay, M.; Faris, H.; Obeid, N. (2018). Improving Extreme Learning Machine by Competitive Swarm Optimization and its application for medical diagnosis problems. *Expert Systems with Applications*, 104, 134–152.

Esmaeilzadeh, S.; Yang, Y.; Adeli, E. (2018). End-to-end Parkinson disease diagnosis using brain MR images by 3D-CNN. *arXiv*, arXiv:1806.05233.

Fitriyani, N.L.; Syafrudin, M.; Alfian, G.; Rhee, J. (2019). Development of disease prediction model based on ensemble learning approach for diabetes and hypertension. *IEEE Access*, 7, 144777–144789.

Fernández-Edreira, D.; Liñares-Blanco, J.; Fernandez-Lozano, C. (2021). Machine Learning analysis of the human infant gut microbiome identifies influential species in type 1 diabetes. *Expert Systems with Applications*, 185, 115648.

Gala, Dhir, et al. (2024). The Role of Artificial Intelligence in Improving Patient Outcomes and Future of Healthcare Delivery in Cardiology: A Narrative Review of the Literature. *Healthcare*, Vol. 12, No. 4, MDPI.

Hamilton-Craig, C.R.; Friedman, D.; Achenbach, S. (2012). Cardiac computed tomography—Evidence, limitations and clinical application. *Heart Lung and Circulation*, 21(2), 70–81.

# Machine Learning-Aided Breast Cancer Detection: Towards Reducing Mortality Rates

**Dhruba Jyoti Borah and Sangita Baruah**

**Abstract** Breast cancer is regarded as the primary cause of death in women. Therefore, timely diagnosis is essential for the patient to recover from it. In this study, a novel supervised machine-learning based-approach is developed to detect breast cancer. A feature-selection algorithm is used to extract the optimal feature from the Wisconsin Breast Cancer (WBC) dataset and then some well-known cutting-edge supervised machine learning models are used to detect cancerous cell nuclei. Finally, a voting ensemble algorithm is used to enhance the detection accuracy of the proposed approach. The approach delivers 98.07% accuracy, 98.09% precision, 96.7% recall and 97.39% F1 in differentiating malignant and benign tumours in the WBC dataset.

**Keywords** Wisconsin breast cancer · Recursive feature elimination · Machine learning

## 1 Introduction

Breast cancer significantly impacts on women's life causing causality of millions worldwide each year. According to Globocan 2018, breast cancer is one of the most common forms of cancer diagnosed in women globally, approximately one in four female cancer cases (Piñeros et al. 2017). As per the statistics provided by WHO (World Health Organization), around two million new cases were identified in 2018 (Kumari and Verma 2019). This statistic shows the urgent need for detection methods of breast cancer to prevent the mortality rates.

Identifying breast cancer in early stage is crucial for reducing death numbers. Breast cancer can be successfully treatable when it is identified in initial stage.

D. J. Borah (✉)
Majuli University of Culture, Chitadarchuck, Majuli, Assam 785104, India
e-mail: dhrubajyotiborah209@gmail.com

S. Baruah
Cotton University, Guwahati, Assam, India



T. P. Singh et al. (eds.), *Revolutionizing Healthcare: Impact of Artificial Intelligence on Diagnosis, Treatment, and Patient Care*, Studies in Computational Intelligence 1182,
https://doi.org/10.1007/978-3-031-80813-5_6

Unfortunately, there is subtle or no symptoms in the early-stage breast cancer, which causes a challenging task for early detection.

Mammograms have been a real lifesaver for countless women when it comes to catching breast cancer early. It has helped a lot of people to survive this terrible disease. However, mammograms can be less effective at detecting cancer in women with dense breast tissue. Furthermore, mammograms involve exposure to low levels of radiation, which may be a cause of concern for women with a lower likelihood of developing breast cancer.

Researchers are trying to find better, faster ways to detect breast cancer. This leads to the advancements of medical imaging. New techniques like contrast-enhanced spectral mammography and tomosynthesis show a promising picture of breast tissue, which improves the detection process of breast cancer. Numerous experiments have been performed using artificial Intelligence (AI) algorithms in the research area of breast cancer identification to analyse mammograms and other imaging data, which may help the radiologists to identify the subtle abnormalities.

The implementation of supervised machine learning (ML) algorithms offers promising advancements in the breast cancer diagnosis. ML techniques have been utilising to analyse medical data and enhance their decision-making processes (Litjens et al. 2017). This implies the ML to become a significant part of the future of breast cancer detection.

This study analyses the limitations of current breast-cancer identification and detection methods and explores the potentiality of cutting-edge supervised ML algorithms to enhance breast cancer tumour classification. A novel machine learning algorithm-based approach is proposed for isolating malignant and benign breast tumours. The Wisconsin Breast Cancer (WBC) dataset is used for model validation. Recursive Feature Elimination (RFE) is used to optimize feature selection. Subsequently, state-of-the-art machine learning algorithms are implemented for malignant tumour detection. Finally, a voting ensemble technique is incorporated to improve the overall performance of the proposed method.

The remaining of this paper is organised as: Sect. 2 presents a survey and review of recently developed malignant breast cancer tumour detection methodologies, Sect. 3 details the proposed machine learning-based detection approach for breast cancer tumour, Sect. 4 discusses the experiments done and the results obtained. Finally, Sect. 5 concludes the paper by outlining potential future research directions in breast cancer tumour detection research.

## 2 Related Work

Muhammet (AK 2020) explored machine learning for malignant breast tumour detection using logistic regression, k-nearest neighbours, and other algorithms. The Wisconsin breast cancer (WBC) dataset (Wolberg et al.1992) was used for model validation. They investigated the algorithms' performance on datasets with varying feature sets (all features, highly correlated, low correlated). Logistic regression achieved the highest accuracy (98.1%) when all features are used. However, the evaluation relied solely on accuracy, omitting metrics like precision and recall.

In Elsadig et al. (2023), the authors utilized 8 ML algorithms to detect the malignant breast tumour. Five feature-selection algorithms, i.e. ANOVA, ReliefF, Gini index, Chi-square (X2) and Gain ratio are used to reduce the irrelevant features from the WBC dataset, which reduces features to 17 features. KNN, DT, SVM, Random Forest, MLP (neural network), NB, Logistic regression and stack are used to train and test the dataset. According to the author, SVM outperforms other classifiers and attained a classification accuracy of 97.7% However, the evaluation relied solely on accuracy, omitting metrics like precision and recall.

Sri Hari Nallamala et. al use 16 features of the WBC datasets as input (Nallamala et al. 2019). Then, logistic regression, KNN and SVM are used to classify the malign and benign breast tumour. Finally, voting ensemble is used to predict the final output, where the model delivers 98.50% precision. However, the author does not show the 16 features, which are used for classification. Furthermore, other performance metrics such as accuracy, recall, F1 are omitted by the author to validate the performance.

Shubham Sharma et al. uses KNN, Random Forest and Naïve Bayes to classify the malign and benign tumour (Sharma et al. 2018). They also use the WBC dataset for validation. According to their observation, KNN performs the best results delivering 95.9% accuracy, 98.27% precision, 90.47% recall and 94.20% F1.

In Mangukiya et al. (2022), the authors use seven supervised machine learning algorithms: Adaboost, XGboost, Naive Bayes (NB), KNN, Random Forest, Support Vector Machine, and Decision Tree. WBC dataset is used for validation. Initially, data are pre-processed and converted into standard scale. Then the method uses ML algorithms for classification. According to the author, XGboost delivers the best result with 98.24% accuracy. However, the author used only accuracy metric to evaluate the performance of their model omitting other important performance metrices such as, precision, recall, F1.

Chowdhury et al. use five ML algorithms, i.e. KNN, Naive Bayes, Decision Tree, SVM, Logistic Regression to detect malignant breast cancer tumour (Chowdhury 2020). The author uses WBC dataset. After implementing the model, the author observes that, Random Forest classifier delivers the best result with 96.5% accuracy. However, they have not used other performance metrics such as precision, recall and F1 to evaluate the performance of the model.

In Jain and Singla 2023), the author uses WBC dataset, where the model is constructed using several ML algorithms like Decision Tree, SVM, Logistic Regression, Random Forest, KNN, etc. Here data is pre-processed to reduce the noise and then applied into the ML algorithms to classify the malignant and benign breast tumour. According to the author, SVM delivers the best result with 97.2% accuracy and 97% precision, recall and F1. However, the author did not clearly mention the steps to perform the pre-processing, feature extraction and selection.

In Khalid et al. (2023), 6 different categorisation models for the detection of breast cancer, including logistic regression (LR), linear support vector classifier (linear SVC), decision tree (DT), support vector classifier (SVC), k-nearest neighbors (KNN), and random forest (RF) are used. The dataset used to validate the model contains 3002 merged pictures from 1501 individuals who had performed digital mammography between February 2007 and May 2015. For feature selection, 3 distinct algorithms are used: recursive feature elimination, low-variance features, and univariate feature selection. Then the ML algorithms are used to classify the benign and malignant breast tumours. According to the authors, RF algorithm delivers the best accuracy of 96.49%. However, the model is evaluated using only accuracy, omitting precision, recall and F1.

# 3 Methodologies

The proposed method is a three-phase methodology for detecting malignant breast tumours using machine learning algorithms. The methodology leverages the strengths of various algorithms and feature sets to achieve robust classification performance.

The first phase focuses on identifying the most informative features from a relevant breast cancer dataset. Recursive Feature Elimination (RFE) is employed to iteratively remove the least informative features based on a chosen evaluation metric. To optimize RFE performance, a parameter tuning step is implemented. The dataset is parsed multiple times with different RFE parameter settings, resulting in a collection of distinct feature sets.

In the second phase, the tumours are classified as malignant or benign. Here, various machine learning algorithms are applied to each feature set generated in phase 1. This will evaluate different algorithms on diverse feature representations of the data. Each feature set is parsed as input as separate instance of the chosen machine learning algorithms.

Given the use of multiple ML algorithms with distinct feature sets in Phase 2, an ensemble learning approach is employed in Phase 3 to harness the collective capabilities of these models. The outputs of individual algorithms for each feature set are parsed as inputs to the ensemble learning algorithm. The ensemble algorithm then produces a final set of classifications. The final classification for the proposed model is selected on the basis of the best performing ensemble output.

A graphical representation of our methodology is depicted in the Fig. 1.

## *3.1 Feature Selection Using Recursive Feature Elimination (RFE)*

In the proposed approach, Recursive Feature Elimination (RFE) algorithm is used for feature selection. This algorithm is particularly advantageous with datasets that have a high dimensionality (large number of features). RFE algorithm can identify optimal subset of features efficiently, which leads to an efficient and robust model

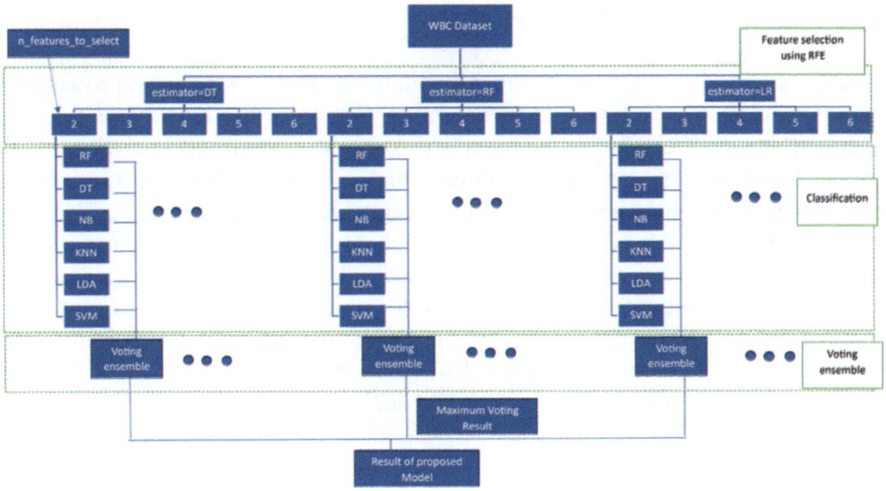


**Fig. 1** Architecture of our proposed method

training and prediction processes. For selecting the most relevant features, RFE algorithm utilizes an iterative feature elimination process (Chen and Jeong 2007). At the beginning, the model is trained using the entire set of available features. An importance score is calculated for each feature, which quantifies its contribution to the model's performance. Features that are least important will score less and are eliminated from the dataset subsequently. This process is repeated with all the remaining features until the targeted number of features is achieved. This helps in retaining the most informative and reduced subset of features useful for creating a robust model (Misra and Yadav 2020).

## 3.2 *Machine-Learning Algorithms Utilized for Classification in Our Proposed Model*

The proposed approach utilises six different machine learning algorithms to differentiate malignant breast tumours: Random Forest, Gaussian Naïve Bayes, Decision Tree, K-Nearest Neighbor (KNN), Support Vector Machine, and Linear Discriminant Analysis.

### 3.2.1 Random Forest (RF)

In machine learning, Random Forest algorithm stands out as a method to achieve precise predictions. This technique combines the strengths of multiple trees by creating Decision Trees during the training process (Venkatesh and Anuradha 2019). Known for its effectiveness in handling complex datasets, this versatile method performs well in scenarios that require accurate forecasts. A Random Forest is constructed by a group of decision trees which are trained using bagging method, on a subset of the training data selected randomly. To enhance diversity among trees only a random subset of features is evaluated for splitting at every node of each decision tree. In classification tasks the ultimate prediction is based on a majority vote, among the individual tree predictions. This ensemble strategy effectively addresses overfitting issues commonly associated with decision trees.

### 3.2.2 Decision Tree (DT)

DTs are learning methods commonly employed in tasks like regression and classification. Their effectiveness, ease of understanding and straightforwardness make them quite popular in scenarios where it is crucial to grasp the logic behind the model. A decision tree is like a flowchart—it has nodes, that represent decisions based on features and leaf nodes indicating the class label denoting the target value (Maimon 2005). The tree construction involves iteratively splitting the training data into homogeneous subsets at each node that better identifies the target classes. Initially the tree construction process begins at the root node. The root node represents the whole dataset. To take a decision and split, attributes of each node are compared with the decision rules, until it reaches to the leaf node that determines the classification results (Mittal et al. 2017). Due to their clear decision-making structure, interpretability of decision trees is very high, which makes them a valuable tool for explaining a model's behaviour.

### 3.2.3 Gaussian Naive Bayes (NB)

Gaussian Naive Bayes (NB) is a supervised ML algorithm, which is basically utilized for its simplicity, speed, and capability to handle high-dimensional data (Chen and Jeong 2007). This method of intuitive classification uses past knowledge of characteristics (conditions) that are thought to be significant to predict the likelihood of an event (class membership). The underlying premise of NB is feature independence, which states that every feature independently affects the total class probability. In actuality, the NB classifier applies the Bayes theorem to get the posterior probability (P(A|B)) for each class (A) given the observed features (B):

$$P(A|B) = \frac{P(B|A)P(A)}{P(B)} \tag{1}$$

where P(B|A) represents the likelihood of features given the class, and P(A) is the prior probability of the class, and P(B) is the probability of observing the features. Notably, Gaussian NB assumes features follow a Gaussian distribution and estimates the mean and standard deviation of every feature and for every class during training.

### 3.2.4 K-Nearest Neighbors (KNN)

KNN is a popular non-parametric supervised learning technique for regression and classification applications (Razavi-Termeh et al. 2024). Its ease of use comes from its ability to categorise data points according to how similar they are to their k-nearest neighbours in the training set. KNN is useful for non-linear and complex data since it works on the premise that related data points are likely to share class labels. KNN focuses its predictions on instance similarity and avoids making significant assumptions about the underlying data distribution because it is a non-parametric approach. To quantify the similarity among data points, it uses distance metrics such as the Euclidean distance (alternatively Manhattan or Minkowski distance). Based on the selected distance metric, KNN classifies a new data by identifying its k closest neighbours in the training dataset and assigning the majority class label (Imandoust and Bolandraftar 2013). This approach has several benefits, such as: it requires less pre-processing of training data, is flexible enough to work with different kinds of data, and can adjust to intricate decision limits. KNN is an invaluable instrument in machine learning and data mining applications such as anomaly detection, recommendation systems, and pattern recognition due to its simplicity to use, interpretability, and efficacy for a variety of tasks.

### 3.2.5 Linear Discriminant Analysis (LDA)

It is used for dimensionality reduction and topic modelling more often, but Linear Discriminant Analysis (LDA) is also applicable to classification tasks (FISHER 1936). LDA projects data points to a lower dimensional space that maximizes the separation between different classes unlike KNN that focuses on nearest neighbours only. As such, LDA can predict the class label of a new query point from its location in the transformed space so as to enable development of an efficient decision boundary that separates classes effectively (Kleinbaum et al. 1998). This method is particularly suitable for those datasets with well-defined classes arranged in a linear hierarchy. Because of this, LDA is useful in facial recognition, biology, finance and other high dimensional data processing applications.

### 3.2.6 Support Vector Machine (SVM)

SVMs are learning algorithms that allow supervised classification and regression. SVMs are known for their versatility and effectiveness especially in relation to manage high-dimensional data and complex non-linear relations (Wang and Hu 2005). One of the main strengths of this algorithm is their ability to locate a perfect hyperplane in the feature space that maximizes the margin among different class labels. Margin is an area from the hyperplane to the nearest points belonging to each class; they are called support vectors as well. By maximizing this margin, SVMs develop strong models that can generalize well on unseen data. One advantage of this algorithm is that it can work with nonlinear data. The way in which it does this is through use of several kernel functions such as radial or polynomial basis functions. These kernels help in converting input space into higher dimensional spaces where linear separations between classes may become possible. Nevertheless, SVM can be quite expensive when running big datasets due optimization process involved in finding maximum-margin hyperplane(s). Additionally, SVM performance depends on choice of kernel function as well as its parameters. However, given their effectiveness, interpretability (in case of linear SVM) and ability to handle high-dimensional data, they have remained popular for many applications such as text classification, image recognition, bioinformatics and finance across different fields or industries over time.

### *3.3 Majority Voting as Ensemble Learning Algorithm*

Ensemble methods, a cornerstone of machine learning, combine predictions from diverse models to achieve superior performance and generalization (Borah and Sarma 2023). Majority voting, a powerful yet straightforward ensemble technique for classification tasks, exemplifies this approach. It leverages predictions from multiple base models, each independently classifying a data point. The ultimate prediction is chosen by the class receiving the highest votes, essentially a democratic choice among the models (Baruah et al. 2023). This method harnesses the collective wisdom of the ensemble, potentially surpassing the accuracy and robustness of any single model. Crucially, diverse base models with complementary strengths and weaknesses are essential for robust majority voting. This diversity mitigates individual model biases and errors, leading to improved generalization performance. Consequently, majority voting is a widely employed technique across various machine learning domains, including both classification and regression tasks.

# 4 Experiments and Results

## *4.1 Dataset*

The Wisconsin Breast Cancer (WBC) dataset (Wolberg et al. 1993) is used for the purpose of validating the proposed methodology. It is a publicly available well-known dataset used for research in diagnosis of breast cancer. The WBC dataset consists of 569 samples, each sample has 30 features and a corresponding class label (malignant or benign). There is an imbalanced class distribution, with 357 samples (62.75%) belong to the benign class and 212 samples (37.25%) belong to the malignant class. Table 1 shows a detailed description of WBC dataset.

**Table 1** Description of dataset

| Sl. No. | Feature name | Sample value | Sl. No. | Feature name | Sample value |
|---|---|---|---|---|---|
| 1 | id | 842,302 | 17 | smoothness_se | 0.0064 |
| 2 | diagnosis | M | 18 | compactness_se | 0.04904 |
| 3 | radius_mean | 17.99 | 19 | concavity_se | 0.05373 |
| 4 | texture_mean | 10.38 | 20 | concave points_se | 0.01587 |
| 5 | perimeter_mean | 122.8 | 21 | symmetry_se | 0.03003 |
| 6 | area_mean | 1001 | 22 | fractal_dimension_se | 0.00619 |
| 7 | smoothness_mean | 0.1184 | 23 | radius_worst | 25.38 |
| 8 | compactness_mean | 0.2776 | 24 | texture_worst | 17.33 |
| 9 | concavity_mean | 0.3001 | 25 | perimeter_worst | 184.6 |
| 10 | concave points_mean | 0.1471 | 26 | area_worst | 2019 |
| 11 | symmetry_mean | 0.2419 | 27 | smoothness_worst | 0.1622 |
| 12 | fractal_dimension_mean | 0.07871 | 28 | compactness_worst | 0.6656 |
| 13 | radius_se | 1.095 | 29 | concavity_worst | 0.7119 |
| 14 | texture_se | 0.9053 | 30 | concave points_worst | 0.2654 |
| 15 | perimeter_se | 8.589 | 31 | symmetry_worst | 0.4601 |
| 16 | area_se | 153.4 | 32 | fractal_dimension_worst | 0.1189 |

## 4.2 Results

### 4.2.1 Phase I: Feature Selection Using RFE Algorithm

The proposed methodology starts with feature selection on the WBC dataset using the Recursive Feature Elimination (RFE) algorithm. Here, two key parameters of RFE are tuned to optimise the feature selection process: estimator and n_features_to_select.

- **estimator**: This parameter determines the supervised learning algorithm used to evaluate feature importance. Three algorithms were employed to perform this: Logistic Regression (LR), Random Forest (RF), and Decision Tree (DT).
- **n_features_to_select**: This parameter specifies the desired number of features to be retained after the RFE process. Here, we investigated a range of values from 2 to 6, aiming to identify the optimal number of features for accurate breast tumour detection.

Table 2 presents the results obtained with the RFE algorithm using DT as the estimator, evaluating different values for n_features_to_select.

Table 3 demonstrates the results of **n_features_to_select** of RFE algorithm when RF is used as estimator.

Table 4 demonstrates the results of **n_features_to_select** of RFE algorithm when LR is used as estimator.

**Table 2** RFE results when estimator = Decision Tree

| Number of features | Features | | | | | |
|---|---|---|---|---|---|---|
| 2 | Mean concave points | Worst perimeter | | | | |
| 3 | Mean concave points | Worst texture | Worst perimeter | | | |
| 4 | Mean concave points | Worst texture | Worst perimeter | Worst concave points | | |
| 5 | Mean concave points | Worst radius | Worst texture | Worst perimeter | Worst concave points | |
| 6 | Mean concave points | Fractal dimension error | Worst radius | Worst texture | Worst perimeter | Worst concave points |

**Table 3** RFE results when estimator = Random Forest

| Number of features | Features | | | | | |
|---|---|---|---|---|---|---|
| 2 | Mean concave points | Worst perimeter | | | | |
| 3 | Mean concave points | Worst perimeter | Worst concave points | | | |
| 4 | Mean concave points | Worst perimeter | Worst area | Worst concave points | | |
| 5 | Mean concave points | Worst radius | Worst perimeter | Worst area | Worst concave points | |
| 6 | Mean concavity | Mean concave points | Worst radius | Worst perimeter | Worst area | Worst concave points |

**Table 4** RFE results when estimator = Logistic Regression

| Number of features | Features | | | | | |
|---|---|---|---|---|---|---|
| 2 | Worst compactness | Worst concavity | | | | |
| 3 | Texture error | Worst compactness | Worst concavity | | | |
| 4 | Texture error | Worst radius | Worst compactness | Worst concavity | | |
| 5 | Mean radius | Texture error | Worst radius | Worst compactness | Worst concavity | |
| 6 | Mean radius | Texture error | Worst radius | Worst compactness | Worst concavity | Worst symmetry |

### 4.2.2 Phase II: Classification of Breast Cancer Tumour

The above feature selection experiment has provided a set of features containing 15 instances. However, it is not clear which instance should be used for getting the best classification result. Therefore, all the 15 instances have been applied separately to the machine learning algorithm in phase II.

In the phase II, 6 supervised ML algorithms are employed for classification of breast cancer benign and malignant tumour namely Decision Tree (DT), Gaussian Naive Bayes (NB), Random Forest (RF), Linear Discriminant Analysis (LDA), Support Vector Machine (SVM) and K-Nearest Neighbors (KNN). Before deploying

**Table 5** Decision Tree parameters to tune, their ranges and the optimal value

| Parameter | Range | Optimal value |
|---|---|---|
| criterion | 'gini', 'entropy' | 'entropy' |
| splitter | 'best', 'random' | 'best' |
| max_depth | None, 5, 10, 15, 20 | 10 |
| min_samples_split | 2, 3, 4, 5, 6, 7, 8, 9, 10 | 7 |
| min_samples_leaf | 1, 2, 3, 4, 5, 6, 7, 8, 9, 10 | 9 |
| max_features | 'auto', 'sqrt', 'log2', None | None |

these ML algorithms, their parameters have been tuned and have utilised optimal parameter of all the algorithms. The Table 5 shows the parameters to tune of DT with their ranges and also shows the optimal values.

Table 6 shows the parameters to tune of NB with their ranges and also shows the optimal values.

Table 7 shows the parameters to tune of KNN with their ranges and also shows the optimal values.

Table 8 shows the parameters to tune of KNN with their ranges and also shows the optimal values.

Table 9 shows the parameters to tune of KNN with their ranges and also shows the optimal values.

Table 10 shows the parameters to tune of KNN with their ranges and also shows the optimal values.

Figures 2, 3, 4, and 5 present the performance metrics (accuracy, precision, recall, and F1-score) of the RFE algorithm using a decision tree (DT) as the estimator. The impact of the n_features_to_select parameter (2, 3, 4, 5, and 6) have been evaluated by varying on these metrics. Since, the DT-based RFE achieved superior performance

**Table 6** NB parameters to tune, their ranges and the optimal value

| Parameter | Range | Optimal value |
|---|---|---|
| 'priors' | None, 0.2, 0.3, 0.5 | None |
| 'var_ smoothing' | 1.00000000e−09, 1.00000000e−08, 1.00000000e−07, 1.00000000e−06,1.00000000e−05, 1.00000000e−04, 1.00000000e−03, 1.00000000e−02, 1.00000000e−01, 1.00000000e+00 | 1.00000000e−09 |

**Table 7** KNN parameters to tune, their ranges and the optimal value

| Parameter | Range | Optimal value |
|---|---|---|
| 'n_neighbors' | 1, 2, 3, 4, 5, 6, 7, 8, 9, 10, 11, 12, 13, 14, 15, 16, 17, 18, 19, 20 | 7 |
| 'weights' | 'uniform', 'distance' | 'distance' |
| 'algorithm' | 'auto', 'ball_tree', 'kd_tree', 'brute' | 'auto' |
| 'p' | 1, 2 | 1 |

**Table 8** LDA parameters to tune, their ranges and the optimal value

| Parameter | Range | Optimal value |
|---|---|---|
| solver | 'svd', 'lsqr', 'eigen' | Eigen |
| shrinkage | None, 'auto', 0.1, 0.5, 0.9 | 0.9 |
| priors | None, 0.2, 0.3, 0.5 | None |
| n_components | None, 1, 2 | None |
| store_covariance | True, False | True |
| tol | 1e−4, 1e−3, 1e−2 | 1e−2 |

**Table 9** RF parameters to tune, their ranges and the optimal value

| Parameter | Range | Optimal value |
|---|---|---|
| n_estimators | 10, 20, 30, 40, 50, 60, 70, 80, 90, 100 | 60 |
| min_samples_split | 2, 3, 4, 5, 6, 7, 8, 9, 10 | 6 |
| min_samples_leaf | 1, 2, 3, 4, 5, 6, 7, 8, 9, 10 | 6 |
| max_features | 'auto', 'sqrt', 'log2', None | sqrt |
| Bootstrap | True, False | True |
| Max_depth | 5, 10, 15, 20, None | 20 |

**Table 10** SVM parameters to tune, their ranges and the optimal value

| Parameter | Range | Optimal value |
|---|---|---|
| C | 1.00000000e−03, 1.00000000e−02, 1.00000000e−01, 1.00000000e+00, 1.00000000e+01, 1.00000000e+02, 1.00000000e+03 | 1.00000000e+00 |
| max_iter | 100, 500, 1000 | 500 |
| class_ weight | None, 'balanced' | Balanced |
| Tol | 1.e−04, 1.e−03, 1.e−02, 1.e−01, 1.e+00 | 1.e−02 |
| Probability | True, False | True |
| Shrinking | True, False | True |
| coef0 | − 1, − 0.9, − 0.8, − 0.7, − 0.6, − 0.5, − 0.4, − 0.3, − 0.2, − 0.1, 0, 0.1, 0.2, 0.3, 0.4, 0.5, 0.6, 0.7, 0.8, 0.9, 1 | − 0.1 |
| Kernel | 'linear', 'poly', 'rbf', 'sigmoid' | Linear |
| Gamma | 'scale', 'auto', 0.001, 0.01, 0.1, 1.0, 10.0, 100.0, 1000.0 | 10.0 |
| degree | 1, 2, 3, 4, 5 | 4 |

compared to other estimators (e.g., Random Forest and Logistic Regression), only these results are presented for brevity.

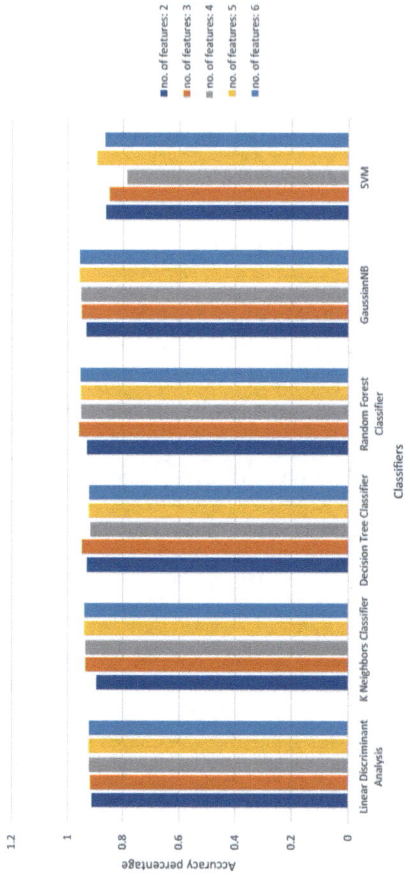


**Fig. 2** Accuracies of RFE algorithm while DT is used as feature selection estimator

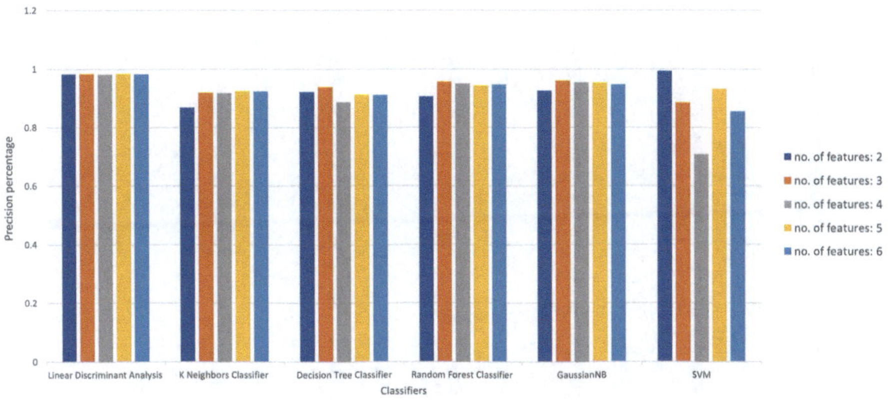


**Fig. 3** Precision of RFE algorithm while DT is used as feature selection estimator

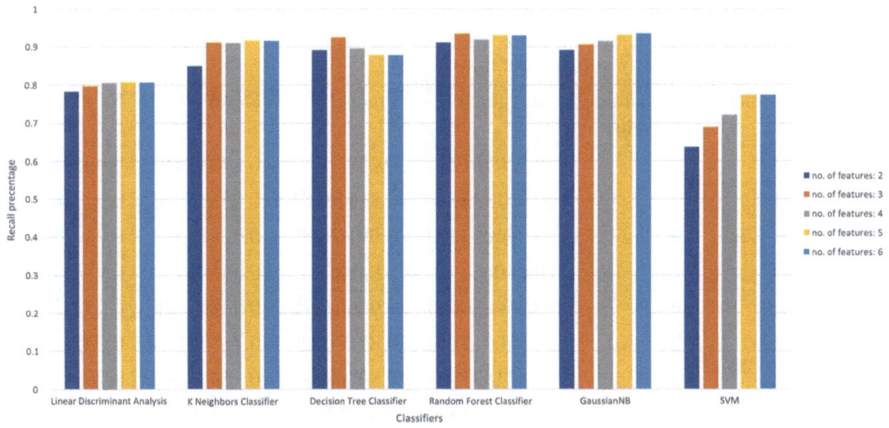


**Fig. 4** Recall of RFE algorithm while DT is used as feature selection estimator

### 4.2.3 Phase III: Ensemble Learning for Performance Enhancement

In order to enhance the predictive performance of the proposed method, the majority voting ensemble algorithm is implemented. This approach leverages the predictions from six machine learning (ML) classifiers, aiming to achieve superior overall performance compared to any single model.

In this phase, the results of the majority voting on all instances were obtained. Table 11 displays the experimental results of majority voting ensemble of all the instances of the proposed method.

Table summarizes the performance of the ensemble model using different base estimators and feature selections (n_features_to_select). As evident from the table, the decision tree (DT) estimator achieved the best overall performance with 5 selected

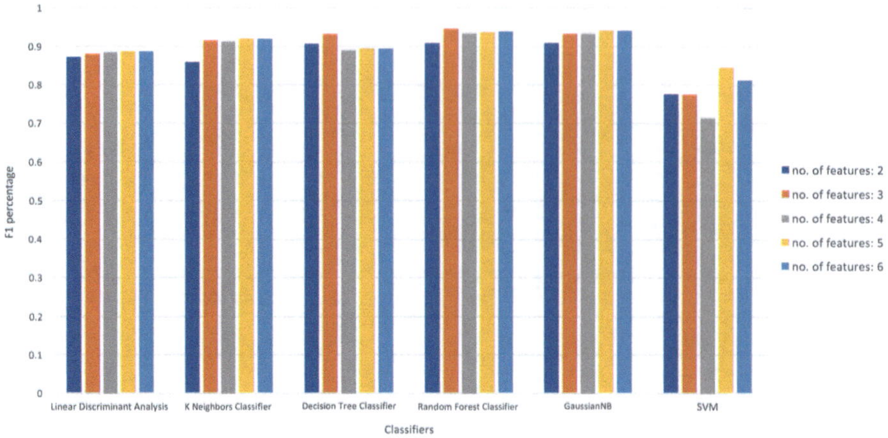


**Fig. 5** F1 of RFE algorithm while DT is used as feature selection estimator

**Table 11** Results of majority voting ensembles

| Sl. No | Estimator | n_features_ to_select | Accuracy | Precision | Recall | F1 |
|---|---|---|---|---|---|---|
| 1 | DT | 2 | 0.92970123 | 0.93 | 0.87735849 | 0.90291262 |
| 2 | DT | 3 | 0.961335677 | 0.96116505 | 0.93396226 | 0.94736842 |
| 3 | DT | 4 | 0.956063269 | 0.95169082 | 0.92924528 | 0.94033413 |
| 4 | DT | 5 | 0.980668 | 0.980861 | 0.966981 | 0.973872 |
| 5 | DT | 6 | 0.964850615 | 0.96601942 | 0.93867925 | 0.95215311 |
| 6 | LR | 2 | 0.850615114 | 0.80676329 | 0.78773585 | 0.79713604 |
| 7 | LR | 3 | 0.841827768 | 0.80808081 | 0.75471698 | 0.7804878 |
| 8 | LR | 4 | 0.947275923 | 0.94174757 | 0.91509434 | 0.92822967 |
| 9 | LR | 5 | 0.949033392 | 0.94634146 | 0.91509434 | 0.93045564 |
| 10 | LR | 6 | 0.964850615 | 0.96601942 | 0.93867925 | 0.95215311 |
| 11 | RF | 2 | 0.92970123 | 0.93 | 0.87735849 | 0.90291262 |
| 12 | RF | 3 | 0.933216169 | 0.92647059 | 0.89150943 | 0.90865385 |
| 13 | RF | 4 | 0.945518453 | 0.94581281 | 0.90566038 | 0.9253012 |
| 14 | RF | 5 | 0.943760984 | 0.94554455 | 0.9009434 | 0.92270531 |
| 15 | RF | 6 | 0.942003515 | 0.93236715 | 0.91037736 | 0.92124105 |

features. In this configuration, the ensemble model attained an accuracy of 98.07%, precision of 98.09%, recall of 96.70%, and F1-score of 97.39%. While other estimators (Logistic Regression—LR, Random Forest—RF) yielded promising results with different feature selections (LR: 6 features, RF: 4 features), the DT-based ensemble consistently surpassed them. Therefore, the model with DT as estimator with 5

**Table 12** Final result of our proposed approach

| Accuracy | Precision | Recall | F1 | TP | TN | FP | FN |
|---|---|---|---|---|---|---|---|
| 98.07% | 98.09% | 96.7% | 97.39% | 205 | 353 | 4 | 7 |

features is considered the optimal configuration for the proposed method, and its detailed performance metrics are displayed in Table 12.

## 5 Conclusion

In this work, the potentiality of supervised ML algorithms for improved breast cancer tumour detection was investigated. The limitations of existing methods were analysed and a novel strategy is proposed for the identification of breast cancer tumour. The proposed strategy starts with utilising RFE algorithm for optimal feature selection, followed by the implementation of a set of cutting-edge machine learning algorithms and a voting ensemble technique to enhance detection accuracy. The findings of this study shows that, the ML algorithms exhibit a significant role for the identification of breast cancer tumour classification. The proposed approach delivers 98.07% accuracy, 98.09% precision, 96.7% recall and 97.39% F1 in differentiating malignant and benign tumours in the WBC dataset.

Further research is necessary to refine and validate the proposed approach in larger datasets. Additionally, experiment of integrating this approach with existing detection techniques could be more beneficial to detect breast cancer tumour.

## References

Ak, M. F. (2020). A comparative analysis of breast cancer detection and diagnosis using data visualization and machine learning applications. In Healthcare (Vol. 8, No. 2, p. 111). MDPI.

Baruah, S., Borah, D. J., & Deka, V. (2023). Detection of peer-to-peer botnet using machine learning techniques and ensemble learning algorithm. International Journal of Information Security and Privacy (IJISP), 17(1), 1–16.

Borah, D. J., & Sarma, A. (2023). DETECTION OF PEER-TO-PEER BOTNETS USING GRAPH MINING. International Journal of Computer Networks & Communications (IJCNC), 15, 105–125.

Chen, X. W., & Jeong, J. C. (2007). Enhanced recursive feature elimination. In Sixth international conference on machine learning and applications (ICMLA 2007) (pp. 429–435). IEEE.

Chowdhury, A. (2020). Breast cancer detection and prediction using machine learning.

Elsadig, M. A., Altigani, A., & Elshoush, H. T. (2023). Breast cancer detection using machine learning approaches: a comparative study. International Journal of Electrical & Computer Engineering (2088–8708), 13(1).

Fisher, R. A. (1936). The use of multiple measurements in taxonomic problems. Annals of eugenics, 7(2), 179–188.

Imandoust, S. B., & Bolandraftar, M. (2013). Application of k-nearest neighbor (knn) approach for predicting economic events: Theoretical background. International journal of engineering research and applications, 3(5), 605–610.

Jain, B., & Singla, N. (2023). Breast Cancer Detection using Machine Learning Algorithms. Journal of Computers, Mechanical and Management, 2(6), 30–35.

Khalid, A., Mehmood, A., Alabrah, A., Alkhamees, B. F., Amin, F., AlSalman, H., & Choi, G. S. (2023). Breast cancer detection and prevention using machine learning. Diagnostics, 13(19), 3113.

Kleinbaum, D. G., Kupper, L. L., Muller, K. E., & Nizam, A. (1998). Applied regression analysis and other multivariable methods (3rd ed.). Thomson Brooks/Cole Publishing Co.

Kumari, N., & Verma, K. (2019). A SURVEY ON VARIOUS MACHINE LEARNING APPROACHES USED FOR BREAST CANCER DETECTION. International Journal of Advanced Research in Computer Science, 10(3).

Litjens, G., Kooi, T., Bejnordi, B. E., Setio, A. A. A., Ciompi, F., Ghafoorian, M., ... & Sánchez, C. I. (2017). A survey on deep learning in medical image analysis. Medical image analysis, 42, 60–88.

Maimon O., Rokach (2005), Data Mining and Knowledge Discovery Handbook.

Mangukiya, M., Vaghani, A., & Savani, M. (2022). Breast cancer detection with machine learning. International Journal for Research in Applied Science and Engineering Technology, 10(2), 141–145.

Misra, P., & Yadav, A. S. (2020). Improving the classification accuracy using recursive feature elimination with cross-validation. Int. J. Emerg. Technol, 11(3), 659–665.

Mittal, K., Khanduja, D., & Tewari, P. C. (2017). An insight into 'decision tree analysis'. World Wide Journal of Multidisciplinary Research and Development, 3(12), 111–115.

Nallamala, S. H., Mishra, P., & Koneru, S. V. (2019). Breast cancer detection using machine learning way. Int J Recent Technol Eng, 8(2–3), 1402–1405.

Piñeros, M., Znaor, A., Mery, L., & Bray, F. (2017). A global cancer surveillance framework within noncommunicable disease surveillance: making the case for population-based cancer registries. *Epidemiologic reviews*, *39*(1), 161–169.

Razavi-Termeh, S. V., Sadeghi-Niaraki, A., Razavi, S., & Choi, S. M. (2024). Enhancing flood-prone area mapping: fine-tuning the K-nearest neighbors (KNN) algorithm for spatial modelling. *International Journal of Digital Earth*, *17*(1), 2311325.

Sharma, S., Aggarwal, A., & Choudhury, T. (2018, December). Breast cancer detection using machine learning algorithms. In *2018 International conference on computational techniques, electronics and mechanical systems (CTEMS)* (pp. 114–118). IEEE.

Venkatesh, B., & Anuradha, J. (2019). A review of feature selection and its methods. *Cybernetics and information technologies*, *19*(1), 3–26.

Wolberg, W. H., Street, W. N., & Mangasarian, O. L. (1993). Breast cancer Wisconsin (diagnostic) data set. *UCI Machine Learning Repository*.

Wang, H., & Hu, D. (2005, October). Comparison of SVM and LS-SVM for regression. In *2005 International conference on neural networks and brain* (Vol. 1, pp. 279–283). IEEE.

# Advancements in Diabetic Retinopathy Detection Using Deep Learning

**Sonia Sarmah, Manasi Hazarika, Pranab Das, Amal Satheesh, Dhritiraj Barman, and Ankit Kumar**

**Abstract** ***Background***: Diabetic Retinopathy (DR) is an advanced complication of diabetes that damages the retinal blood vessels. If DR is not treated for a long time, it might result in irreversible visual loss or permanent blindness. ***Problem statement***: In the initial stage of DR, the lesions are so minute that the patients do not usually exhibit any symptoms, making early detection challenging. Only specialized eye examinations by experts can monitor the presence of diseases. However, these traditional DR screening methods require a considerable amount of time and labor. Thus, to stop irreversible vision loss, effective and precise automated techniques for early DR identification are required. ***Objective***: This study attempts to capture the recent advancements in DR lesion segmentation, classification and grading using deep learning techniques using fundus images. ***Methods***: Artificial intelligence and computer vision have become effective tools for various automation processes in the health sector including automated disease diagnosis. In the field of DR also, AI-based technologies, especially deep learning has demonstrated promising outcomes by achieving high accuracy for DR detection, stage classification and lesion detection.

S. Sarmah (✉) · P. Das · A. Satheesh · D. Barman
Department of Computer Applications, Assam Don Bosco University, Airport Road, Azara, Assam 781017, India
e-mail: sonia.sarmah@dbuniversity.ac.in

P. Das
e-mail: pranab.das@dbuniversity.ac.in

A. Satheesh
e-mail: amalsatheesh001@gmail.com

D. Barman
e-mail: shillron044@gmail.com

M. Hazarika
Department of Computer Science and Engineering, Assam Don Bosco University, Airport Road, Azara, Assam 781017, India
e-mail: manasi.hazarika@dbuniversity.ac.in

A. Kumar
University of Petroleum and Energy Studies, 248007 Dehradun, India
e-mail: anketvit@gmail.com

 
T. P. Singh et al. (eds.), *Revolutionizing Healthcare: Impact of Artificial Intelligence on Diagnosis, Treatment, and Patient Care*, Studies in Computational Intelligence 1182,
https://doi.org/10.1007/978-3-031-80813-5_7

The paper covers the standardized framework for DR diagnosis using DL, publicly available fundus datasets, preprocessing techniques, DL-based segmentation, classification, and grading methods for DR. ***Results***: Automated DR diagnosis with DL approaches has demonstrated encouraging outcomes. Segmentation approaches such as Mask R-CNN, SegNet, U-Net, and DeepLabv3 have shown high accuracy in DR lesion segmentation. High accuracy rates are also achieved by DL- based classification techniques that employ ensemble models, and transfer learning; with accuracies reported up to 95% and 97% for multi-class and binary classification respectively. ***Conclusion***: Deep learning presents a viable approach to enhance the effectiveness, precision, and accessibility of DR screening. But issues like data availability and model interpretability still exist.

**Keywords** Diabetic retinopathy · Deep learning · Diabetic lesions · DR segmentation · DR classification

## 1 Introduction

Diabetes mellitus is a chronic condition caused by raised sugar levels in the blood. It has grown to be a major global health issue and each year an increasing number of adults are getting affected by it (International Diabetes Federation). Persistent elevation of the blood sugar in diabetes mellitus affects the tiny blood vessels in the retina, the light sensitive region at the rear of the eye. This state, known as Diabetic Retinopathy (DR), is an escalating microvascular issue that may lead to irreversible vision loss or permanent blindness. In fact, DR has been identified as the primary cause of blindness in diabetic patients and making it crucial to detect the condition in early stages (Fisher et al. 2016).

DR is characterized by the existence of different retinal lesions, such as-hard exudates (HEs), soft exudates (SEs), hemorrhages (HMs), and microaneurysms (MAs). It may also be distinguished by alterations and anomalies in the optical disc and retinal vessels (E. T. D. R. S. R. Group 1991; Scanlon et al. 2009; Bhadra et al. 2024). These lesions are important markers of the severity and course of the disease. The visual representation of a normal retina next to a retina with different types of DR lesions is shown in Fig. 1.

- MAs are tiny red spots on the retina that occur due to the microvascular damage caused by diabetes. These are usually the first indicators of DR. However, these lesions are so small that the patients often do not exhibit any discomfort or visual symptoms.
- The presence of HEs indicates more advanced stages of DR, which causes further damage to the blood vessels. These lesions, mainly formed due to the accumulation of fats and proteins that leak from the blood vessels, are typically found in the outer retinal layer, having a hard or waxy appearance with yellowish color.

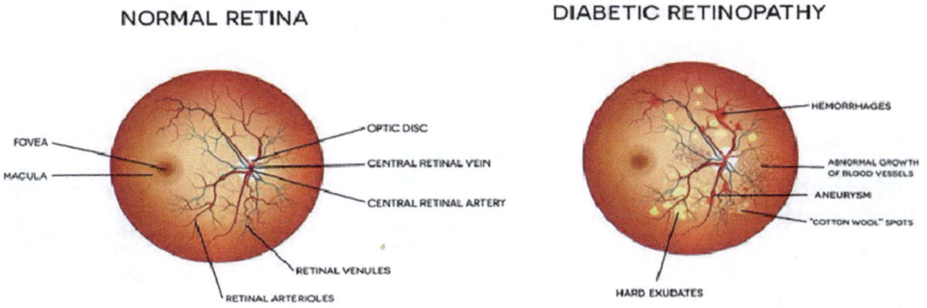


**Fig. 1** Normal retina and retina with various DR lesions (*Source* https://neoretina.com/blog/diabetic-retinopathy-can-it-be-reversed/)

- SEs are fluffy white or greyish retinal lesions, also known as cotton wool spots. These represent the areas of retinal ischemia resulting due to inadequate blood supply. They are mainly caused by the blockages in the tiny retinal blood vessels, triggering localized swelling of nerve fibers.
- HMs in the retina are characterized by the leakage of blood from tiny retinal blood vessels onto the retinal surface. They can vary in size from small dot-like hemorrhages to larger blotches. The size and the severity of the HMs are important markers for DR progression.

DR progresses through various stages that are crucial for its classification and management. Figure 2 presents the various stages of DR progression which can broadly be divided into two classes—(a) non-proliferative DR (NPDR) and (b) proliferative DR (PDR) (Yang et al. 2022). NPDR is the initial stage of DR which is characterized by the presence of various microvascular lesions in the retinas such as MAs, HMs, HEs and SEs. Based on the severity, NPDR can further be classified into three substages namely mild, moderate, and severe NPDR. In mild NPDR, only MAs are present. As the condition advances to the moderate stage, there is a notable increase in MAs and HMs, along with the appearance of intraretinal microvascular abnormalities (IRMAs) that indicate early signs of retinal ischemia. The sever NPDR is marked by the presence of extensive retinal hemorrhages across all retinal quadrants along with venous beading and prominent IRMAs (Wu et al. 2013). PDR stage is characterized by neovascularization, the process of formation of abnormal blood vessels in the retina. These new blood vessels are very weak, delicate and prone to leakage. Blood from these vessels may seep into the vitreous cavity, causing vitreous hemorrhage and retinal detachment. This is a serious condition that exhibits symptoms like—floaters, blurred vision, and in severe cases, complete loss of vision (Ak et al. 2016; Ting et al. 2019). A detailed description of the symptoms associated with different stages of DR is outlined in Table 1.

In the conventional method, DR is diagnosed by a trained ophthalmologist or eyecare specialist by performing dilated fundus exams (Hadziahmetovic et al. 2019). These eye exams also utilize advanced imaging techniques to aid the eyecare specialists. The ophthalmologists or eyecare specialists meticulously examine each image

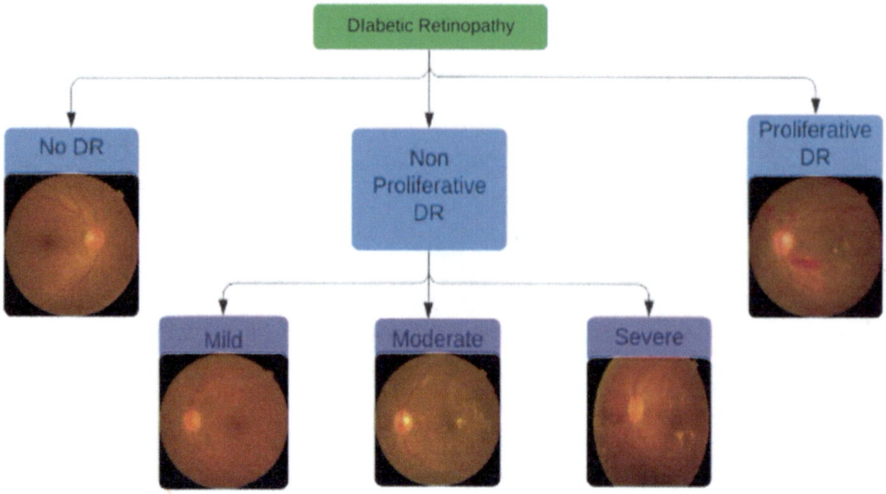


**Fig. 2** Different stages of diabetic retinopathy

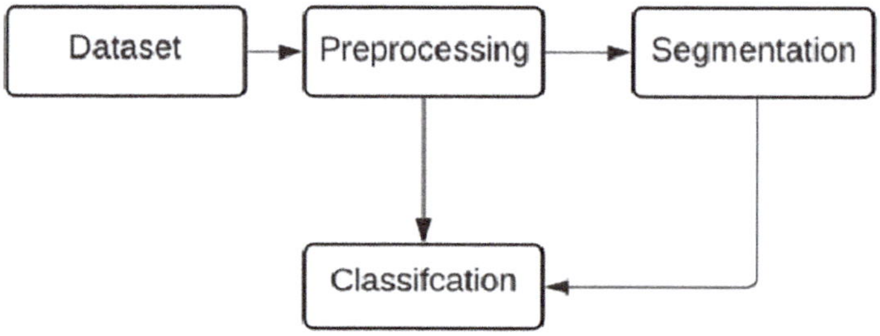


**Fig. 3** Block diagram for data processing pipeline for DR detection

**Table 1** Diabetic retinopathy severity and features

| DR severity | Key features |
|---|---|
| No DR | Absence of microvascular abnormalities |
| Mild NPDR | Presence of MAs only |
| Severe NPDR | Presence of Mas, dot and blot HMs, HEs, SEs; but not severe NPDR |
| PDR | Presence of intraretinal hemorrhages ($\geq$ 20 in each of 4 quadrants), definite venous beading (in at least 2 quadrants), or apparent IRMA (in at least 1 quadrant), but not PDR |
| PDR | Presence of neovascularization, and/or vitreous hemorrhage |

to diagnose and assess the stage of DR, if detected. Thus, regular eye examination is very crucial for the diabetic patients for timely detection of DR. However, this traditional method of manual DR detection is very time-consuming and may cause delay between analysis and treatment initiation (Harper and Keeffe 2012; Zabihollahy et al. 2019). The method is also prone to human error and can be laborious for the eyecare professionals, considering the high volume of patients (Das et al. 2022). Artificial intelligence (AI) has become a game-changing force in healthcare, transforming many facets of the industry. From diagnosis to personalized treatment planning and continuous health monitoring, AI technologies play a vital role in overall healthcare management (Chen et al. 2021; Bajwa et al. 2021). Computer vision, natural language processing, machine learning, and deep learning, are a few of these technologies. In diagnostics, AI-powered systems are used for analyzing and detecting abnormalities in medical images. This helps the healthcare professionals to make accurate and timely diagnoses (Zabihollahy et al. 2019; Keel et al. 2019; Scotland et al. 2010). In the field of DR screening and management also AI-based methods have shown promising results. Several studies have explored the application of AI in diagnosing and managing DR, researchers have increasingly turned to deep learning (DL) techniques to automate the detection and grading of DR (Ronneberger et al. 2015; Yi et al. 2022; Quellec et al. 2017). Deep learning is a subfield of artificial intelligence that focuses on developing algorithms inspired by the human brain's neural networks. These algorithms use multiple layers of nonlinear processing to extract complex features and categorize patterns. DL is widely used in computer-aided medical diagnosis, specifically in medical image analysis including classification, segmentation, detection, retrieval, and registration of medical images (Ker et al. 2017). This study encapsulates the latest advancements in the diagnosis of DR using fundus images, specifically emphasizing the application of DL methods. The rest of the paper includes two more sections. Section 2 presents an overview of the standard framework for DR diagnosis using DL, encompassing data collection, preprocessing, segmentation, and classification showcasing its potential to revolutionize the field of ophthalmology. This section also covers information about some publicly available datasets utilized for DR detection, discusses preprocessing techniques for improving retinal fundus image quality, reviews deep learning-based segmentation methods, and explores DL based classification approaches for DR severity classification. Section 3 summarizes the advancements in DR detection and classification with DL, while also suggesting future research directions to tackle these challenges.

## 2 Diagnosis of DR Using DL

As shown in Fig. 3, a standard framework for diagnosis of DR using DL can be divided into four stages: (i) Input dataset, (ii) Preprocessing, (iii) Segmentation and (iv) Classification (Nadeem et al. 2022). A diverse dataset of retinal fundus images representing various DR stages is collected in the first stage. Many datasets related to DR are also publicly available and can be utilized by researchers in this stage. In the

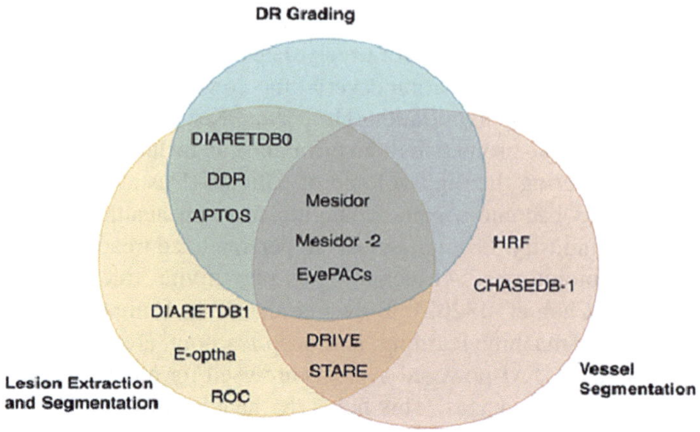


**Fig. 4** Utilization of publicly available DR datasets discussed in this study

second stage, the dataset undergoes preprocessing to ensure consistency and quality, which includes resizing, normalization, and augmentation (Rahim et al. 2023). In a subsequent step, though optional, segmentation can be performed over the preprocessed images to focus on specific region of interest. In this step lesions, blood vessels and optic discs can be extracted from the original image which may be useful for improving the effectiveness of the model (Mustafa et al. 2022). In the last phase, the DL model classifies the images into various classes. The classification can either be binary or multi class. In binary classification only the presence or absence of DR is checked, whereas in multi class classification the model categories the images according to the different stages of severity. Planning a diagnosis and course of treatment is aided by this. We go over each of these phases in brief in this section, with an emphasis on the deep learning-based methods used for classification and segmentation.

## 2.1 Datasets

Many datasets related to DR are publicly available (Kumar and Das 2021). The images in the datasets are primarily of two types of images—fundus color images or Optical Coherence Tomography (OCT) scans. Fundus images are two dimensional images of retina, captured by a special digital Fundus camera. Fundus image is a very valuable tool to document the changes in the retina and to identify any abnormalities (Abràmoff et al. 2010). OCT images, on the other hand, offer both 2D and 3D retinal view. These images, captured through low-coherence light, reveal the deeper layers of the retina, where many disease processes begin, often detecting conditions before a patient's vision has been adversely affected (Zeppieri et al. 2023). A wide variety of datasets related to DR are available to support different research objectives. Some datasets contain images of different stages of DR while some datasets focus solely on different types of lesions. Some datasets provide vessel segmentation masks which can be utilised in segmentation process. We outline some popular publicly available fundus datasets utilised for DR detection research in this subsection. Detailed information of the discussed datasets is presented in Table 2.

(a) **EyePACs**: This dataset is a collection of 88,702 retinal images with high resolution, each annotated with one of five DR stages, are included in this dataset. The dataset is split into 35,126 and 53,576 training and testing images respectively. Ground truths are publicly available only for the training images. It must be noted that some images on Kaggle may exhibit poor quality and incorrect labelling.
(b) **APTOS 2019 Blindness Detection**: This dataset consists of 3662 samples and is publicly available in Kaggle. It contains images from all the five stages of DR.
(c) **Mesidor**: There are1200 fundus color images in this dataset. The images are divided into four classes-Normal (no DR) and three DR stages. Stage-1 class contains images of Mild NPDR, stage-2 consists of images from moderate and severe NPDR, and stage-3 contains images of PDR.
(d) **Mesidor-2**: This dataset includes 1748 DR images; however, they are not labelled.
(e) **DIARETDB0**: A total of 130 DR images are there in this dataset, out of which 20 images belong to the normal class and the other 110 images belong to different stages of DR.
(f) **E-optha**: This dataset, provided by the telecommunications network OPHDIAT, contains two sub-datasets—E-ophtha EX and E-ophtha MA. The E-ophtha EX dataset has 82 images in total, 35 of which are normal and 47 of which are affected by exudates (EX). In contrast, 233 healthy images and 148 images with MAs are present in E-ophtha MA.
(g) **DIARETDB1**: This dataset is a collection of 89 images, out of which 5 are normal, and the other 84 images have some lesions related to DR. The images are accompanied by ground truth (annotated by experts) for prominent lesions, including HEs, SEs, MAs, and HMs.
(h) **Retinopathy Online Challenge (ROC)**: This dataset has100 retina images, split equally into 50 training and 50 testing images. Annotated for MA detection, only training ground truths are available.
(i) **STARE**: This dataset includes expert-designed vessel masks to aid in training and evaluating segmentation algorithms. The dataset, particularly useful for segmentation of blood vessels from retinal images, is composed of 20 images, out of which 10 are for normal retina.
(j) **Digital Retinal Images for Vessel Extraction (DRIVE)**: The images in this dataset are accompanied by manually designed vessel masks by an expert for training and evaluation. Out of the 40 fundus photos in the collection, 7 contain DR lesions (MAs, HMs, Exs, etc.). These seven photos are further split into two sets: three for the testing set and four for the training set.

**Table 2** Details of various publicly available datasets of fundus images

| Dataset | #Images | #Classes | Usage | | | Source |
|---|---|---|---|---|---|---|
| | | | Lesion extraction and segmentation | Vessel segmentation | DR grading | |
| EyePACs | 35,216 | Normal-25810 Mild NDPR-2443 Moderate NDPR-5292 Severe NDPR-873 PDR-798 | Yes | Yes | Yes | https://www.kaggle.com/competitions/diabetic-retinopathy-detection/data |
| APTOS 2019 | 3662 | Normal-1805 Mild NDPR-370 Moderate NDPR-999 Severe NDPR-193 PDR-295 | Yes | No | Yes | https://www.kaggle.com/datasets/mariaherrerot/aptos2019 |
| Mesidor | 1200 | Normal-548 Stage 1 (Mild NDPR)-152 Stage 2(Moderate + Severe NDPR)-246 Stage 3(PDR)-254 | Yes | Yes | Yes | https://www.adcis.net/en/third-party/messidor/ |
| Mesidor-2 | 1748 | – | Yes | Yes | Yes | https://www.medicmind.tech/retinal-image-databases |
| DDR | 12,522 | Normal-6266 Mild NDPR-630 Moderate NDPR-4477 Severe NDPR-236 PDR-913 | Yes | No | Yes | https://www.kaggle.com/datasets/mariaherrerot/ddrdataset |
| DIARETDB0 | 130 | Normal-20 Mild NDPR-50 Moderate NDPR-35 Severe NDPR-15 PDR-10 | Yes | No | Yes | https://www.medicmind.tech/retinal-image-databases |

(continued)

**Table 2** (continued)

| Dataset | #Images | #Classes | Usage | | | Source |
|---|---|---|---|---|---|---|
| | | | Lesion extraction and segmentation | Vessel segmentation | DR grading | |
| E-optha | E-ophtha EX (82) | with EXs-47<br>35-Normal | Yes | No | No | https://www.adcis.net/en/third-party/e-ophtha/ |
| | E-ophtha MA | 148-with MAs<br>233-Normal | | | | |
| DIARETDB1 | 89 | Normal-5<br>With DR lesions- 84 | Yes | No | No | https://www.kaggle.com/datasets/nguyenhung1903/diaretdb1-standard-diabetic-retinopathy-database |
| STARE | 20 | Normal-10<br>DR-10 | Yes | Yes | No | https://www.kaggle.com/datasets/vidheeshnacode/stare-dataset |
| DRIVE | 40 | Normal-33<br>DR-7 | Yes | Yes | No | https://www.kaggle.com/datasets/andrewmvd/drive-digital-retinal-images-for-vessel-extraction |
| CHASE-DB1 | 28 | – | No | Yes | No | https://www.kaggle.com/datasets/khoongweihao/chasedb1 |

(continued)

**Table 2** (continued)

| Dataset | #Images | #Classes | Usage | | | Source |
|---|---|---|---|---|---|---|
| | | | Lesion extraction and segmentation | Vessel segmentation | DR grading | |
| HRF | 45 | Normal-15<br>DR-15<br>Glucoma-15 | No | Yes | No | https://www5.cs.fau.de/research/data/fundus-images/ |
| ROC | 100 | MA only | Yes | No | No | http://webeye.ophth.uiowa.edu/ROC/ |

(k) **CHASE-DB1**: There are a total of 28 images in this dataset, each accompanied by handcrafted vessel segmentation masks, designed by two independent experts. Pixel-level semantic segmentation annotations are applied to images in the CHASE DB1 dataset, meaning that each pixel is given a unique class label.

(l) **HRF (High-Resolution Fundus)**: This dataset consists of 45 images with 15 images each from healthy, DR and glaucoma patients. Each image is accompanied by expertly annotated vessel segmentation images, serving as valuable reference data for vessel segmentation algorithms.

## *2.2 Preprocessing*

In DR detection, data preprocessing is performed for improving the quality of the retinal images and preparing the data for analysis. Image resizing, cropping, and normalization are some typical preprocessing steps. Data augmentation is also performed in this stage by applying different techniques like- rotation, flipping, and zooming etc. (Alwakid et al. 2023). As the green channel provides high contrast (Sharif et al. 2021), some studies also focus on extracting the green channel in the preprocessing stage. The Contrast Limited Adaptive Histogram Equalization (CLAHE) technique has been effectively applied in retinal image preprocessing to enhance contrast (Reza 2004; Alwakid et al. 2023). Researchers also employ data augmentation methods, such as translation, rotation, and contrast scaling, to address class imbalances and increase dataset size (Liu et al. 2022).

## 2.3 Segmentation

DL-based methods are widely used by researchers to segment DR lesions and blood vessels from retinal images. One such method is the U-Net, which is particularly effective in localizing DR lesion such as- HMs, Exs, and MAs within retinal images (Yan et al. 2019; Laibacher et al. 2019; Jin et al. 2019). In the year 2015, Olaf Ronneberger designed a convolutional neural network (CNN) architecture, called U-Net model for segmentation of biomedical images (2015). The U-Net mode has a symmetric encoder and decoder architectures, where the encoder captures contextual information, and the decoder reconstructs the image for pixel-level segmentation. To reduce overfitting and boost network robustness, authors in Guo et al. (2019) presented a Structured Dropout U-Net (SD-UNet) that made use of structured dropout convolutional blocks. The authors utilized the model to retinal vessels segmentation and were able to achieve impressive results. They obtained the highest accuracy (ACC) values of 0.9674 for DRIVE, 0.9738 for CHASE DB1 and 0.9725 for STARE. They also achieved the highest area under the curve (AUC) values of 0.9850 (STARE), 0.9872 (CHASE DB1), and 0.9836 (DRIVE).

Yi et al. presented a Bi-directional ConvLSTM residual U-Net (BCR-UNet), intended for segmentation of blood vessels in the retina (2022). For further improvement in robustness and segmentation accuracy of the network, authors utilised a Structured Dropout Residual Block (SDRB). The BCR Net obtained excellent results with AUC scores of 0.9866 (DRIVE), 0.8246 (CHASEDB1), 0.7015 (STARE), and 0.8075 (IOSTAR) respectively. Qualitative assessment also indicated that BCR-Net, as compared to its counterparts, regularly shown greater accuracy in predicting both thick and microscopic vessels. In another study of retinal blood vessel segmentation, a Dense U-Net architecture was integrated with a patch-based learning method (Wang et al. 2019). The Dense U-Net are widely known for its dense connections that facilitate gradient flow and feature reuse (Cai et al. 2020). The integrated model improved vessel delineation by utilizing localized information. During the training phase, patches were extracted randomly from the training images. For the testing phase also, patches were generated from the test images. These patches were then fetched to a trained model, and the segmentation results were reconstructed applying a sequential reconstruction strategy on the overlapping-patches (Wang et al. 2019). The proposed method achieved ACC values of 0.9511 and 0.9538 for DRIVE and STARE datasets respectively. The AUC values achieved were 0.9740 (DRIVE) and 0.9704 (STARE).

A Deeply Supervised Multiscale Attention U-Net (Mult-Attn-U-Net) was utilized for DR lesion segmentation in Basu and Mitra (2021). The attention mechanism enables selective focus on crucial information while filtering out irrelevant data (Vaswani 2017). It has shown exceptional performance in segmentation and classification of medical images (Zhang et al. 2019; Liu et al. 2021; Li et al. 2023). The model attained the highest segmentation ACC for MA, HE, and SE at 0.65, 0.70, and 0.72, respectively. SegNet is another popular segmentation model based on encoder-decoder architecture, capable of pixel-wise semantic segmentation. The encoder in SegNet extracts the high-order features and the decoder reconstructs the detailed information. The model also includes a pixelwise classifier for prediction (Badrinarayanan et al. 2017). The implementation of SegNet in DR segmentation has shown encouraging results by offering accurate segmentation and localization information of DR lesions. SegNet was first driven by scene understanding applications. Ananda et al. effectively used modified SegNets in Ananda et al. (2019) to segment the optic disc, HE, SE, HMs, and MAs. The method obtained noteworthy Dice Coefficient (DC) scores: 0.9880 for the optic disc, 0.5560 for HEs, 0.8852 for SEs, 0.6988 for HMs, and 0.7520 for MAs. In another study (Pappu et al. 2022), authors presented a novel approach termed deeply supervised maximum response text on-based SegNet (MRT-SegNet) for concurrent segmentation of multiple retinal lesions. Evaluated on the IDRiD dataset, the suggested approach produced a mean area under the precision and recall curve (mAUC_PR) of 0.698. Furthermore, for MA, HMs, HEs, and SEs, specific Area Under the Precision and Recall Curve (AUC_PR) values of approximately 0.50, 0.71, 0.82, and 0.77 were obtained.

Google Research team developed DeepLabv3, a deep learning-based advanced model for semantic segmentation of images, by enhancing the original DeepLab (Chen et al. 2017a, b) architecture. The original DeepLab architecture contains atrous convolution at multiple scales and employs a feature pyramid network (FPN) for improved feature representation (Chen et al. 2017a, b). Atrous convolution, or dilated convolution, allows for multi-scale feature extraction without requiring parameter adjustments. For DR lesion segmentation, a segmentation model that included DeepLabv3 and Xception was used in (Shaukat et al. 2022). With mean Intersection over Union (mIoU) scores of 0.94 (EX), 0.87 (HM), 0.71 (HE), 0.87 (MA), and 0.86 (SEs) on the DIARETTB1 dataset, the suggested method—which consists of 205 layers—showed encouraging results. The approach produced mIoU scores of 0.86 for HM, 0.88 for HE, 0.71 for MA, 0.86 for OD, and 0.84 for SEs on the IDRiD dataset. R-CNN (Region-Based CNN) is an framework for object detection that was introduced by Ross Girshick in 2014. It uses a CNN model to classify regions of interest (ROIs) within pictures (Girshick et al. 2014). Mask R-CNN, a modified version of R-CNN was utilized by the authors for the purpose of segmenting DR lesion (He et al. 2020; Santos et al. 2023). The proposed Mask R-CNN model was an extended version of the Faster R-CNN (He et al. 2020), where a parallel framework was added for predicting the object segmentation masks for extracting both segmentation and localization information. Compared to other methods, like DeepLab v3, the Mask R-CNN model performed better, especially with a notable improvement in precision for MAs.

Deep learning models require a huge volume of data for proper trading of the model. However, in many cases data gathering can be difficult and expensive. Also, there may be cases where datasets are imbalanced having many images for some classes but inadequate number of images for the remaining classes. In these cases, a biased outcome inclined towards the classes with a greater number of images. In such situations, generative adversarial networks (GANs) (Goodfellow et al. 2014) can be useful for unsupervised feature learning and data augmentation (Iqbal and Ali 2018). In the case of DR also, researchers have experimented by incorporating GANs for data augmentation to improve model performance. Paolo Andreini et al. (2021) proposed a two-phased approach for generating retinal images using GAN. In the first phase, GAN was utilized to generate the blood vessel structure by learning from a small set of training samples and corresponding label maps. In the second stage, these structures are then translated into realistic retinal images by using an image-to-image translation algorithm. The authors utilized the Segmentation Multiscale Attention Network (SMANet) for segmentation (Bonechi et al. 2020). The application of the model over the combined original and synthesized datasets resulted in superior results, achieving an AUC of 98.65% for DRIVE and 99.16% for CHASE-DB1. Table 3 includes a summary of the various DL-based approaches utilized in diabetic retinopathy segmentation discussed in this study.

## *2.4 Classification*

Deep learning is extensively applied by researchers for classifying DR into various stages. Based on the classification approach, the methods are categorized into binary and multilevel classifications. Binary classifiers divide the fundus images into one two classes: DR and no DR (normal). Multi-class classifiers, on the other hand, categorize the fundus images into multiple classes like normal, mild NDPR, moderate NDPR, severe NDPR, and PDR. Additionally, other types of classification also exist, such as lesion-based classification, where models classify the fundus images according to the presence or absence of specific DR lesions. This section presents some of the deep learning techniques employed by researchers for binary and multilevel classification. A summary of the various DL-based approaches utilized in diabetic retinopathy classification, discussed in this study, is included in Table 3.

### 2.4.1 Classification Using CNN

Anoop (2022) proposed a CNN-based binary classifier for DR. They conducted comparisons among different optimizers and utilized the APTOS-2019 Blindness Detection dataset from Kaggle for training. Additionally, they developed a web-based interface for convenient interaction with the model. Nasir, N. et al. also developed a CNN model for binary classification of DR stages and achieved an impressive accuracy of reaching 96% (Nasir et al. 2022). Furthermore, Quellec et al. proposed generating heatmaps from ConvNets to illuminate local patterns influencing image-level predictions (2017). Their approach offered insights into the underlying features driving the model's predictions, enhancing interpretability, and potentially aiding in refining diagnostic criteria for DR.

**Table 3** An overview of a few DL-based methods for diabetic retinopathy

| Paper | Technique used | Objective | Dataset used | Outcome | Results |
|---|---|---|---|---|---|
| Yi et al. (2022) | BCR-UNET | Blood Vessel Segmentation | DRIVE, CHASE DB1, STARE, IOSTAR | Tiny blood vessels in the peripheral areas with poor contrast could be preserved using the suggested technique | AUC Scores: DRIVE: 0.9866, CHASE DB1: 0.8246, STARE: 0.7015, IOSTAR0.8075, |
| Yan et al. (2019) | UNET | Lesion Segmentation (MA, EX, HE and SE) | IDRId | The proposed network outperformed both local-only and global-only networks for MA and EX | Area Under Precision Recall curve (AUPR): MA: 0.889, EX:0.525, HE:0.703, SE:0.679 |
| Laibacher et al. (2019) | M2U-NET | Blood vessel segmentation in retina | DRIVE, CHASE DB1, HRF | This proposed model bridged the gap between supervised and unsupervised approaches in this domain, operating in real-time on high-resolution images with low demands on memory and processing | DC: CHASE DB1: 0.8006 |
| Jin et al. (2019) | Deformable U-NET (DU-Net) | Blood vessel segmentation in retina | DRIVE, STARE, CHASE_DB1 and HRF | DUNet offered high accuracy, detailed vessel extraction, and adaptability to diverse vessel shapes and scales | The method achieved the highest accuracy of 0.9690 for STARE dataset |
| Guo et al. (2019) | Structured Dropout U-Net | Blood vessel segmentation in retina | DRIVE, STARE and CHASE-DB1 | The outcomes of the method were compared with some other standard method and found to yield improved results | AUC: DRIVE: 0.9836, STARE: 0.9850, CHASE DB1: 0.9872 |

(continued)

**Table 3** (continued)

| Paper | Technique used | Objective | Dataset used | Outcome | Results |
|---|---|---|---|---|---|
| Wang et al. (2019) | Dense U-net based on patch-based learning | Blood vessel segmentation in retina | DRIVE and STARE | The proposed method showed competitive accuracy, surpassing specialists in segmentation accuracy, highlighting its clinical applicability | AUC: DRIVE: 0.9740, STARE: 0.9704, |
| Basu and Mitra (2021) | Deeply Supervised Multiscale Attention U-Net | Lesion segmentation (MA, HM, and EX) | IDRId | The method achieved high accuracy in segmenting various DR lesions | ACC: MA-0.65, HM-0.0.70, EX-0.72, |
| Ananda et al. (2019) | SegNets | Lesions (HE, SE, HMs, and MAs) and optic disc segmentation | IDRiD and MESSIDOR | The study addressed fundus image segmentation for DR diagnosis using deep learning, tackling the challenge of imbalanced data with individual CNNs for each disease type and employing data augmentation | DC scores for MESSIDOR: optic disc: 0.9880, HEs: 0.5560, SEs: 0.8852, HMs: 0.6988, MAs: 0.7520 |
| Pappu et al. (2022) | Deeply supervised maximum response texton-based SegNet (MRT-SegNet) | Lesion segmentation (MA, HMs, HEs, and SEs) | IDRiD | The proposed MRT-SegNet model delivered promising results in automatically segmenting various retinal lesions | AUC_PR score: 0.495, 0.706, 0.823, and 0.769 are reported for MA, HMs, HEs, and SEs |

(continued)

**Table 3** (continued)

| Paper | Technique used | Objective | Dataset used | Outcome | Results |
|---|---|---|---|---|---|
| Shaukat et al. (2022) | DeepLabv3 and Exception | Semantic segmentation of DR lesions | DIARETDB1, E-ophtha-EX, IDRiD, and Messidor | The method successfully segmented the various DR lesions with high accuracy | mIoU scores: DIARETDB1 Dataset-EX = 0.94, HM = 0.87, HE = 0.71, MA = 0.87, and SEs = 0.86 IDRiD dataset: HM = 0.86, HE = 0.88, MA = 0.71, OD = 0.86, and SEs = 0.84 |
| Santos et al. (2023) | Mask R-CNN | Lesion segmentation (MA, SE, HE and HMs) | DDR | Promising results were obtained by the suggested method, especially in MA segmentation with a significant improvement in precision by approximately 16% | Mean Average Precision of 0.2941 for MA, SE, HE and HMs |
| Andreini et al. (2021) | GAN and SMANet | Synthetic image generation and Retinal Vessel Segmentation | DRiVE, Chase DB1 | The proposed approach enabled the generation of high-resolution and highly realistic images, showcasing its potential for training deep segmentation networks effectively | AUC of 98.65% for DRIVE and 99.16% for CHASE-DB1 |

(continued)

**Table 3** (continued)

| Paper | Technique used | Objective | Dataset used | Outcome | Results |
|---|---|---|---|---|---|
| Anoop (2022) | CNN | Classification (binary) | APTOS-2019 | The paper introduced a CNN model for DR classification and was evaluated against different optimizers. A user-friendly web interface complements the model for easy interaction | Training Accuracy up to 99% |
| Nasir et al. (2022) | CNN | Classification (binary) | IDRID | A CNN structure was proposed in this study for fundus images' pixel classification for assisting the doctors in faster, more accurate diagnoses without direct intervention | Accuracy up to 96% |
| Quellec et al. (2017) | ConvNet | Classification (binary and lesion level) | Kaggle Diabetic Retinopathy, DiaretDB1, E-ophtha | This paper introduced a method to create heatmaps from ConvNets, revealing which pixels influence image-level predictions, aiding in lesion detection without expert input | Az = 0.949 in e-ophtha and an Area Under the ROC Curve Score of 0.954 in Kaggle's dataset |

(continued)

**Table 3** (continued)

| Paper | Technique used | Objective | Dataset used | Outcome | Results |
|---|---|---|---|---|---|
| Esfahani et al. (2018) | Resent | Classification (binary) | Kaggle Dataset | The paper achieved satisfactory accuracy using transfer learning | Accuracy up to 85% |
| Wan et al. (2018) | TL: AlexNet, VggNet, GoogleNet, ResNet | Classification (multi class) | Kaggle dataset | The work demonstrated the utility of transfer learning and CNNs for DR image classification | Accuracy up to 95.68% using VGG Net |
| Salvi et al. (2021) | TL: VGG16, ResNet50 V2, and EfficientNetB0 | Classification (multi class) | MESSIDOR | The research findings indicated that employing Transfer Learning with CNN architectures facilitated accurate predictive analysis of DR from fundus images | Highest accuracy of 95% using VGG16 |

(continued)

**Table 3** (continued)

| Paper | Technique used | Objective | Dataset used | Outcome | Results |
|---|---|---|---|---|---|
| Anita et al. (2023) | TL: MobileNetV2 | Classification (multi class) | APTOS 2019 | The study proposed a lightweight deep learning approach using MobileNetV2 for efficient detection and classification of DR requiring fewer computational resources | Accuracy 85.28% |
| Jiang et al. (2019) | Ensemble model (EM): Inception V3, ResNet152, and Inception-ResNet-V2 | Classification (binary) | Own dataset | With the help of an image augmentation module and an Adaboost ensemble approach, the paper presented an interpretable DR screening method that demonstrated satisfactory performance | Accuracy: 88.21% |
| Nazir et al. (2021) | EM: Densenet-100 and CenterNet | Classification (multi class) | APTOS-2019 and IDRiD | The study showcased accurate classification of DR and DME lesions from low-intensity, noisy data images | Accuracy Score: APTOS-2019 = 97.93% and IDRiD = 98.10% |

(continued)

**Table 3** (continued)

| Paper | Technique used | Objective | Dataset used | Outcome | Results |
|---|---|---|---|---|---|
| Butt et al. (2022) | EM: GoogleNet and ResNet-18 | Classification (binary and multi class) | APTOS-19 | The study proposed a hybrid feature extraction method for DR detection, achieving high accuracy rates and also highlights the potential of exploring customized CNNs for enhanced interclass similarity capture | Accuracy Score: Binary and multi class classification yielded accuracy scores of 97.8% and 89.29% respectively |
| Attota et al. (2022) | EM EfficientNetB7, ResNet50V2, and Inception V3 | Classification (multi class) | EyePACs dataset | Achieving 97.43% accuracy on the EyePacs Dataset, this method demonstrates high performance in detecting DR subtypes | Accuracy upto 97.43% |

(continued)

**Table 3** (continued)

| Paper | Technique used | Objective | Dataset used | Outcome | Results |
|---|---|---|---|---|---|
| Mustafa et al. (2022) | EM: ResNet-50 and DenseNet-121 | Classification (multi class) | EyePACs and Messidor-2 Dataset | With high accuracy the proposed model showed the potential of ensemble model in classification of DR | Accuracy Scores: Messidor2 = 95.58%, EyePACS = 89.20%, APTOS = 89%, and DDR = 76.81% |
| Tri Handoyo and Kusuma (2022) | EM: EfficientNet, EfficientNetV2, LCNet, MobileNetV3, TinyNet, and FBNetV3 | Classification (multi class) | APTOS 2019 | To increase the accuracy of DR classification, the study investigated the integration of different deep learning models and ensemble stacking techniques and achieved moderate accuracy | Testing Accuracy of 84.17% |

### 2.4.2 Transfer Learning Based Classification

Many researchers use pre-trained models in their approaches for feature extraction or classification. This approach, known as transfer learning, not only saves time and computer resources, but also enhances the performance on a task by utilizing the pre-trained model's knowledge on related tasks (Shin et al. 2016). Esfahani et al. (2018) performed binary classification using ResNet (He et al. 2016) over a Kaggle dataset and achieved 85% accuracy. In another study, AlexNet (Krizhevsky et al. 2012), VGG (Simonyan and Zisserman 2014), GoogleNet (Szegedy et al. 2015), and ResNet were evaluated on the Kaggle dataset and an accuracy of up to 95.68% was achieved (Wan et al. 2018). Salvi et al. (2021), explored VGG16, ResNet50 V2, and EfficientNetB0 (Tan and Le 2019) for predictive analysis of DR. VGG16 immersed as the model with highest accuracy, achieving a notable 95% accuracy rate in classifying DR. Resource-intensive models like ResNet and DenseNet (Huang et al. 2017) pose challenges in the detection and classification of DR, particularly due to their high computational requirements. To tackle this issue, a deep learning strategy using MobileNetV2 (Sandler et al. 2018), known for its computational efficiency, was investigated in the study by Anita et al. (2023). The authors employed a large dataset for training and testing the model and achieved an impressive accuracy of 85.28%

for DR classification, demonstrating its effectiveness in automating the diagnosis process while overcoming computational constraints (Table 3).

### 2.4.3 Classification Using Ensembles Models

In recent years, ensembles learning approach has gained popularity in various domains. To improve the prediction ability of deep learning models, researchers have also investigated several ensemble methodologies in the DR domain. In a study Jiang et al. (2019) ensembled multiple deep learning models, including Inception V3 (Szegedy et al. 2015), ResNet152 (He et al. 2016), and Inception-ResNet-V2 (Baldassarre et al. 2017), using the Adaboost algorithm. Their approach aimed to mitigate the bias inherent in individual models and yielded promising results, achieving an accuracy of 88.21%. Nazir et al. also used an ensemble approach to automatically identify eye diseases related to diabetes from retinal images (2021). Their approach integrated DenseNet-100 and a custom CenterNet (Duan et al. 2019). DenseNet-100 was utilised for dataset preparation and feature extraction, while CenterNet was used for lesion localization, classification of DR and diabetic macular oedema (DME). Experiments over APTOS-2019 and IDRiD datasets yielded average accuracies of 97.93% and 98.10% respectively. Highest accuracy scores of 97.8% and 89.29% binary and multi class classification respectively were achieved by another hybrid deep learning approach, combining GoogleNet and ResNet-18, proposed by Butt et al. (2022). Attota et al. (2022) combined EfficientNetB7, ResNet50V2, and Inception V3 to classify DR subtypes and achieved an impressive accuracy of 97.43% on the EyePACs dataset. In addition, a multi-stream ensemble deep network to classify DR severity was proposed in (Mustafa et al. 2022). The method, achieving up to 95.58% accuracy, utilized models like ResNet-50 and DenseNet-121 for feature extraction. Furthermore, the use of ensemble stacking techniques with different models such as EfficientNet, LCNet (Yi and Ma 2022), MobileNetV3, and others, combined with meta-learners, was explored to provide better accuracy in classifying the severity of DR (Tri Handoyo and Kusuma 2022). The model achieved satisfactory classification results with training, validation, testing accuracies, and an F1 score of 96.56%, 95.33%, 84.17%, and 70.16%, respectively for the APTOS 2019 dataset.

## 3 Conclusion and Future Scope

This article presents some recent advancements in the diagnosis of DR highlighting the utilization of DL. Some common publicly available fundus DR datasets are also outlined in the study. Significant advancements have been made in DR diagnosis using DL based algorithms, applied at different phases of data processing including data preprocessing, data augmentation, segmentation, and classification. These advancements pave the way for early detection preventing irreversible vision loss. For data augmentation, GANs are applied to create synthetic images. This is very useful when enough samples are not available for a particular class. In lesion and blood vessel segmentation, deep learning models such as U-Net, DeepLab have demonstrated high accuracy. In binary and multilevel DR classification, DL-based techniques have been remarkably accurate. The researchers have extensively used different CNN models, transfer learning and ensemble models to enhance the classification accuracy. Table 3 summarizes the different DL-based methods used in this study's segmentation and categorization of diabetic retinopathy.

Although a significant advancement has been achieved in DL-based DR detection and classification, there are still some challenges to be addressed.

- DL models are often viewed as black boxes, making it challenging for the healthcare professionals to trust the models' decision. Future studies may explore the development of systems that would explain and justify model's decision.
- DL models are required to be trained with a high volume of labelled data for optimal performance. This is quite challenging because it takes a lot of time and effort to annotate the images.
- Imbalanced dataset is a major problem in medical imaging data. For DR datasets also, this problem persists as there are fewer positive samples (with DR) compared to normal images (without DR). For multi-class datasets, there are even fewer samples for PDR class. Data imbalance affects the outcome of DL models, leading to biased predictions and reduced performance. Advanced data augmentation techniques, GANs for synthetic dataset generation, hybrid ML-DL methods are some potential and open areas which the researchers may explore to address this challenge.
- Additionally, developing a low-cost automated standalone system capable of detecting multiple eye disorders including DR, glaucoma, diabetic macular oedema etc. is essential for further advancements in the eye-care sector.

## References

Abràmoff, M. D., Garvin, M. K., & Sonka, M. (2010). *Retinal imaging and image analysis*. IEEE reviews in biomedical engineering, 3, 169–208.

AK, Fujita H, Hagiwara Y, Chua CK, Laude A (2016) *Automated screening system for retinal health using bi-dimensional empirical mode decomposition and integrated index*. Comput Biol Med 75:54–62.

Alwakid, G., Gouda, W., & Humayun, M. (2023, March). *Deep Learning-based prediction of Diabetic Retinopathy using CLAHE and ESRGAN for Enhancement*. In Healthcare (Vol. 11, No. 6, p. 863). MDPI.

Ananda, S., Kitahara, D., Hirabayashi, A., & Reddy, K. U. K. (2019, November). *Automatic fundus image segmentation for diabetic retinopathy diagnosis by multiple modified U-Nets and SegNets*. In 2019 Asia-Pacific Signal and Information Processing Association Annual Summit and Conference (APSIPA ASC) (pp. 1582–1588). IEEE.

Andreini, P., Ciano, G., Bonechi, S., Graziani, C., Lachi, V., Mecocci, A., & Bianchini, M. (2021). *A two-stage GAN for high-resolution retinal image generation and segmentation*. Electronics, 11(1), 60.

Anita, K., Nivedha., Sreelakshmi., T. (2023). *Detection and Classification of Diabetic Retinopathy Using Deep Learning*. International Journal For Science Technology And Engineering, https://doi.org/10.22214/ijraset.2023.51626.
Anoop, B. K. (2022). Binary classification of DR-diabetic retinopathy using CNN with fundus colour images. Materials Today: Proceedings, 58, 212–216.
Attota, D., Tadikonda, D. N., Pethe, S., & Khan, M. A. A. H. (2022, June). *An ensembled method for diabetic retinopathy classification using transfer learning*. In 2022 IEEE 46th Annual Computers, Software, and Applications Conference (COMPSAC) (pp. 1444–1449). IEEE.
Badrinarayanan, V., Kendall, A., & Cipolla, R. (2017). *Segnet: A deep convolutional encoder-decoder architecture for image segmentation*. IEEE transactions on pattern analysis and machine intelligence, 39(12), 2481–2495.
Bajwa, J., Munir, U., Nori, A., & Williams, B. (2021). *Artificial intelligence in healthcare: transforming the practice of medicine*. Future healthcare journal, 8(2), e188.
Baldassarre, F., Morín, D. G., & Rodés-Guirao, L. (2017). *Deep koalarization: Image colorization using cnns and inception-resnet-v2*. arXiv preprint arXiv:1712.03400.
Basu, S., & Mitra, S. (2021, November). *Segmentation in diabetic retinopathy using deeply-supervised multiscalar attention*. In 2021 43rd Annual International Conference of the IEEE Engineering in Medicine & Biology Society (EMBC) (pp. 2614–2617). IEEE.
Bhadra, S.; Kumar, C.J.; Bhattacharyya, D.K. (2024). Multiview EEG signal analysis for diagnosis of schizophrenia: an optimized deep learning approach. *Multimedia Tools and Applications*. https://doi.org/10.1007/s11042-024-20205-y.
Bonechi, S.; Bianchini, M.; Scarselli, F.; Andreini, P. *Weak supervision for generating pixel–level annotations in scene text segmentation*. Pattern Recognit. Lett. 2020, 138, 1–7.
Butt, M. M., Iskandar, D. A., Abdelhamid, S. E., Latif, G., & Alghazo, R. (2022). *Diabetic retinopathy detection from fundus images of the eye using hybrid deep learning features*. Diagnostics, 12(7), 1607.
Cai, S., Tian, Y., Lui, H., Zeng, H., Wu, Y., & Chen, G. (2020). *Dense-UNet: a novel multiphoton in vivo cellular image segmentation model based on a convolutional neural network. Quantitative imaging in medicine and surgery, 10*(6), 1275.
Chen, L. C., Papandreou, G., Kokkinos, I., Murphy, K., & Yuille, A. L. (2017a). Deeplab: *Semantic image segmentation with deep convolutional nets, atrous convolution, and fully connected crfs*. IEEE transactions on pattern analysis and machine intelligence, 40(4), 834–848.
Chen, L. C., Papandreou, G., Schroff, F., & Adam, H. (2017b). *Rethinking atrous convolution for semantic image segmentation*. arXiv preprint arXiv:1706.05587.
Chen, Z. H., Lin, L., Wu, C. F., Li, C. F., Xu, R. H., & Sun, Y. (2021). *Artificial intelligence for assisting cancer diagnosis and treatment in the era of precision medicine*. Cancer Communications, 41(11), 1100–1115.
Das, Dolly, Saroj Kr Biswas, and Sivaji Bandyopadhyay. *A critical review on diagnosis of diabetic retinopathy using machine learning and deep learning. Multimedia Tools and Applications* 81, no. 18 (2022): 25613–25655.
Duan, K., Bai, S., Xie, L., Qi, H., Huang, Q., & Tian, Q. (2019). *Centernet: Keypoint triplets for object detection*. In Proceedings of the IEEE/CVF international conference on computer vision (pp. 6569–6578).
E. T. D. R. S. R. GROUP. *Grading diabetic retinopa thy from stereoscopic color fundus photographs-an extension of the modified* Airlie House classification. Ophthalmology 1991; 98(5):786–806.
Esfahani, M. T., Ghaderi, M., & Kafiyeh, R. (2018). *Classification of diabetic and normal fundus images using new deep learning method*. Leonardo Electron. J. Pract. Technol, 17(32), 233–248.
Fisher DE, Jonasson F, Klein R, Jonsson PV, Eiriksdottir G, Launer LJ, Gudnason V, Cotch MF (2016) *Mortality in older persons with retinopathy and concomitant health conditions: the age, gene/environment susceptibility-reykjavik study*. Ophthal- mology 123(7):1570–1580.
Girshick, R., Donahue, J., Darrell, T., & Malik, J. (2014). *Rich feature hierarchies for accurate object detection and semantic segmentation*. In Proceedings of the IEEE conference on computer vision and pattern recognition (pp. 580–587).

Goodfellow I, Pouget-Abadie J, Mirza M, Xu B, Warde-Farley D, Ozair S, Courville A, Bengio Y (2014) *Generative adversarial nets*. Advances in neural information processing systems, vol 27.

Guo, C., Szemenyei, M., Pei, Y., Yi, Y., & Zhou, W. (2019, October). SD-UNet: *A structured dropout U-Net for retinal vessel segmentation*. In 2019 IEEE 19th international conference on bioinformatics and bioengineering (BIBE) (pp. 439–444). IEEE.

Hadziahmetovic, M., Nicholas, P., Jindal, S., Mettu, P. S., & Cousins, S. W. (2019). *Evaluation of a remote diagnosis imaging model vs dilated eye examination in referable macular degeneration. JAMA ophthalmology, 137*(7), 802–808.

Harper CA, Keeffe JE. *Diabetic retinopathy management guidelines*. Expet Rev Ophthalmol 2012;7(5):417–39.

He, K., Zhang, X., Ren, S., & Sun, J. (2016). *Deep residual learning for image recognition*. In Proceedings of the IEEE conference on computer vision and pattern recognition (pp. 770–778).

K. He, G. Gkioxari, P. Dollár, and R. Girshick, *Mask R-CN*." IEEE Trans. Pattern Anal. Mach. Intell., vol. 42, no. 2, pp. 386–397, Jun. 2020.

Huang, G., Liu, Z., Van Der Maaten, L., & Weinberger, K. Q. (2017). *Densely connected convolutional networks*. In Proceedings of the IEEE conference on computer vision and pattern recognition (pp. 4700–4708).

International diabetes federation - *what is diabetes [Online]. Available,* www.idf.org/aboutdiabetes/what-is-diabetes.html.

Iqbal, T., & Ali, H. (2018*). Generative adversarial network for medical images (MI-GAN).* Journal of medical systems, 42(11), 231.

Jiang, H., Yang, K., Gao, M., Zhang, D., Ma, H., & Qian, W. (2019, July*). An interpretable ensemble deep learning model for diabetic retinopathy disease classification.* In 2019 41st annual international conference of the IEEE engineering in medicine and biology society (EMBC) (pp. 2045–2048). IEEE.

Jin Q, Meng Z, Pham TD, Chen Q, Wei L, Su R (2019) *DUNet: A deformable network for retinal vessel segmentation*. Knowl- Based Syst 178:149–162.

Keel S, Wu J, Lee PY, Scheetz J, He M (2019*) Visualizing deep learning models for the detection of referable diabetic retinopa- thy and glaucoma*. JAMA Ophthalmol 137(3):288–292.

Ker, J., Wang, L., Rao, J., & Lim, T. (2017). *Deep learning applications in medical image analysis*. IEEE Access, 6, 9375–9389.

Krizhevsky, A., Sutskever, I., & Hinton, G. E. (2012). *Imagenet classification with deep convolutional neural networks*. Advances in neural information processing systems, 25.

Kumar, C.J, Das PR.(2021) The diagnosis of ASD using multiple machine learning techniques. International Journal of Developmental Disabilities. 68(6):973-983. https://doi.org/10.1080/20473869.2021.1933730.

Laibacher T, Weyde T, Jalali S (2019) *M2u-net: effective and efficient retinal vessel segmentation for real-world applications*. In: Proceedings of the IEEE/CVF conference on computer vision and pattern recognition workshops.

Li, X., Li, M., Yan, P., Li, G., Jiang, Y., Luo, H., & Yin, S. (2023). *Deep learning attention mechanism in medical image analysis: Basics and beyonds. International Journal of Network Dynamics and Intelligence*, 93–116.

Liu, R., Gao, S., Zhang, H., Wang, S., Zhou, L., & Liu, J. (2022). MTNet: *A combined diagnosis algorithm of vessel segmentation and diabetic retinopathy for retinal images*. Plos one, 17(11), e0278126.

Liu, X.; Zhang, L.; Li, T.; Wang, D.; Wang, Z. *Dual attention guided multi-scale CNN for fine-grained image classification*. Inf. Sci. 2021, 573.

Mustafa, H., Ali, S. F., Bilal, M., & Hanif, M. S. (2022). *Multi-stream deep neural network for diabetic retinopathy severity classification under a boosting framework*. IEEE Access, 10, 113172–113183.

Nadeem, Muhammad Waqas, Hock Guan Goh, Muzammil Hussain, Soung-Yue Liew, Ivan Andonovic, and Muhammad Adnan Khan. *Deep learning for diabetic retinopathy analysis: a review, research challenges, and future directions. Sensors* 22, no. 18 (2022): 6780.

Nasir, N., Oswald, P., Alshaltone, O., Barneih, F., Al Shabi, M., & Al-Shammaa, A. (2022, February). *Deep dr: Detection of diabetic retinopathy using a convolutional neural network.* In 2022 Advances in Science and Engineering Technology International Conferences (ASET) (pp. 1–5). IEEE.

Nazir, T., Nawaz, M., Rashid, J., Mahum, R., Masood, M., Mehmood, A., ... & Hussain, A. (2021). *Detection of diabetic eye disease from retinal images using a deep learning based CenterNet model.* Sensors, 21(16), 5283.

Pappu, G. P., Krishna, T., Biswal, B., Karn, P. K., Biswal, P. K., Hasan, S., & Nayak, D. (2022*). A deeply supervised maximum response texton based SegNet for simultaneous multi retinal lesion segmentation.* International Journal of Imaging Systems and Technology, 32(5), 1709–1726.

Quellec, G., Charriere, K., Boudi, Y., Cochener, B., & Lamard, M. (2017). *Deep image mining for diabetic retinopathy screening.* Medical image analysis, 39, 178–193.

Rahim, S., Sabri, K., Ells, A., Wassyng, A., Lawford, M., Chu, L., & He, W. (2023). *Novel Fundus image preprocessing for Retcam images to improve deep learning classification of retinopathy of prematurity. arXiv preprint* arXiv:2302.02524.

Reza, A. M. (2004). *Realization of the contrast limited adaptive histogram equalization (CLAHE) for real-time image enhancement.* Journal of VLSI signal processing systems for signal, image and video technology, 38, 35–44.

Ronneberger, O., Fischer, P., & Brox, T. (2015). *U-net: Convolutional networks for biomedical image segmentation. In Medical image computing and computer-assisted intervention*–MICCAI 2015: 18th international conference, Munich, Germany, October 5–9, 2015, proceedings, part III 18 (pp. 234–241). Springer International Publishing.

Salvi, R. S., Labhsetwar, S. R., Kolte, P. A., Venkatesh, V. S., & Baretto, A. M. (2021, January). *Predictive analysis of diabetic retinopathy with transfer learning.* In 2021 4th Biennial International Conference on Nascent Technologies in Engineering (ICNTE) (pp. 1–6). IEEE.

Sandler, M., Howard, A., Zhu, M., Zhmoginov, A., & Chen, L. C. (2018). Mobilenetv2: *Inverted residuals and linear bottlenecks.* In Proceedings of the IEEE conference on computer vision and pattern recognition (pp. 4510–4520).

Santos, Carlos, Marilton Aguiar, Daniel Welfer, Marcelo Dias, Alejandro Pereira, Marcelo Ribeiro, and Bruno Belloni. *A new approach for fundus lesions instance segmentation based on mask r-cnn x101-fpn pre-trained architecture.* IEEE Access (2023).

Scanlon PH, Wilkinson CP, Aldington SJ, Matthews DR. *A Practical manual of diabetic retinopathy management.* first ed. Wiley-Blackwell; 2009.

Scotland GS, et al. *Costs and consequences of automated algorithms versus manual grading for the detection of referable diabetic retinopathy.* Br J Ophthalmol 2010; 94(6):712–9.

Sharif, N. A. M., Azhar, A. S. M., Harun, N. H., Bakar, J. A., Abdullah, A. A., & Chong, Y. F. (2021, August). *Green channel and top hat-based image enhancement for diabetic retinopathy screening.* In Journal of Physics: Conference Series (Vol. 1997, No. 1, p. 012002). IOP Publishing.

Shaukat, N., Amin, J., Sharif, M., Azam, F., Kadry, S., & Krishnamoorthy, S. (2022). *Three-dimensional semantic segmentation of diabetic retinopathy lesions and grading using transfer learning.* Journal of Personalized Medicine, 12(9), 1454.

Shin, H. C., Roth, H. R., Gao, M., Lu, L., Xu, Z., Nogues, I.,& Summers, R. M. (2016). *Deep convolutional neural networks for computer-aided detection: CNN architectures, dataset characteristics and transfer learning.* IEEE transactions on medical imaging, 35(5), 1285–1298.

Simonyan, K., & Zisserman, A. (2014). *Very deep convolutional networks for large-scale image recognition.* arXiv preprint arXiv:1409.1556.

Szegedy, C., Liu, W., Jia, Y., Sermanet, P., Reed, S., Anguelov, D., & Rabinovich, A. (2015). *Going deeper with convolutions.* In Proceedings of the IEEE conference on computer vision and pattern recognition (pp. 1–9).

Tan, M., & Le, Q. (2019, May). *Efficientnet: Rethinking model scaling for convolutional neural networks.* In International conference on machine learning (pp. 6105–6114). PMLR.

Ting DS, Wu WC, Toth C (2019) *Deep learning for retinopathy of prematurity screening.* Br J Ophthalmol 103(5):577–579.

Tri Handoyo, A., & Kusuma, G. P. (2022). *Severity Classification of Diabetic Retinopathy Using Ensemble Stacking Method.* Revue d'Intelligence Artificielle, 36(6).

Vaswani, A. (2017). *Attention is all you need.* Advances in Neural Information Processing Systems.

Wan, S., Liang, Y., & Zhang, Y. (2018). *Deep convolutional neural networks for diabetic retinopathy detection by image classification.* Computers & Electrical Engineering, 72, 274–282.

Wang, C., Zhao, Z., Ren, Q., Xu, Y., & Yu, Y. (2019). *Dense U-net based on patch-based learning for retinal vessel segmentation.* Entropy, 21(2), 168.

Wu, L., Fernandez-Loaiza, P., Sauma, J., Hernandez-Bogantes, E., & Masis, M. (2013). *Classification of diabetic retinopathy and diabetic macular edema. World journal of diabetes, 4*(6), 290.

Yan, Z., Han, X., Wang, C., Qiu, Y., Xiong, Z., & Cui, S. (2019, April). *Learning mutually local-global u-nets for high-resolution retinal lesion segmentation in fundus images.* In 2019 IEEE 16th International Symposium on Biomedical Imaging (ISBI 2019) (pp. 597–600). IEEE.

Yang, Z., Tan, T. E., Shao, Y., Wong, T. Y., & Li, X. (2022). *Classification of diabetic retinopathy: Past, present and future. Frontiers in Endocrinology, 13,* 1079217.

Yi, X., & Ma, B. (2022, February). LCNet: *Location Combination for Object Detection.* In Proceedings of the 6th International Conference on Digital Signal Processing (pp. 152–158).

Yi, Y., Guo, C., Hu, Y., Zhou, W., & Wang, W. (2022). BCR-UNet: *Bi-directional ConvLSTM residual U-Net for retinal blood vessel segmentation.* Frontiers in Public Health, 10, 1056226.

Zabihollahy F, Lochbihler A, Ukwatta E (2019) *Deep learning-based approach for fully automated detection and segmentation of hard exudate from retinal images.* In: Medical imaging 2019: biomedical applications in molecular, structural, and functional imaging, vol 10953. International Society for Optics and Photonics, p 1095308.

Zeppieri, M., Marsili, S., Enaholo, E. S., Shuaibu, A. O., Uwagboe, N., Salati, C., ... & Musa, M. (2023). *Optical coherence tomography (OCT): a brief look at the uses and technological evolution of ophthalmology. Medicina, 59*(12), 2114.

Zhang, T.; Li, A.; Wang, M.; Wu, X.; Qiu, B. *Multiple attention fully convolutional network for automated ventricle segmentation in cardiac magnetic resonance imaging.* J. Med. Imaging Health Inform. 2019, 9, 1037–1045

# Advancements in Computer-Aided Diagnosis Systems for Mammographic Mass Detection: A Comprehensive Review

**Manasi Hazarika, Sonia Sarmah, Pranab Das, and Lipi B. Mahanta**

**Abstract** **Background**: Mammographic-mass segmentation and classification are crucial for early diagnosis of breast cancer. Traditional machine learning approaches have shown promising results for both mass segmentation and classification but deep learning approaches have demonstrated superior performance. **Objective**: This review focuses on both traditional and deep learning-based machine learning techniques employed in mammographic-mass segmentation, localization, and classification techniques. It traces the evolution from classical machine learning algorithms to deep learning for mass detection and classification tasks. **Methods**: The mass segmentation and detection methods discussed in this review are built on the principles such as region-growing, contour-based approaches, and thresholding, as well as modern deep learning models like U-Net, YOLO, and R-CNN. For mass classification, traditional classifiers like Artificial Neural Network, Support Vector Machine, and their deep learning counter parts are discussed. **Results**: Traditional segmentation methods often struggle with complex mammograms with dense tissues or indistinct mass boundaries. In contrast deep learning has brought remarkable improvement in segmentation, localization of masses as well mass in classification. **Conclusion**: Deep learning-based approaches are proven to be more efficient in segmentation and classification of mammographic mass compared to the traditional approaches. Deep learning can efficiently extract features from mammograms and may fully integrate into the breast cancer diagnosis.

M. Hazarika (✉)
Computer Science and Engineering, Assam Don Bosco University, Guwahati, India
e-mail: manasi.hazarika@dbuniversity.ac.in

S. Sarmah · P. Das
Computer Applications, Assam Don Bosco University, Guwahati, Assam, India
e-mail: sonia.sarmah@dbuniversity.ac.in

P. Das
e-mail: pranab.das@dbuniversity.ac.in

L. B. Mahanta
Mathematical and Computational Sciences Division, Institute of Advanced Study in Science and Technology, Guwahati, Assam, India
e-mail: lbmahanta@iasst.gov.in



T. P. Singh et al. (eds.), *Revolutionizing Healthcare: Impact of Artificial Intelligence on Diagnosis, Treatment, and Patient Care*, Studies in Computational Intelligence 1182,
https://doi.org/10.1007/978-3-031-80813-5_8

**Keywords** Mammogram · Breast cancer · Computer-aided diagnosis · Machine learning · Deep learning

## 1 Introduction

Breast cancer has ascended to become the foremost prevalent form of cancer, outpacing lung cancer, according to a report disclosed by the International Agency for Research on Cancer (IARC) in December 2020. It now constitutes 12.5% of all newly diagnosed cancer cases across genders. Particularly among females, breast cancer contributes a staggering 25.8% of all new cases worldwide (WCRF 2024). As per the latest report of WHO, in the year 2022, breast cancer diagnoses among women worldwide totaled 2.3 million, with 670,000 fatalities recorded because of the disease (WHO 2024). All the tumors found in the breast are not cancerous. Benign tumors are non-cancerous and non-invasive in nature which can be removed easily (NBCF 2024) while malignant tumors are carcinogenic and invasive. Malignant tumors can easily invade other body parts including the lymphatic system of liver and bone. Presence of a mass is one of the most noticed signs of breast carcinoma. Usually, a malignant mass is painless, hard, and irregular in shape. But lumps with soft tissues and regular shapes can also be cancerous. They can even be painful. Despite extensive research, the precise causes of these abnormalities remain elusive, rendering prevention strategies impractical. Consequently, early detection emerges as the linchpin for effective treatment (Lu et al. 2009). While mammography (La Vecchia et al. 1993) stands as a cornerstone for early breast cancer detection, its efficacy is hampered by inherent challenges in interpreting mammogram images, leading to a suboptimal detection rate. Even with the expertise of seasoned radiologists, the potential for human error looms large. Around 10–30% of the signs of breast cancer are missed by the radiologist during the time of screening (Ekpo et al. 2018). Consequently, seeking a second opinion (Garcia et al. 2018) becomes imperative to enhance detection accuracy. However, the burden on radiologists often complicates the process of securing a second reader. Considering these challenges, leveraging Computer Aided Diagnostic System (CADx) (Doi 2007) as secondary readers present a viable solution, offering a reliable means to augment detection rates and alleviate the strain on radiologists. With improvements in machine intelligence, radiologists can rely on computer systems to improve their diagnosis (Eadie et al. 2012). With the assistance CADx readers sensitivity can be raised by 10% on average (Masud et al. 2019). An automated system that can classify abnormalities is highly desirable; this can be designed based on knowledge-based anatomy. Therefore, the integration of CADx systems will enhance the accuracy of mass detection, complementing the skills of radiologists.

Deep learning techniques have revolutionized medical image analysis, specifically detection and classification of abnormalities like tumors in various parts of the body including breast. Deep learning models have demonstrated remarkable accuracy in automatically detecting mammographic masses. These models are often based on convolutional neural networks (CNNs). CNNs have the capacity to learn complex patterns and features from mammograms, enabling them to identify subtle abnormalities which are suggestive of breast cancer. By utilizing deep learning based CADx systems, radiologists can benefit from improved efficiency and accuracy in mass detection, potentially reducing interpretation time and aiding in early diagnosis. Integrating deep learning algorithms into CADx systems can enhance screening programs, aiding healthcare professionals in identifying and characterizing suspicious lesions, ultimately improving patient outcomes in breast cancer management.

Despite these advances, there are still significant gaps in the application of deep learning-based approaches for breast cancer detection. Most of the current systems, while solving the task at hand, have struggled to meet results due to data insufficiency data and class imbalance. In addition, deep learning models are not well understood, which makes it difficult for their application in practice. There is clearly a requirement for further studies to create strong, interpretable, and generalizable such models which can be organized in the context of clinical practice rather than in research conditions. Further, even though most of the studies are particularly aimed at such single tasks as detection or classification, no integrated systems only, which include and cover all processes of breast mass diagnosis, including segmentation and classification, are available. This review seeks to address these gaps by providing a thorough examination of both classical and deep learning-based machine learning algorithms for the automated diagnosis of mammographic masses.

The objective of this study is to deliver a comprehensive review of both traditional and deep learning-based machine learning approaches for the automated diagnosis of mammographic masses. The study includes methods used for breast mass segmentation, b detection (localization) and classification. The rest of the paper is organized as follows: Sect. 2 highlights the major abnormalities found in breast, Sect. 3 presents a list of most used benchmark datasets of mammogram images, Sect. 4 provides a review on traditional machine learning-based breast mass diagnosis methods and Sect. 5 presents a review on deep-learning based breast mass diagnosis methods. Finally, Sect. 6 presents the summary and conclusion of the review.

## 2 Abnormalities Seen in Breast

On mammograms, various abnormalities manifest prominently, including microcalcifications, masses, asymmetry, and architectural distortion (McKenna 1994). Among these, masses and calcifications are the most frequently observed anomalies. A mass typically presents as a lesion occupying space, characterized by curved or concave boundaries (Agarwal et al. 2009). The mass classification into malignant or benign depends on factors such as morphology (shape, margin, size), texture and density. Benign masses often exhibit round, oval, or slightly lobular shapes, while malignant masses tend to display multilobular or irregular contours. Additionally, a mass with a well-defined boundary typically suggests a benign nature, whereas those with spiculated margins raise suspicion for malignancy (Guray and Sahin 2006; Thomassin-Naggara et al. 2014).

Calcifications are calcium deposits appearing as minuscule dots scattered throughout the mammary gland or clustered together. They form because of cellular division during breast growth. Microcalcifications commonly emerge in areas where cells are actively dividing and growing. These deposits can vary in size, with larger ones termed as macrocalcifications, typically present as benign with smooth, round surfaces. Conversely, smaller calcifications, known as microcalcifications, often signal malignancy, although not all microcalcifications are indicative of cancer. While most macrocalcifications are benign, they can occasionally act as an initial signs of cancer in breast (Breastcancer.org 2024; Wilkinson et al. 2017). Additionally, malignant masses may appear along with clusters of microcalcifications, leading to rapid growth or enlargement over time, and exhibit architectural distortion or nipple retraction. These characteristics can help radiologists differentiate between benign and malignant mammographic masses and guide appropriate management decisions, such as further imaging or biopsy. Mammographic masses can vary in their characteristics, but there are certain features that helps to distinguish between malignant and benign tumors. Figure 1 illustrates mammograms with different mammograms which include normal mammogram, mammogram with benign mass, mammogram with malignant mass, mammogram with benign calcification and mammogram with malignant calcification.

## 3 Mammographic Datasets

### *3.1 Mini MIAS*

The mini-Mammographic Image Analysis Society Digital Mammogram Database (MIAS) (Suckling et al. 1994) comprises of 322 mammograms (mediolateral oblique) images collected from 161 patients, with two images captured per patient: one for the left breast and one for the right. The dataset encompasses cases of benign, malignant, and normal conditions. Images are organized in pairs, with odd numbered filenames assigned to right mammograms and even-numbered filenames assigned to left mammograms for each patient. Each image has a spatial resolution of 200 microns per pixel, with a total size of 1024 × 1024 pixels and an 8-bit gray-scale depth. Ground truths include information on centers, radii of regions of interest, abnormality types, mass category, and pathology. It also provides information related to tissue type (dense, fatty, glandular), abnormality type (masses, calcifications, asymmetry, architectural distortion).

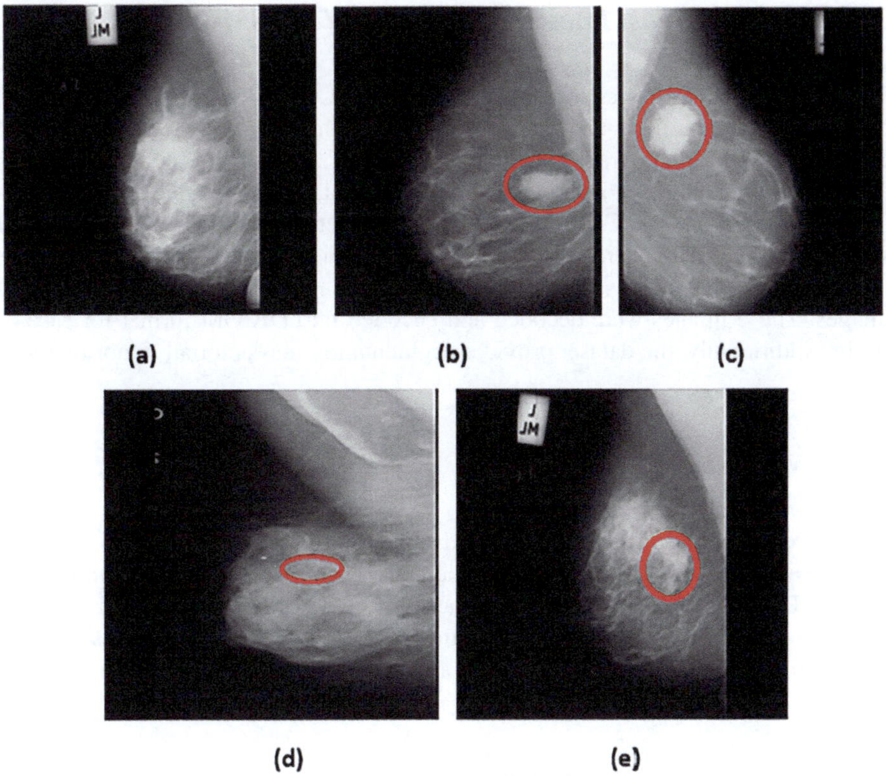


**Fig. 1** Sample mammogram images (*Source* MIAS dataset). The sub-images **a** indicate a normal Mammogram. The sub-images from **b–e** indicates the presence of (b) circular benign mass, **c** spiculated malignant mass, **d** benign microcalcification and **e** Malignant microcalcification

## *3.2 Digital Database for Screening Mammography*

The digital database for screening mammography (DDSM) (Heath et al. 2001) is an extensive repository which consists more than 2600 studies of scanned film mammography. Within this collection lies a subset known as the curated breast imaging subset of the DDSM (CBIS-DDSM), comprising meticulously annotated and labelled images. This dataset offers extensive details, including bounding boxes for regions of interest (ROIs) and comprehensive pathological information like breast mass type, grade, and stage of the tumor (Das et al., 2024). It is mainly limited to scanned film-screen mammography and lags behind more sophisticated imaging methods such as digital breast tomosynthesis (DBT) and full-field digital mammography (FFDM).

### 3.3 Curated Breast Images of a Subset of DDSM

The Curated Breast Images of a Subset of DDSM (CBIS-DDSM) (Lee et al. 2017) dataset is a standardized and adapted version of the DDSM. Originally, DDSM comprises 2620 mammography studies conducted on scanned films, featuring images depicting mild, benign, and malignant cases. Each data entry includes verified pathological information. The CBIS-DDSM series is a meticulously selected and curated subset of DDSM data sourced from professional mammograms. It consists of 1318 images showcasing masses, comprising 711 MLO images and 607 craniocaudal (CC) images. These images were decoded and encoded into DICOM format for ease of study. Additionally, the dataset provides segmentation masks for all tumor regions.

### 3.4 INbreast

The INbreast (Moreira et al. 2012) public dataset comprises full-field digital mammograms (FFDMs) obtained using the Siemens MammoNovation mammography system. These FFDMs include both CC and MLO views collected from 115 cases, totalling of 410 DICOM images. From these FFDMS, a total of 116 masses were identified from 107 samples. The masses are categorized into benign and malignant.

### 3.5 Breast Cancer Digital Repository

The Breast Cancer Digital Repository (BCDR) (Pereira and Papa 2011) is a publicly accessible dataset comprising of mammograms and breast ultrasound images. The dataset maintains to repositories, one with film mammography (BCDR-FM) and one with FFDMs (BCDR-DM). The BCDR-FM contains 3703 mammograms with 1044 clinically detected lesions. The images are in TIFF format, with a spatial resolution of $720 \times 1168$ and 8-bit gray-scales. The BCDR-DM contains 3612 mammography images, with 452 clinically diagnosed lesions. These images are also in TIFF format, but with a higher spatial resolution ($4084 \times 3328$ pixels) and gray-scales (14-bit), stored in TIFF format.

## 4 Computer Aided Diagnosis System for Mass Detection and Classification

CADx systems have been playing an important role in breast cancer detection in aiding radiologists by providing additional assistance and interpretation of mammographic images. These systems typically apply a variety of image processing and machine learning approaches to analyze mammograms and detect potential signs of breast cancer. They typically undergo several key phases to analyze mammographic images effectively as indicated in Fig. 2. Firstly, there is a preprocessing phase where the mammographic images are enhanced and normalized to improve contrast and remove noise, ensuring better quality for subsequent analysis. In the next phase, the regions of interests (ROI) are segmented from the whole mammogram images. Accurate segmentation could facilitate precise lesion detection. But mammographic mass segmentation often suffers from challenges as masses come with different sizes, different shapes, irregular boundaries, low contrast, overlapped dense tissues. Following this, the system moves into a feature extraction phase where various features such as texture, geometry, and density are extracted from ROIs. These features serve as input for the classification phase, where machine learning algorithms or other classifiers are trained to discriminate between benign and malignant masses. Finally, in classification phase, the trained model evaluates new mammograms and provides diagnostic insights, such as highlighting suspicious regions or providing probability scores for malignancy. The performance of the CADx system is often evaluated through rigorous testing and validation procedures to ensure its reliability and effectiveness in clinical settings.

### 4.1 Mass Detection and Segmentation Using Traditional Machine Learning Approaches

In past few decades, various studies on mass segmentation have been made. Segmentation methods can be roughly categorized into a region, clustering, contour, threshold and model based (Oliver et al. 2010). In region-based approaches, spatially connected pixels having similar intensities form groups called regions. These approaches are either based on region growing or on split and merge techniques. In past decades, region growing methods were widely used for segmentation of mammographic masses. Guliato et al. (1998) developed a fuzzy region growing approach to segment mammographic masses. In this, uncertainty present around the tumor boundaries was considered so that the shift from mass to normal tissue can be preserved. Petrick et al. (1999) used gradient information for region growing to segment. In this technique, the initial seed was selected using local maxima and Gaussian filtering is used to create a gradient image. From the gradient image, threshold values of different regions were calculated. The contour-based image segmentation methods relies on the edge information of the ROIs. Petrick et al. (1995, 1996) used Gaussian Laplacian edge detector to localize the edges of mammographic masses. Subsequently, an adaptive density-weighted contrast enhancement filter was used for the enhancement the images.

Gabor filters were used by Parr et al. (1994) to identify the edges of the spiculated masses. Hough transform was applied in (Groshong and Kegelmeyer 1996) to detect circumscribed masses while the same was used in Zhang and Giger (1995) to model

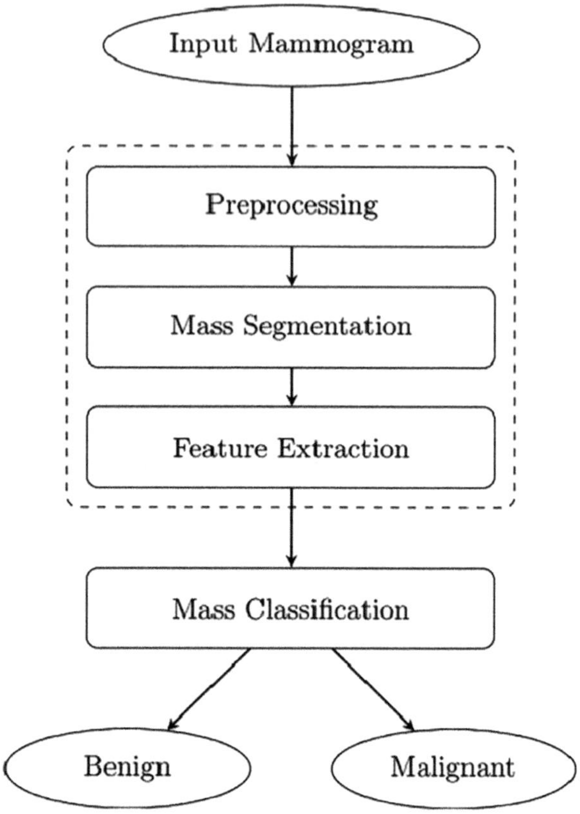


**Fig. 2** Generic framework for computer aided mammographic mass detection system

tissue texture change close to spiculated masses. Few attempts of contour-based image enhancement methods without using edge filtering have been evident in some of the works. One such attempt was made by Hong and Brady (2003) in which a topographic representation of mammograms was used to detect salient regions like pectoral muscle or masses by applying thresholding at different topographic levels. Yin et al. (2003) used a similar approach where an intelligent mesh was used for finding lesions. In some approaches, contour information was used for the refinement of initial segmentation. Timp and Karssemeijer (2004) used dynamic programming-based optimization technique to retrieve the best contour of the masses.

Another most widely used image segmentation techniques is global-thresholding. Masses in mammograms usually exhibit higher intensity values than the surrounding tissues that make global thresholding methods a very popular choice for mass segmentation. However, in most cases, dense tissues are superimposed over masses and in such a scenario global thresholding is not a good choice for mass segmentation (Bhadra et al., 2024). In this case, local thresholding is a better option than global thresholding. Matsubara (1996) developed a histogram-based thresholding algorithm to classify mammogram images into three categories based on the tissue type. Based on the tissue type, suspicious masses are segmented using multiple thresholds. Dominguez and Nandi (2007) proposed a segmentation method using multiple threshold levels. An adaptive thresholding technique to detect potential masses was presented in Varela et al. (2007). Kom et al. (2007) presented a local thresholding method to segment regions containing potential masses. This technique employed two windows of different sizes around the neighborhood of a pixel. Next, a local threshold for the region is adaptively calculated on the basis of the moving average of the smaller window and the intensity difference between the maximum and minimum values in the larger window.

Another popular approach for image segmentation is clustering techniques. These techniques can be grouped into hard clustering and soft clustering. Sahiner et al. (1998) used the K-means clustering method to perform the initial segmentation followed by an edge based refinement. Model-based segmentation follows the supervised paradigm of machine learning. In this, a model is trained to learn the patterns of the ROIs. One such method is pattern matching. Constantinidis et al. (2001) used a pattern matching algorithm for mammographic mass detection using normalized cross-correlation distance as the similarity. Dhungel et al. (2015) developed a mass detection technique using a multiscale deep belief network and Gaussian mixture model. Xu et al. (2005) developed two mass models to segment various types of mass segmentation in which they used iterative thresholding to identify the suspicious regions and then applied a discrete wavelet transform approach to locate the masses. In this, Canny edge detection was used to extract rough mass regions which were further refined by using an active contour model. Yuan et al. (2007) designed an active contour model-based technique for mass segmentation based on geometry. Kupinski and Giger (1998) proposed two mammographic mass segmentation methods: one using a radial gradient index and one using a probabilistic model.

Ball and Bruce (2007a) developed a mass detection method in the polar domain. In this, an adaptive level-set segmentation technique was employed to fine-tune the threshold at each angle. This was further extended in Ball and Bruce (2007b), in which they combined Dixon and Taylor line operators with a level set to segment spiculated lesions. Hassanien and Ali (2004) proposed a spiculated mass segmentation method integrating pulse coupled neural network and fuzzy set theory.

### *4.2 Traditional Machine Learning Approaches on Feature Extraction and Mass Classification*

Feature extraction and mass classification is the final step of a CADx system. The feature space of mammographic masses is very huge and complex due to the wide variety of masses as well as diverged types of tissues. However, all the features are not equally significant; some features are very important while some are not. For an instance, shape and margin play a very important role in classifying masses, as masses with smooth boundaries and regular shapes like circular are mostly found as benign while masses with ill-defined boundaries and irregular shapes are found mostly malignant. Some features may not be useful while used alone for classification but can be very useful if used in combination with other features. The feature selection phase selects the best features that will describe the ROIs. The feature space is categorized into intensity features (contrast measure, mean gray level of ROIs, standard deviation or variance inside ROIs, skewness, kurtosis), geometric features, and texture features.

For mass classification, a variety of methods have been used, out of which the most popular ones are the Artificial Neural Network (ANN), random forest (RF), linear discriminate analysis (LDA), and Support Vector Machines (SVM). Bruce and Adhami (1999) designed an LDA based classifier to classify 60 mammographic lesions into three classes viz., stellate, nodular and round. In this study, the classifier used both conventional and three multiscale shape features. They observed an increase in overall classification accuracy from 72 to 83% after the use of multiscale features. Another LDA based classifier is seen in (Chan et al. 1995). They have used texture features for their classifier; eight features were extracted from each of ROIs. The study used a total of 672 ROIs out of which 168 ROIs were masses and 504 ROIs were normal tissues. An average area under the ROC curve (AUC) of 0.84 and 0.82 is achieved during the training and testing phase respectively. In (Datta 2008), a classification method was presented to classify breast masses from mass spectrometry profiles using LDA. For feature selection, they used the Random Forest method. A texture classification method for texture analysis of masses was presented in (Mavroforakis et al. 2002).

Bovis et al. (2000) used two different classifiers: three-layer backpropagation neural network (BNN) and Radial Basis Network (RBN) to classify suspicious mammographic masses. When compared the results of both the classifiers, it was seen that the recognition rate of the classifiers with BNN and RBN was 72% and 65% respectively. The average Az value of the BNN also showed better results than RBN. Pratiwi et al. (2015) presented a Radial Basis Function Neural Network (RBFNN) based mass classifier. The result showed accuracies of 93.98% and 94.29% for classification of normal and abnormal and classification of benign and malignant respectively. In Jaleel et al. (2014) an RBFNN was applied for benign and malignant classification with an accuracy of 93.7%.

Tan et al. (2004a) used an SVM-based recursive feature elimination (SVM-RFE) approach with a normalized mutual information feature selection (NMIFS) process to assess mammography masses. LDA and SVM classifiers were used to analyze the selected feature sets. Another SVM based feature selection method integrating a normalized mutual information feature selection (NMIFS) was proposed in Moayedi et al. (2010). Compared to existing methods this technique significantly improves the accuracy and efficacy of classifying breast masses as benign or malignant. In one of the early attempts, Liu et al. (2011) evaluated RF to classify masses based on shape and gray level co-occurrence matrix (GLCM) based texture features. Singh et al. (2016) proposed an RF based method to classify whole mammograms into normal and abnormal with an accuracy of 92.24%. Another RF based mammogram classification is presented in Berks et al. (2011). Ramos et al. (2012) made a study on texture features analysis to discriminate normal breast tissues from tissues containing lesion using RF algorithm.

# 5 Deep Learning in Diagnosis of Breast Masses

When it comes to processing medical images, deep learning has several benefits over conventional techniques. Its ability to extract pertinent features from data without the requirement for human feature engineering is a major benefit. Traditional approaches frequently call for expertly created handmade features, which might take time and could not capture all pertinent data. Deep learning methods like convolutional neural networks (CNNs), can automatically extract hierarchical representations of data, which allows them to adjust to different visual complexity and attributes. Deep learning approaches can perform better on a variety of imaging modalities and tasks because of this versatility. Large datasets may also be used to train such models, which enhances their resilience and allows them to generalize effectively to previously unknown data. Moreover, deep learning models can continuously improve with additional data and refinement, offering scalability and adaptability to evolving medical imaging challenges. Overall, these advantages position deep learning as a powerful tool in medical image processing, facilitating more accurate and efficient diagnosis and disease monitoring.

## *5.1 Deep Learning Based Breast Mass Detection and Segmentation Methods*

Deep learning-based approaches gained immense popularity for object detection and localization in recent time. One of such popular deep neural network based object detection method is Region-based Convolutional Neural Network (R-CNN) (Girshick et al. 2014). It generates a large set of proposed regions containing probable object of interest with the help of a moving window. These proposed regions are subsequently classified by a CNN model. Despite showing good performance, the method was very slow and to overcome this problem several variants of it such as Fast R-CNN (Girshick 2015), Faster R-CNN (Ren et al. 2015), Mask R-CNN (He et al. 2017) etc. have been proposed. Another popular sliding window-based object detection method is YOLO (Redmon et al. 2016). It showed very overwhelming results and takes short computation time. But it fails to detect smaller objects.

CNN based object detection algorithms (R-CNN, YOLO) have achieved state-of-the-art performance in medical image analysis. An R-CNN based mass detection approach was presented in Wu et al. (2018). Initially, a hierarchical technique was employed to identify probable mass regions followed by an improved version of Faster R-CNN for the final detection. Min et al. (2020) presented a Mask R-CNN based for automatic detection and segmentation of masses. In another attempt, Agarwal et al. (2020) developed a mass detection framework using faster R-CNN and transfer learning. The framework obtained a sensitivity of $0.91 \pm 0.06$ at 1.69 false positive per image (FPPI) for INbreast dataset.

In Zhang et al. (2022) introduces an anchor-free YOLOv3 to detect the masses in mammograms, addressing challenges in deep learning-based detection networks. Key improvements include a new anchor-free detection approach, the use of generalized intersection over union (GIoU) loss for refining bounding boxes, and the implementation of focal loss for object. Additionally, a new feature aggregation method is proposed. Comparative experiments on INbreast and TXMD datasets show superior performance, achieving 0.95 TPR with 1.7 FPPI on INbreast and 0.94 TPR with 5.75 FPPI on TXMD, outperforming YOLOv3 and existing methods. In Coşkun et al. (2023), a comparative analysis was made to investigate the performance of YOLO based techniques and a transformer-based YOLOv5 for the detection of mammographic masses. In this, Swin Transformers were integrated with popular versions of YOLO. The transformer-based YOLO model of YOLOv5 showed the best performance when trained with $832 \times 832$ image size.

Encoder-Decoder architectures like U-NET (Ronneberger et al. 2015), fully convolution network (FCN) (Long et al. 2015) and their variants are frequently investigated for medical image segmentation. Unlike the sliding window approaches like R-CNN, these networks accept whole images like the conventional segmentation methods rather than image patches. A deep U- Net with a hybrid deep supervision scheme was proposed in Zhang et al. (2018) to segment and classify of masses. In order to enhance the performance of the standard U-Net, Li et al. (2018) combined residual learning to address gradient vanishing and explosion problems, especially when deepening the CNN architecture. Li et al. (2019) presented a new encoder-decoder architecture system for mammographic mass segmentation. In this study, an asymmetrical encoder-decoder architecture was incorporate with a proposed attention guided dense upsampling block to perform accurate mass segmentation. Baccouche et al. (2021) introduces Connected-UNets by connecting two standard UNets using skip connection to segment breast masses. A double shot model to detect and subsequently segment out mammographic masses was presented in Su et al. (2022) by integrating YOLOV5-L6 and Local Global (LOGO). YOLOV5-L6 initially identifies tumors in breast cancer images, followed by LOGO for segmentation. Achieving F1-Score and IoU of 74.52% and 69.37%, respectively, on the CBIS-DDSM dataset, and 69.37% and 61.09%, respectively, on the INbreast dataset. The model, in Tiryaki and Tutkun (2024), proposed U-net++ uses pretrained Xception weights in the encoder region for mass segmentation. The model produced quite promising results with an AUC of 0.7829 for mass segmentation.

Zhu et al. (2018) proposed an adversarial network using FCN and conditional random field for the segmentation of mammographic lesions. They used adversarial training to reduce the problem of over-fitting associated with small dataset size. Another attempt of mass segmentation using FCN is seen in Hang et al. (2017). A multichannel, multiscale FCN was proposed (Xu et al. 2020) to segment masses by incorporating pre-processed images. The technique utilized multiscale feature for global mass localization and fine-border delineation. The evaluation results showed good dice similarity coefficient and Hausdorff distance demonstrating reliable mass segmentation capabilities.

A deep learning model was developed in Shen t al. (2019) to accurately identify breast cancer on screening mammograms. This technique optimally utilizes training datasets that include either thorough clinical annotations or just the image's overall cancer state. The algorithm was tested independently on CBIS-DDSM and INbreast datsets. The algorithm achieved an AUC of 0.91 and 0.98 per image respectively for CBIS-DDSM and INbreast. The technique proposed in Zahoor et al. (2023) employed the Skipping Dilated semantic segmentation approach for mass abnormalities segmentation in mammogram images. It addresses class imbalance issues by incorporating class weights and enhances contextual awareness through dilation convolution. Evaluation on INbreast and CBIS-DDSM datasets shows impressive performance, with a weighted Intersection over Union (WIOU) of 98.51% for INbreast and 94.82% for CBIS-DDSM achieved by the Skipping Dilated Semantic CNN.

## 5.2 *Deep Learning Based Methods for Classification of Mammographic Masses*

Recent research reports that the application of deep learning has significantly improved the performance of CADx systems. Jiao et al. (2016) proposed a deep convolutional network for breast mass classification. The method showed an accuracy of 96.7% when evaluated on 600 images of DDSM dataset. A cross-model deep learning scheme was presented in Hadad et al. (2017) for the detection of breast lesion. Jadoon et al. (2017) presented a descriptive CNN features based classification technique to classify the mammograms into normal, benign and malignant. Liu et al. (2018), embedded fully-connected layers before the first convolutional layer of CNN model. The experimental result showed this approach outperforms both conventional MLP and conventional CNN based classifiers for the same datasets. Dhungel et al. (2017) presented an integrated framework to detect, segment, and classify masses using deep learning. In this, a CNN model trained for estimating handcrafted features which are eventually fine-tuned to perform mass classification. In Suh et al. (2020), a breast lesion detection technique using Dense-Net-169 and Efficient-Net-B5 was proposed. This approach was tested on a private dataset of 301 images with very good accuracies of 95.2% and 95.4% for Dense-Net-169 and Efficient-Net-B5 respectively. Additionally, Matthews Correlation Coefficient (MCC) values for both networks decline for both Dense-Net-169 (66.5%) and Efficient-Net-B5 (76%). Yan et al. (2021) presented a method to enhance breast cancer detection by integrating information from both CC and MLO views. In their study (Kavitha 2021), the authors introduced a breast lesion segmentation method by integrating multilevel thresholding with Shell Game Optimization algorithms. The method showed accuracies of 98.5% and 97.55% respectively when evaluated on Mini-MIAS and DDSM datasets.

Hassan et al. (2020) presented a deep CNN based classification method for breast masses using by using pretrained AlexNet and GoogleNet. Experimental evaluation on CBIS-DDSM, INbreast, Egyptian National Cancer Institute (NCI), and MIAS, demonstrates the superior performance of the proposed AlexNet model. Achieving 100% accuracy on CBIS-DDSM and INbreast databases, and high accuracy and AUC scores on NCI and MIAS databases, the AlexNet model outperforms GoogleNet. These results highlight the robustness and efficacy of the proposed AlexNet model, surpassing previous methods and achieving state-of-the-art performance on various datasets. Ayana et al. (2022) proposed a transfer learning-based approach to classify masses into benign or malignant. The model used for transfer learning was pretrained on ImageNet as well as with cancer cell line images. The model was evaluated on three public datasets and a mixed dataset, the model achieves highest AUC scores of 1 for DDSM dataset. Compared to patch-based methods, the proposed approach demonstrates statistically significant performance improvement (p-value = 0.0029). Additionally, it outperforms patch-based and full image-based methods, achieving a 91.41% test accuracy, reducing computational burden, and enhancing early breast cancer diagnosis. In (Escorcia-Gutierrez et al. 2022), an automated DL based method was proposed to detect breast cancers using mammograms. Here, images were pre-processed using Gaussian filtering and the ROIs were segmented using Tsallis entropy. The features were extracted using Residual Network (ResNet 34), following this a wavelet neural network (WNN) was employed for the classification of the segmented ROIs. Testing the method on 322 images from the mini-MIAS dataset resulted in an accuracy of 96.07%.

Ma and Peng (2024) presented a cross-view variational autoencoder (CV-VAE) based framework to segment and classify masses. The method achieves impressive results, with Dice similarity coefficients of 92.46% and 93.70% for segmentation, and AUC scores of 93.20% and 95.01% for classification on public datasets. Jyoti et al. (2023) introduced a CAD system to segment and classify masses. The method uses ResNet50 for segmentation of ROIs and SVM for classification. Li et al. (2024) utilized a knowledge transfer-based approach is to achieve improvements in CNN by transferring attention maps from binary masks onto regions of interest (ROIs). In experiments conducted on the BCDR dataset, it is shown that application of the DenseNet121 model increased mass classification accuracy up to 71% comparatively raised from the results where no knowledge transfer technique was conducted and with mass classification accuracy merely achieving 58%. Laishram and Rabidas (2024) introduces a novel framework hybrid, namely, deep and machine learning methodology, which combines the application of both methodologies for mammographic mass classification. The models achieve a sensitivity of 93.8% on the mini-MIAS and 92.3% on DDSM.

## 6 Discussion and Conclusion

This paper explores various techniques within a CADx system for diagnosing breast cancer using digital mammography images. CADx systems have significantly transformed the landscape of breast cancer detection, providing crucial support to radiologists in interpreting mammographic images. These systems integrate different pattern recognition and machine learning algorithms to perform analysis of mammograms and identify potential signs of breast cancer. The process typically involves phases such as preprocessing, segmentation, feature extraction, and classification. The job of the preprocessing techniques enhances image quality for subsequent analysis. Segmentation methods then identify potential regions of interest (ROIs) containing suspicious masses. Feature extraction techniques quantify characteristics like texture, shape, and density within these ROIs. Lastly, classification algorithms distinguish between benign and malignant masses using their features. A summary of the important methods for mammographic mass segmentation and detection is presented in Table 1, while Table 2 summarizes some important mass classification methods discussed in this paper. These tables include traditional as well as deep learning-based techniques.

This paper reviewed both traditional machine learning and deep learning approaches for mass detection, segmentation, and classification. Incorporating traditional machine learning models provides a comprehensive perspective of the landscape of breast cancer detection methodologies. It allows for a comparison between the performance, strengths, and weaknesses of both traditional and deep learning approaches, enabling researchers and practitioners to make informed decisions about which methods are most suitable for their specific needs. For many years, traditional machine learning models have played a significant role in the field of medical image analysis. Incorporating these models into the review provides readers with a deeper insight into the historical background and evolution of breast cancer detection techniques. This contextual understanding is valuable for appreciating the progression and advancements in the field.

For ROI segmentation various methods various techniques related to region-based, contour-based, clustering-based, thresholding-based, model-based and CNN-based have been explored. Classical segmentation methods offer interpretability, computational efficiency, and may be appropriate with limited labelled data. In contrast, deep learning-based segmentation approaches deliver scalability, generalization, and end-to-end learning capabilities, making them particularly effective for complex segmentation tasks involving large datasets.

In traditional approaches to breast mass classification, the efficiency lies in the meticulous crafting of features derived from intensity, geometry, and texture attributes of mammographic images. These handcrafted features which are carefully engineered to encapsulate domain-specific knowledge and insights serve as crucial indicators to differentiate between benign and malignant masses. The classification process predominantly employs established techniques such as ANN, SVM,

**Table 1** An overview of the key methods developed for mammographic mass detection and segmentation

| Techniques used | Datasets used | Number of images/ ROIs used | Results |
|---|---|---|---|
| Faster RCNN (Wu et al. 2018) | INbreast | 107 | TPR of 0.88 at 0.75 FFPI |
| Mask R-CNN (Min et al. 2020) | INbreast | 410 | TPR of 0.90 at 0.9 FPPI |
| R-CNN (Agarwal et al. 2020) | INbreast | 410 | 0.91 ± 0.06 at 1.69 |
| Region-growing (Petrick et al. 1999) | Private | 253 | TPR of 90% at 4.2 FPPI, TPR of 80% at 2 FPPI |
| Edge detection algorithms (Petrick et al. 1995) | Private | 84 | TPR of 80% at 2.4 FPPI |
| Edge detection algorithms (Petrick et al. 1996) | Private | 85 | TPR of 90% at 4.4 FPPI, TPR of 80% at 2.3 FPPI |
| Contour-based (Zhang and Giger 1995) | MIAS | 48 | 95% segmentation accuracy |
| Dynamic programming (Hong et al. 2003) | Private | 1210 | 69% Mean Overlap with manually segmented masses |
| Adaptive thresholding (Matsubara 1996) | MIAS | 57 | TPR 80% at 2.3 FPPI |
| Adaptive thresholding (Dominguez et al. 2007) | Private | 198 | TPR of 88% and 94% at 1.02 FPPI for lesion-based and case-based evaluation |
| Adaptive thresholding (Varela et al. 2007) | Private | 61 | TPR of 95.91% |
| Deep learning and random forests (Dhungel et al. 2015) | INbreast, DDSM-BCRP | | TPR 0.96 ± 0.03 at 1.2FPPI on INbreast and TPR of 0.75 at 4.8FPPI on DDSM-BCRP |
| Model-based (Xu et al. 2005) | Private | 107 masses | Mean precision 95.1% |
| Active contour model (Yuan et al. 2007) | Private | 739 | 85% accuracy at overlap threshold 0.4 |
| Model-based (Kupinski et al. 1998) | Private | 118 | 96% accuracy at overlap threshold 0.3 |

(continued)

**Table 1** (continued)

| Techniques used | Datasets used | Number of images/ ROIs used | Results |
|---|---|---|---|
| UNET (Li et al. 2019) | INbreast, DDSM-BCRP | 344 ROIs | Dice index: 93.66% and 91.14% for INbreast and DDSM-BCRP respectively |
| UNET, Attention Gates (Zahoor et al. 2023) | DDSM | 2500 | F1-socre: 82.24%, sensitivity: 77.89% and accuracy:78.38% |
| UNET, Atrous Spatial Pyramid Pooling (Su et al. 2022) | CBIS-DDSM, INbreast, Private | 1963 | Dice scores: (89.52%, 95.28%, 95.88%), IoU: (80.02%, 91.03%, 92.27%) respectively for CBIS-DDSM, INbreast, Private |
| YOLOV5-L6 and Local Global (LOGO) (Zhu et al. 2018) | | 1574 | F1-Score: 74.52%, IoU: 69.37%, for CBIS-DDSM dataset, and F1-Score: 69.37%, IoU: 61.09%, for INbreast dataset |
| FCN and conditional random field (Hang et al. 2017) | INbreast, DDSM-BCRP | 1783 | Dice Score of 90.97% and 91.30% for INbreast and DDSM-BCRP respectively |
| CNN (Baccouche et al. 2021) | DDSM, INbreast | 2500 | Dice coefficient: 0.915 ± 0.031, Hausdorff distance: 6.257 ± 3.380, for DDSM and Dice coefficient: 0.918 ± 0.038, Hausdorff distance: 2.572 ± 0.956, for INbreast |
| UNET (Xu et al. 2020) | DDSM | 400 | F1-socre: 82.24%, sensitivity: 77.89% and accuracy: 78.38% |
| GlimpseNet (Li et al. 2018) | DDSM | 3800 | 66.2% accuracy |
| U-Net (Bruce and Adhami 1999) | INbreast | 116ROIs | Dice coefficient: 0.85, classification accuracy: 89% |

The terms TPR, FPPI, ROC, AUC, IoU indicate True Positive Rates, False Positives Per Image, Receiver's Operative Characteristics Curve, Area Under the Curve and Intersection over Union respectively

RF, among others. The efficacy of traditional classification methods hinges significantly upon the quality and relevance of these handcrafted features. Engineers and researchers meticulously design algorithms to extract features that encapsulate key characteristics of breast masses, including their shape, texture patterns, and intensity distributions.

These features are carefully chosen to capture meaningful information that aids in the accurate categorization of masses. However, this approach can be labour-intensive and may not fully exploit the wealth of information present in complex medical images. As a result, the performance of traditional methods may plateau when faced with highly nuanced or heterogeneous datasets. In this regard deep

**Table 2** An overview of the key methods developed for mammographic mass classification

| Techniques used | Datasets used | Number of Images/ ROIs used | Results |
|---|---|---|---|
| Wavelet transform, LDA (Ball 2007) | Private | 60 | Classification accuracy of 83% and 80% with multiresolution and uniresolution shape features |
| Adaptive level set segmentation (Chan et al. 1995) | DDSM | 60 | 87% overall accuracy, AZ = 0.9687 w |
| (Xu et al. 2020) | Private | 672 ROIs | AUC of ROC curve of 0.82 |
| LDA (Bovis et al. 2000) | Private | 153 | Sensitivity: 90.8%, specificity: 84.4% |
| Backpropagation Neural, Network and Radial Basis Network (Pratiwi et al. 2015) | MIAS | 72 | Accuracy: 77%, Az of ROC:0.74 |
| GLCM, Radial Basis Function Neural Network (Tan et al. 2014) | MIAS | 322 | Classification Accuracy: 93.98% for Normal-Abnormal and 94.29% for benign and malignant |
| GLCM, Radial Basis Function Neural Network (Liu and Tang 2013) | MIAS | 148 | Accuracy: 93.7% accuracy |
| SVM (Moayedi et al. 2010) | Private | 1200 | AUC of 0.805 ± 0.012 |
| SVM (Liu et al. 2011) | DDSM | 826 | 93% accuracy |
| SVM, support vector-based fuzzy neural network (SVFNN), and kernel SVM (Singh et al. 2016) | MIAS | 322 | 96.6%, 91.5% and 82.1% accuracy with SVM, SVFNN and kernel SVM, |
| RF, SVM (Berks et al. 2011) | DDSM | 236 ROIs | Az value of 0.86 with SVM and 0.83 with random forest |
| RF (Hadad et al. 2017) | MIAS | 322 | 92.24% |
| CNN, transfer learning (Jadoon et al. 2017) | DDSM | 600 | 96.7% accuracy |
| CNN, DW, curvelet transform (Liu et al. 2018) | IRMA | 2796 | 83.74% accuracy |

(continued)

**Table 2** (continued)

| Techniques used | Datasets used | Number of Images/ ROIs used | Results |
|---|---|---|---|
| CNN (Suh et al. 2020) | WDBC and WBCD | | Accuracy: 99.28%, sensitivity: 98.65%, and specificity: 99.57% for WDBC, and accuracy: 98.71%, sensitivity: 97.60%, and specificity: 99.43% for WBCD |
| Active contour model, CNN (Kavitha 2021) | INbreast | 410 | Segmentation: TPR 90% at 1 FFPI, classification: sensitivity of 0.98 and specificity of 0.7 |
| Siamese network, YOLO (Hassan et al. 2020) | INbreast, CBIS-DDSM | 621 | 94.78% as AUC |
| Optimal Multi-Level Thresholding, Capsule Network (Yan et al. 2021) | MIAS, DDSM | 13,450 | Accuracy: 98.50% for Mini-MIAS and 97.55% for DDSM dataset, respectively |
| DenseNet-169, EfficientNet-B5 (Suh et al. 2020) | Private | 3002 | Accuracy: 95.2% ± 0.005, 95.4% ± 0.020 for DenseNet-169 and EfficientNet-B5 respectively |
| ResNet34, chimp, optimization algorithm, wavelet neural network (Jiao et al. 2016) | MIAS | 322 | 96.07% accuracy |
| Skipping Dilated semantic segmentation (Zahoor et al. 2023) | INbreast, CBIS-DDSM | 1874 | Accuracy of 98.51% and 94.82% are achieved for INbreast and CBIS-DDSM respectively |
| EfficientNetB2, Transfer Learning (Ayana et al. 2022) | DDSM, INbreast, MIAS | 11,212 | AUC:1 for DDSM, 0.9994 for INbreast, 0.9993, and 0.9998 for mixed datasets |

learning offers several advantages over traditional methods. Recent developments in deep learning have transformed breast cancer diagnosis, delivering enhanced performance compared to traditional methods. Convolutional neural networks (CNNs) are deep learning models that can automatically learn pertinent features straight from data, doing away with the requirement for human feature engineering. Beyond classification, CNN-based approaches have demonstrated exceptional performance in both mass detection and segmentation. Techniques such as R-CNN, Fast R-CNN, YOLO, and auto-encoders (e.g., UNET, FCN) have achieved state-of-the-art results, pushing the boundaries of accuracy and efficiency in these tasks. Additionally, deep learning models can utilize transfer learning techniques, where models pretrained on large-scale datasets (such as ImageNet) are adapted and fine-tuned for specific medical imaging. This approach can mitigate the need for larger labelled medical data and accelerate model training.

Despite of having huge success, the deep learning models suffer from the various challenges which include:

- Data dependency: For training, deep learning models need huge amounts of labelled data. Limited or imbalanced datasets (where one class, like malignant masses, is much rarer) can lead to biases and hinder generalizability.
- Interpretability: The decision-making process of deep learning models is often opaque, as they are commonly referred to as "black boxes," making it hard to interpret how they arrive at their conclusions. There may be issues with this lack of interpretability in important medical applications.
- Computational cost: The process of training deep learning models demands significant computational resources, which can result in significant costs and extended processing durations.
- False positives and false negatives: While deep learning models can achieve high accuracy, they can still generate false positives (identifying benign masses as malignant) and false negatives (missing malignant masses). This can lead to unnecessary biopsies and missed diagnoses, respectively.

Researchers are actively exploring ways to address these challenges and further enhance deep learning for breast mass detection, including techniques like data augmentation, transfer learning, explainable AI, and multimodal learning. In conclusion, CADx systems that combine traditional machine learning with deep learning methodologies have made substantial strides in breast cancer detection and diagnosis. By integrating advanced image processing techniques with robust machine learning algorithms, these systems provide invaluable support to radiologists, significantly enhancing diagnostic accuracy and efficiency in clinical practice. Ongoing research and innovation in this field hold great promise for further improving CADx systems, paving the way for better patient outcomes and a more effective approach to combating breast cancer.

## References

Agarwal, R., Diaz, O., Yap, M. H., Lladó, X., & Martí, R. (2020). Deep learning for mass detection in full-field digital mammograms. *Computers in Biology and Medicine*, 103774. https://doi.org/10.1016/j.compbiomed.2020.103774

Agrawal, G., Su, M.-Y., Nalcioglu, O., Feig, S. A., & Chen, J.-H. (2009). Significance of breast lesion descriptors in the ACR BI-RADS MRI lexicon. *Cancer: Interdisciplinary International Journal of the American Cancer Society, 115*(7), 1363–1380. https://doi.org/10.1002/cncr.24167

Ayana, G., Park, J., & Choe, S.-W. (2022). Patchless multi-stage transfer learning for improved mammographic breast mass classification. *Cancers, 14*(5), 1280. https://doi.org/10.3390/cancers14051280

Baccouche, A., Garcia-Zapirain, B., Castillo Olea, C., & Elmaghraby, A. S. (2021). Connected-UNets: A deep learning architecture for breast mass segmentation. *NPJ Breast Cancer, 7*(1), 151. https://doi.org/10.1038/s41523-021-00331-z

Ball, J. E., & Bruce, L. M. (2007a). Digital mammogram spiculated mass detection and spicule segmentation using level sets. In *2007 29th Annual International Conference of the IEEE Engineering in Medicine and Biology Society* (pp. 4979–4984). IEEE. https://doi.org/10.1109/IEMBS.2007.4353431

Ball, J. E., & Bruce, L. M. (2007b). Digital mammographic computer-aided diagnosis (CAD) using adaptive level set segmentation. *29th Annual International Conference of the IEEE Engineering in Medicine and Biology Society*, 4973–4978. https://doi.org/10.1109/IEMBS.2007.4353430

Berks, M., Chen, Z., Astley, S., & Taylor, C. (2011). Detecting and classifying linear structures in mammograms using random forests. In *Biennial International Conference on Information Processing in Medical Imaging* (pp. 510–524). Springer. https://doi.org/10.1007/978-3-642-22092-0_42

Bhadra, S.; Kumar, C.J.; Bhattacharyya, D.K. (2024). Multiview EEG signal analysis for diagnosis of schizophrenia: an optimized deep learning approach. *Multimedia Tools and Applications.* https://doi.org/10.1007/s11042-024-20205-y

Bovis, K., Singh, S., Fieldsend, J., & Pinder, C. (2000). Identification of masses in digital mammograms with MLP and RBF nets. *Proceedings of the IEEE-INNS-ENNS International Joint Conference on Neural Networks*, vol. 1, pp. 342–347. IEEE. https://doi.org/10.1109/IJCNN.2000.857822

Breastcancer.org. (2024). Calcifications on mammograms. Retrieved April 10, 2024, from https://www.breastcancer.org/screening-testing/mammograms/calcifications

Bruce, L. M., & Adhami, R. R. (1999). Classifying mammographic mass shapes using the wavelet transform modulus-maxima method. *IEEE Transactions on Medical Imaging, 18*(12), 1170–1177. https://doi.org/10.1109/42.819330

Chan, H.-P., Wei, D., Helvie, M. A., Sahiner, B., Adler, D. D., Goodsitt, M. M., & Petrick, N. (1995). Computer-aided classification of mammographic masses and normal tissue: Linear discriminant analysis in texture feature space. *Physics in Medicine & Biology, 40*(5), 857. https://doi.org/10.1088/0031-9155/40/5/003

Constantinidis, A., Fairhurst, M. C., & Rahman, A. F. R. (2001). A new multi-expert decision combination algorithm and its application to the detection of circumscribed masses in digital mammograms. *Pattern Recognition, 34*(8), 1527–1537. https://doi.org/10.1016/S0031-3203(00)00101-7

Coşkun, D., Karaboğa, D., Baştürk, A., Akay, B., Nalbantoğlu, Ö. U., Doğan, S., Pacal, I., & Karagöz, M. A. (2023). A comparative study of YOLO models and a transformer-based YOLOv5 model for mass detection in mammograms. *Turkish Journal of Electrical Engineering and Computer Sciences, 31*(7), 1294–1313. https://doi.org/10.55730/1300-0632.4106

Das, P. R., Talukdar, R. R., & Kumar, C. J. (2024). Exploring the interplay of abstinence self-efficacy, locus of control, and perceived social support in substance use disorder recovery. *Current Medical Research and Opinion, 40*(9), 1625–1635. https://doi.org/10.1080/03007995.2024.2390046

Datta, S. (2008). Classification of breast cancer versus normal samples from mass spectrometry profiles using linear discriminant analysis of important features selected by random forest. *Statistical Applications in Genetics and Molecular Biology, 7*(2). https://doi.org/10.2202/1544-6115.1376

Dhungel, N., Carneiro, G., & Bradley, A. P. (2015). Automated mass detection in mammograms using cascaded deep learning and random forests. *2015 International Conference on Digital Image Computing: Techniques and Applications (DICTA)*, 1–8. https://doi.org/10.1109/DICTA.2015.7371270

Dhungel, N., Carneiro, G., & Bradley, A. P. (2017). A deep learning approach for the analysis of masses in mammograms with minimal user intervention. *Medical Image Analysis, 37*, 114–128. https://doi.org/10.1016/j.media.2017.01.009

Doi, K. (2007). Computer-aided diagnosis in medical imaging: Historical review, current status, and future potential. *Computerized Medical Imaging and Graphics, 31*(4-5), 198–211. https://doi.org/10.1016/j.compmedimag.2007.02.002

Dominguez, A. R., & Nandi, A. K. (2007). Enhanced multi-level thresholding segmentation and rank-based region selection for detection of masses in mammograms. In *2007 IEEE International Conference on Acoustics, Speech and Signal Processing* (ICASSP '07), 449. IEEE. https://doi.org/10.1109/ICASSP.2007.366912

Eadie, L. H., Taylor, P., & Gibson, A. P. (2012). A systematic review of computer-assisted diagnosis in diagnostic cancer imaging. *European Journal of Radiology, 81*(1), 70–76. https://doi.org/10.1016/j.ejrad.2010.12.026

Ekpo, E. U., Alakhras, M., & Brennan, P. (2018). Errors in mammography cannot be solved through technology alone. *Asian Pacific Journal of Cancer Prevention (APJCP), 19*(2), 291–296. https://doi.org/10.22034/APJCP.2018.19.2.291

Escorcia-Gutierrez, J., Mansour, R. F., Beleno, K., Jimenez-Cabas, J., Pérez, M., Madera, N., & Velasquez, K. (2022). Automated deep learning empowered breast cancer diagnosis using biomedical mammogram images. *Computers, Materials and Continua, 71*(3), 4221–4241. https://doi.org/10.32604/cmc.2022.020517

Garcia, D., Spruill, L. S., Irshad, A., Wood, J., Kepecs, D., & Klauber-DeMore, N. (2018). The value of a second opinion for breast cancer patients referred to a national cancer institute (NCI)-designated cancer center with a multidisciplinary breast tumor board. *Annals of Surgical Oncology, 25*(10), 2953–2957. https://doi.org/10.1245/s10434-018-6645-2

Girshick, R. (2015). Fast R-CNN. In *Proceedings of the IEEE International Conference on Computer Vision* (pp. 1440–1448). https://doi.org/10.1109/ICCV.2015.169

Girshick, R., Donahue, J., Darrell, T., & Malik, J. (2014). Rich feature hierarchies for accurate object detection and semantic segmentation. In *Proceedings of the IEEE Conference on Computer Vision and Pattern Recognition* (pp. 580–587). https://doi.org/10.1109/CVPR.2014.81

Groshong, B. R., & Kegelmeyer, W. P. Jr. (1996). Detecting circumscribed lesions with the Hough transform. *Medical Imaging 1996: Image Processing, 2710*, 59–70. International Society for Optics and Photonics. https://doi.org/10.1117/12.240772

Guliato, D., Rangayyan, R. M., Carnielli, W. A., Zuffo, J., & Desautels, J. (1998). Segmentation of breast tumors in mammograms by fuzzy region growing. In *Proceedings of the 20th Annual International Conference of the IEEE Engineering in Medicine and Biology Society, vol. 2*, 1002–1005. IEEE. https://doi.org/10.1109/IEMBS.1998.745638

Guray, M., & Sahin, A. A. (2006). Benign breast diseases: Classification, diagnosis, and management. *The Oncologist, 11*(5), 435–449. https://doi.org/10.1634/theoncologist.11-5-435

Hadad, O., Bakalo, R., Ben-Ari, R., Hashoul, S., & Amit, G. (2017). Classification of breast lesions using cross-modal deep learning. In *2017 IEEE 14th International Symposium on Biomedical Imaging* (ISBI 2017) (pp. 109–112). IEEE. https://doi.org/10.1109/ISBI.2017.7950480

Hang, W., Liu, Z., & Hannun, A. (2017). GlimpseNet: Attentional methods for full-image mammogram diagnosis. *Stanford University Reports*. Retrieved from http://cs231n.stanford.edu/reports/2017/pdfs/517.pdf

Hassan, S. A., Sayed, M. S., Abdalla, M. I., & Rashwan, M. A. (2020). Breast cancer masses classification using deep convolutional neural networks and transfer learning. *Multimedia Tools and Applications, 79*(41), 30735–30768. https://doi.org/10.1007/s11042-020-08967-6

Hassanien, A. E., & Ali, J. M. (2004). Digital mammogram segmentation algorithm using pulse coupled neural networks. In *Third International Conference on Image and Graphics (ICIG'04)* (pp. 92–95). IEEE. https://doi.org/10.1109/ICIG.2004.47

He, K., Gkioxari, G., Dollár, P., & Girshick, R. (2017). Mask R-CNN. In *Proceedings of the IEEE International Conference on Computer Vision* (pp. 2961–2969). https://doi.org/10.1109/ICCV.2017.322

Heath, M., Bowyer, K., Kopans, D., Moore, R., & Kegelmeyer, W. (2001). The digital database for screening mammography. *Proceedings of the Fifth International Workshop on Digital Mammography*, 212–218. https://doi.org/10.1007/978-94-017-2384-2_30

Hong, B.-W., & Brady, M. (2003). A topographic representation for mammogram segmentation. *Medical Image Computing and Computer-Assisted Intervention - MICCAI 2003: 6th International Conference*, Montréal, Canada, November 15–18, 2003. *Proceedings, Vol. 6*, 730–737. Springer.

Jadoon, M. M., Zhang, Q., Haq, I. U., Butt, S., & Jadoon, A. (2017). Three-class mammogram classification based on descriptive CNN features. *BioMed Research International*, 2017. https://doi.org/10.1155/2017/7941428

Jaleel, J. A., Salim, S., & Archana, S. (2014). Textural features-based computer-aided diagnostic system for mammogram mass classification. In *2014 International Conference on Control, Instrumentation, Communication and Computational Technologies (ICCICCT)* (pp. 806–811). IEEE.

Jiao, Z., Gao, X., Wang, Y., & Li, J. (2016). A deep feature-based framework for breast masses classification. *Neurocomputing*, 197, 221–231. https://doi.org/10.1016/j.neucom.2016.03.059

Kavitha, T., Mathai, P. P., Karthikeyan, C., Ashok, M., Kohar, R., Avanija, J., & Neelakandan, S. (2021). Deep learning based capsule neural network model for breast cancer diagnosis using mammogram images. *Interdisciplinary Sciences: Computational Life Sciences*, 1–17. https://doi.org/10.1007/s12539-021-00442-y

Kom, G., Tiedeu, A., & Kom, M. (2007). Automated detection of masses in mammograms by local adaptive thresholding. *Computers in Biology and Medicine*, 37(1), 37–48. https://doi.org/10.1016/j.compbiomed.2006.03.005

Kupinski, M. A., & Giger, M. L. (1998). Automated seeded lesion segmentation on digital mammograms. *IEEE Transactions on Medical Imaging*, 17(4), 510–517. https://doi.org/10.1109/42.688174

La Vecchia, C., Negri, E., Bruzzi, P., Franceschi, S., Bucchi, L., & Parazzini, F. (1993). The impact of mammography on breast cancer detection. *Annals of Oncology*, 4(1), 41–44. https://doi.org/10.1093/oxfordjournals.annonc.a058870

Laishram, R., & Rabidas, R. (2024). An optimized ensemble classifier for mammographic mass classification. *Computers and Electrical Engineering*, 119, 109488. https://doi.org/10.1016/j.compeleceng.2024.109488

Lee, R. S., Gimenez, F., Hoogi, A., & Rubin, D. L. (2017). Curated breast imaging subset of DDSM. *The Cancer Imaging Archive*. https://doi.org/10.7937/K9/TCIA.2016.7O02S9CY

Li, G., & Zwiggelaar, R. (2024, May). Improving the CNNs performance of mammography mass classification via binary mask knowledge transfer. In *17th International Workshop on Breast Imaging (IWBI 2024)* (Vol. 13174, pp. 493–499). SPIE. https://doi.org/10.1117/12.2608455

Li, H., Chen, D., Nailon, W. H., Davies, M. E., & Laurenson, D. (2018). Improved breast mass segmentation in mammograms with conditional residual U-net. In *Image Analysis for Moving Organ, Breast, and Thoracic Images: Third International Workshop, RAMBO 2018, Fourth International Workshop, BIA 2018, and First International Workshop, TIA 2018, Held in Conjunction with MICCAI 2018, Granada, Spain, September 16 and 20, 2018* (pp. 81–89). Springer. https://doi.org/10.1007/978-3-030-04632-6_9

Li, S., Dong, M., Du, G., & Mu, X. (2019). Attention dense-U-net for automatic breast mass segmentation in digital mammogram. *IEEE Access*, 7, 59037–59047. https://doi.org/10.1109/ACCESS.2019.2891021

Liu, J., Chen, J., Liu, X., & Tang, J. (2011). An investigation of mass diagnosis in mammogram with random forest. In *The Fourth International Workshop on Advanced Computational Intelligence* (pp. 638–641). IEEE. https://doi.org/10.1109/WACI.2011.5960351

Liu, K., Kang, G., Zhang, N., & Hou, B. (2018). Breast cancer classification based on fully-connected layer first convolutional neural networks. *IEEE Access*, 6, 23722–23732. https://doi.org/10.1109/ACCESS.2018.2821561

Liu, X., & Tang, J. (2013). Mass classification in mammograms using selected geometry and texture features, and a new SVM-based feature selection method. *IEEE Systems Journal*, 8(3), 910–920. https://doi.org/10.1109/JSYST.2013.2248442

Long, J., Shelhamer, E., & Darrell, T. (2015). Fully convolutional networks for semantic segmentation. In *Proceedings of the IEEE Conference on Computer Vision and Pattern Recognition* (pp. 3431–3440). https://doi.org/10.1109/CVPR.2015.7298965

Lu, W., Jansen, L., Post, W., Bonnema, J., Velde, J., & De Bock, G. (2009). Impact on survival of early detection of isolated breast recurrences after the primary treatment for breast cancer: A meta-analysis. *Breast Cancer Research and Treatment*, 114, 403–412. https://doi.org/10.1007/s10549-008-0222-5

Ma, Y., & Peng, Y. (2024). Mammogram mass segmentation and classification based on cross-view VAE and spatial hidden factor disentanglement. *Physical and Engineering Sciences in Medicine*, 47(1), 223–238. https://doi.org/10.1007/s13246-024-01320-6

Masud, R., Al-Rei, M., Lokker, C., et al. (2019). Computer-aided detection for breast cancer screening in clinical settings: Scoping review. *JMIR Medical Informatics*, 7(3), e12660. https://doi.org/10.2196/12660

Matsubara, T. (1996). Development of mass detection algorithm based on adaptive thresholding technique in digital mammograms. *Digital Mammography '96*, 391–396

Mavroforakis, M. E., Georgiou, H. V., Cavouras, D., Dimitropoulos, N., & Theodoridis, S. (2002). Mammographic mass classification using textural features and descriptive diagnostic data. *2002 14th International Conference on Digital Signal Processing Proceedings. DSP 2002*, Vol. 1, 461–464. IEEE. https://doi.org/10.1109/ICDSP.2002.1020783

McKenna, R. J. Sr. (1994). The abnormal mammogram: Radiographic findings, diagnostic options, pathology, and stage of cancer diagnosis. *Cancer*, 74(S1), 244–255. https://doi.org/10.1002/1097-0142(19940701)74:1+<244::AID-CNCR2820741309>3.0.CO;2-Q

Min, H., Wilson, D., Huang, Y., Liu, S., Crozier, S., Bradley, A. P., & Chandra, S. S. (2020). Fully automatic computer-aided mass detection and segmentation via pseudo-color mammograms and mask R-CNN. In *2020 IEEE 17th International Symposium on Biomedical Imaging (ISBI)* (pp. 1111–1115). IEEE. https://doi.org/10.1109/ISBI45719.2020.9098266

Moayedi, F., Azimifar, Z., Boostani, R., & Katebi, S. (2010). Contourlet-based mammography mass classification using the SVM family. *Computers in Biology and Medicine*, 40(4), 373–383. https://doi.org/10.1016/j.compbiomed.2009.12.006

Moreira, I. C., Amaral, I., Domingues, I., Cardoso, A., Cardoso, M. J., & Cardoso, J. S. (2012). INbreast: Toward a full-field digital mammographic database. *Academic Radiology, 19*(2), 236–248. https://doi.org/10.1016/j.acra.2011.09.014

NBCF (National Breast Cancer Foundation). (2024). Breast tumors. Retrieved April 10, 2024, from https://www.nationalbreastcancer.org/breast-tumors/

Oliver, A., Freixenet, J., Marti, J., Perez, E., Pont, J., Denton, E. R., & Zwiggelaar, R. (2010). A review of automatic mass detection and segmentation in mammographic images. *Medical Image Analysis, 14*(2), 87–110. https://doi.org/10.1016/j.media.2009.12.001

Parr, T., Astley, S., Boggis, C., & others. (1994). The detection of stellate lesions in digital mammograms. *International Workshop on Digital Mammography*, 231–239

Pereira, G. P., & Papa, J. P. (2011). Breast Cancer Digital Repository (BCDR): A mammographic database for CAD systems evaluation. In *Proceedings of the 25th IEEE International Symposium on Computer-Based Medical Systems* (pp. 1–6). https://doi.org/10.1109/CBMS.2011.5999147

Petrick, N., Chan, H.-P., Sahiner, B., & Helvie, M. A. (1999). Combined adaptive enhancement and region-growing segmentation of breast masses on digitized mammograms. *Medical Physics, 26*(8), 1642–1654. https://doi.org/10.1118/1.598656

Petrick, N., Chan, H.-P., Sahiner, B., Wei, D., Helvie, M. A., Goodsitt, M. M., & Adler, D. D. (1995). Automated detection of breast masses on digital mammograms using adaptive density-weighted contrast-enhancement filtering. *Medical Imaging 1995: Image Processing, 2434*, 590–597. https://doi.org/10.1117/12.208832

Petrick, N., Chan, H.-P., Wei, D., Sahiner, B., Helvie, M. A., & Adler, D. D. (1996). Automated detection of breast masses on mammograms using adaptive contrast enhancement and texture classification. *Medical Physics, 23*(10), 1685–1696. https://doi.org/10.1118/1.597822

Pratiwi, M., Harefa, J., Nanda, S., & others. (2015). Mammograms classification using gray-level co-occurrence matrix and radial basis function neural network. *Procedia Computer Science, 59*, 83–91. https://doi.org/10.1016/j.procs.2015.07.486

Ramos, R. P., Nascimento, M. Z., & Pereira, D. C. (2012). Texture extraction: An evaluation of ridgelet, wavelet, and co-occurrence based methods applied to mammograms. *Expert Systems with Applications, 39*(12), 11036–11047. https://doi.org/10.1016/j.eswa.2012.03.022

Redmon, J., Divvala, S., Girshick, R., & Farhadi, A. (2016). You only look once: Unified, real-time object detection. In *Proceedings of the IEEE Conference on Computer Vision and Pattern Recognition* (pp. 779–788). https://doi.org/10.1109/CVPR.2016.91

Ren, S., He, K., Girshick, R., & Sun, J. (2015). Faster R-CNN: Towards real-time object detection with region proposal networks. In *Advances in Neural Information Processing Systems* (pp. 91–99). https://doi.org/10.5555/2969239.2969250

Ronneberger, O., Fischer, P., & Brox, T. (2015). U-net: Convolutional networks for biomedical image segmentation. In *International Conference on Medical Image Computing and Computer-Assisted Intervention* (pp. 234–241). Springer. https://doi.org/10.1007/978-3-319-24574-4_28

Sahiner, B., Chan, H.-P., Petrick, N., & others. (1998). Design of a high-sensitivity classifier based on a genetic algorithm: Application to computer-aided diagnosis. *Physics in Medicine & Biology, 43*(10), 2853. https://doi.org/10.1088/0031-9155/43/10/015

Shen, L., Margolies, L. R., Rothstein, J. H., Fluder, E., McBride, R., & Sieh, W. (2019). Deep learning to improve breast cancer detection on screening mammography. *Scientific Reports, 9*(1), 12495. https://doi.org/10.1038/s41598-019-48995-4

Singh, V. P., Srivastava, A., Kulshreshtha, D., Chaudhary, A., & Srivastava, R. (2016). Mammogram classification using selected GLCM features and random forest classifier. *International Journal of Computer Science and Information Security, 14*(6), 82–87

Su, Y., Liu, Q., Xie, W., & Hu, P. (2022). YOLO-Logo: A transformer-based YOLO segmentation model for breast mass detection and segmentation in digital mammograms. *Computer Methods and Programs in Biomedicine, 221*, 106903. https://doi.org/10.1016/j.cmpb.2022.106903

Suckling, J., Parker, J., Dance, D. R., Astley, S. M., Hutt, I., Breatnach, E., & Ricketts, I. (1994). The Mammographic Image Analysis Society Digital Mammogram Database. *Exerpta Medica. International Congress Series*, 1069, 375–378.

Suh, Y. J., Jung, J., & Cho, B.-J. (2020). Automated breast cancer detection in digital mammograms of various densities via deep learning. *Journal of Personalized Medicine*, 10(4), 211. https://doi.org/10.3390/jpm10040211

Tan, M., Pu, J., & Zheng, B. (2014). Optimization of breast mass classification using sequential forward floating selection (SFFS) and a support vector machine (SVM) model. *International Journal of Computer Assisted Radiology and Surgery*, 9(6), 1005–1020. https://doi.org/10.1007/s11548-014-1028-6

Thomassin-Naggara, I., Tardivon, A., & Chopier, J. (2014). Standardized diagnosis and reporting of breast cancer. *Diagnostic and Interventional Imaging*, 95(7-8), 759–766. https://doi.org/10.1016/j.diii.2014.04.011

Timp, S., & Karssemeijer, N. (2004). A new 2D segmentation method based on dynamic programming applied to computer-aided detection in mammography. *Medical Physics*, 31(5), 958–971. https://doi.org/10.1118/1.1661740

Tiryaki, V. M., & Tutkun, N. (2024). Breast cancer mass classification using machine learning, binary-coded genetic algorithms and an ensemble of deep transfer learning. *The Computer Journal*, 67(3), 1111–1125. https://doi.org/10.1093/comjnl/bxv143

Varela, C., Tahoces, P. G., Mendez, A. J., Souto, M., & Vidal, J. J. (2007). Computerized detection of breast masses in digitized mammograms. *Computers in Biology and Medicine*, 37(2), 214–226. https://doi.org/10.1016/j.compbiomed.2006.04.001

Wilkinson, L., Thomas, V., & Sharma, N. (2017). Microcalcification on mammography: Approaches to interpretation and biopsy. *The British Journal of Radiology*, 90(1069), 20160594. https://doi.org/10.1259/bjr.20160594

WCRF (World Cancer Research Fund). (2024). Worldwide cancer data: Cancer trends. Retrieved from https://www.wcrf.org/cancer-trends/worldwide-cancer-data/
WHO (World Health Organization). (2024). Breast cancer. https://www.who.int/news-room/fact-sheets/detail/breast-cancer#:~:text=In%202022%2C%20there%20were%202.3,increasing%20rates%20in%20later%20life
Wu, Y., Shi, W., Cui, L., Wang, H., Bu, Q., & Feng, J. (2018). Automatic mass detection from mammograms with region-based convolutional neural network. In *Chinese Conference on Image and Graphics Technologies* (pp. 442–450). Springer. https://doi.org/10.1007/978-3-030-04633-3_51
Xu, S., Adeli, E., Cheng, J.-Z., Xiang, L., Li, Y., Lee, S.-W., & Shen, D. (2020). Mammographic mass segmentation using multichannel and multiscale fully convolutional networks. *International Journal of Imaging Systems and Technology*, 30(4), 1095–1107. https://doi.org/10.1002/ima.22382
Xu, W., Xia, S., Xiao, M., & Duan, H. (2005). A model-based algorithm for mass segmentation in mammograms. *2005 IEEE Engineering in Medicine and Biology 27th Annual Conference*, 2543–2546. https://doi.org/10.1109/IEMBS.2005.1616362
Yan, Y., Conze, P.-H., Lamard, M., Quellec, G., Cochener, B., & Coatrieux, G. (2021). Towards improved breast mass detection using dual-view mammogram matching. *Medical Image Analysis*, 71, 102083. https://doi.org/10.1016/j.media.2021.102083
Yin, L., Deshpande, S., & Chang, J. K. (2003). Automatic lesion/tumor detection using intelligent mesh-based active contour. In *Proceedings of the 15th IEEE International Conference on Tools with Artificial Intelligence* (pp. 390–397). IEEE. https://doi.org/10.1109/ICTAI.2003.1250052
Yuan, Y., Giger, M. L., Li, H., Suzuki, K., & Sennett, C. (2007). A dual-stage method for lesion segmentation on digital mammograms. *Medical Physics*, 34(11), 4180–4193. https://doi.org/10.1118/1.2780552
Zahoor, S., Shoaib, U., & Lali, I. U. (2023). Breast masses segmentation: A framework of skip dilated semantic network and machine learning. *International Journal on Artificial Intelligence Tools*, 32(3), 2340012. https://doi.org/10.1142/S0218213023400121
Zhang, L., Li, Y., Chen, H., Wu, W., Chen, K., & Wang, S. (2022). Anchor-free YOLOv3 for mass detection in mammogram. *Expert Systems with Applications*, 191, 116273. https://doi.org/10.1016/j.eswa.2021.116273
Zhang, M., & Giger, M. L. (1995). Automated detection of spiculated lesions and architectural distortions in digitized mammograms. *Medical Imaging 1995: Image Processing* (Vol. 2434, pp. 846–854). SPIE. https://doi.org/10.1117/12.205027
Zhang, R., Zhang, H., & Chung, A. C. (2018). A unified mammogram analysis method via hybrid deep supervision. In *Image Analysis for Moving Organ, Breast, and Thoracic Images: Third International Workshop, RAMBO 2018, Fourth International Workshop, BIA 2018, and First International Workshop, TIA 2018, Held in Conjunction with MICCAI 2018, Granada, Spain, September 16 and 20, 2018, Proceedings 3* (pp. 107–115). Springer. https://doi.org/10.1007/978-3-030-04632-6_12
Zhu, W., Xiang, X., Tran, T. D., Hager, G. D., & s, X. (2018). Adversarial deep structured nets for mass segmentation from mammograms. In *2018 IEEE 15th International Symposium on Biomedical Imaging (ISBI 2018)* (pp. 847–850). IEEE. https://doi.org/10.1109/ISBI.2018.8363635

# Automated Detection of Oral Cancer Through Deep Learning: A Histopathological Approach

**Reviricha Bora**

**Abstract** Oral Cancer is the cancerous tissue growth which occurs in the head and neck regions, primarily affecting key sub-regions such as the lips and the mouth. It is the 6th most revealed oncological disease happening in the developing countries, particularly in India. Oral cancer, which frequently develops in the oral cavity, is also known as Squamous Cell Carcinoma (SCC). It is identified by the differentiation of epithelial squamous tissue and rapid tumor growth, which has an impact on the basal membrane of the inner cheek area. This paper outlines the various steps including obtaining histopathological images using Leica ICC 50 HD digital microscope which is connected to advanced software and a high performing computer, followed by image augmentation, classification and assessment of accuracy of predictions. The primary motive of this chapter is to provide an understanding to the aspiring researchers who wish to apply deep learning methods in oral cancer research. It also focuses on enabling automated detection in the identification of benign cells and abnormal cells through classification through early diagnosis at an affordable cost.

**Keywords** Oral cancer · Deep learning · Histopathological images · Augmentation · Classification

## 1 Introduction

One of the deadliest diseases globally is the oral cancer which affects the neck and head regions. It ranks as one of the most prevalent cancers with 377,713 new cases and 177,757 deaths according to a report in 2020. It is one of the leading cancers worldwide and is considered by late diagnoses, higher morbidity, and mortality rates (Alanazi et al. 2022). Majority of the cases are detected at a later stage leading to delay in referring to a cancer specialist. The early symptoms along with the warning signs are undetectable which results in the disease spreading to the carcinoma stage

R. Bora (✉)
Cotton University, Assam, India
e-mail: borareviricha@gmail.com

 
T. P. Singh et al. (eds.), *Revolutionizing Healthcare: Impact of Artificial Intelligence on Diagnosis, Treatment, and Patient Care*, Studies in Computational Intelligence 1182,
https://doi.org/10.1007/978-3-031-80813-5_9

within a short span of time which makes it life-threatening (Kumar et al. 2019). Oral Cancer has a higher rate of reappearance as compared to any other cancers making it mandatory to conduct a thorough analysis of both its staging and grading. Oral Squamous Cell Carcinoma (OSCC) predominates among all oral cancer cases (84–97%) (Ananthakrishnan et al. 2023).

In India, the incidence is 16.1% in men and 10.4% in women (Sankaranarayanan et al. 2005). While it is intimidating in the region under study, the figures are notably higher making it 44.5% for men and 16.7% for women (Asthana et al. 2016). Cancer registries in India reported 13,500 cases from the period 1990–1996 in comparison to 1272 cases recorded globally (Centre, Disease, Cancer and Programme 2013).

Squamous Cell Carcinoma is the most widespread form of oral cancer found in the oral cavity. They are divided into two types-Benign Cells and Abnormal Cells (namely MDSCC and WDSCC). Both of these cells are differentiated based on Broder's System of histopathological grading. The quantitative analysis of cell morphology highlights the tumor growth which exhibits minute histological differences and separates the different classes that is difficult to be viewed through the human eye (Das et al. 2020). Here an attempt has been made to classify Oral Squamous Cell Carcinoma according to two different categories followed by the evaluation of the results. Augmentation techniques are used to find the accuracy for better training, fine-tuning and evaluation of results. A GPU is required for training in deep learning. Utilizing the data assembled for the study, deep learning-based computer vision techniques were evaluated for the automated oral cancer detection and its classification (Keser et al. 2024).

Progression in computer vision and deep learning fields have led to the growth of powerful techniques have enabled screening of the cancerous region in the oral cavity and these techniques make provisions for feedback to healthcare professionals as well as to the patients itself for proper self-assessment and diagnosis (Marc Aubreville et al. 2017).

Automating the detection of benign cells and abnormal cells in the oral cavity has the potential to facilitate cost-effective and early identification of the disease. The primary motive of this study is to single-out oral cancer in its early stage and develop important essential assistive tools for oral cancer screening. It aims to investigate the various applications of computer vision techniques utilizing photographic images and explore the potential of a deep learning-based approach for screening of oral cancer. Table 1 presents some of the interrelated works in this domain.

Literature study reveals that there are certain limitations of current deep learning models in accurately detecting oral cancer from histopathological images such as challenges related to data scarcity or biasedness in existing datasets (Bhadra and Kumar 2022, 2023). There is a need for exploration of novel deep learning architectures specifically designed to address the unique challenges of oral cancer detection from histopathological images such as the varying tissue textures and structures (Basavegowda and Dagnew 2020). There is also a need for development of interpretability methods for deep learning models in oral cancer detection (Choudhury et al. 2022). Hence, exploring certain techniques to explain the decisions made by these models could improve trust and adoption in clinical settings.

**Table 1** ML/DL Techniques used for carcinoma identification

| Author | Year | Image samples in the dataset | Classifier used | Feature set used | Performance |
|---|---|---|---|---|---|
| Krishnan | 2012 | Total-158 | SVM | HOS<br>LBP | Accuracy—92.41%<br>Accuracy—83.55% |
| Aubreville | 2017 | Total-7894 | SVM<br>RF | LBP<br>GLCM | 88.3% |
| Rana | 2017 | Total-405 | CNN classifier | | Accuracy—78.93% |
| Ananthar-aman | 2017 | Total-75<br>**For training**:<br>Dataset-75<br>**For testing**:<br>Dataset-6 | Clarifai's visual recognition algorithm<br>Inception v3 model | | Training accuracy—100%<br>Validation accuracy—90%<br>Testing accuracy (Clarifai)—100%<br>Testing accuracy (retrained inception)—66% |
| Folmsbee | 2018 | Total-143 | CNN classifier | | 96.44% |
| Uthoff | 2018 | Normal-86<br>Suspicious (OPML + malignant lesions)-84<br>Total-170 | CNN classifier | | 86.88% |
| Gupta | 2019 | Normal-663<br>Mild Dysplasia-681<br>Moderate Dysplasia-675<br>Severe Dysplasia-669<br>Total-2688 | CNN classifier | | Training accuracy—91.65%<br>Testing accuracy—89.3% |
| Jeyaraj | 2019 | Total-100<br>Total-500 | SVM<br>DBN | HSI | 91.4%<br>94.5% |
| Lipi B Mahanta | 2020 | Total-156<br>Benign-53<br>Well differentiated-45<br>Moderately differentiated-32<br>Poorly differentiated-26 | SVM | | 96.6% |

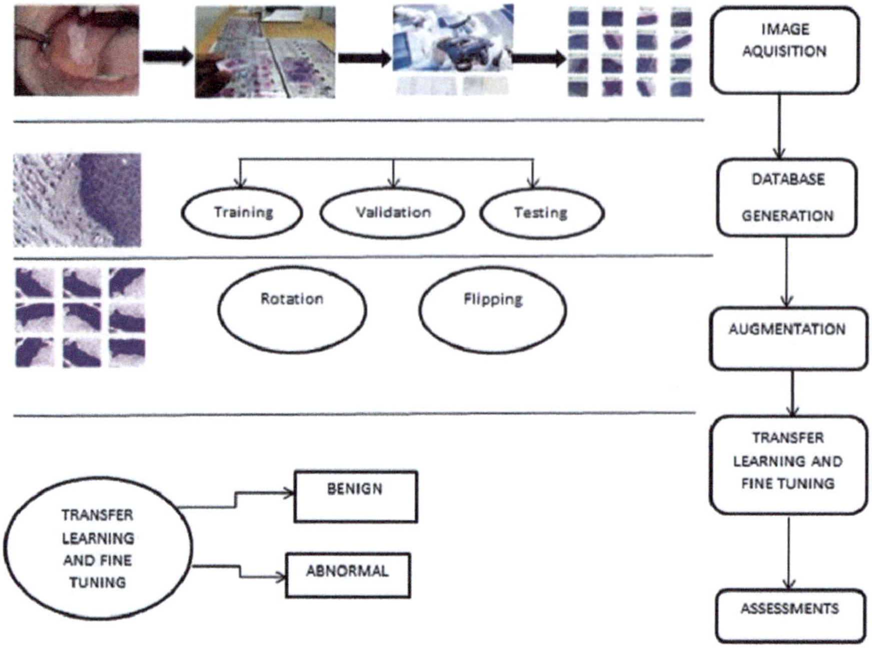


**Fig. 1** Graphical Representation of the different phases of the proposed methodology

# 2 Materials and Methods

## *2.1 Proposed Methodology*

The study on biopsy images was conducted through the application of transfer learning using MobileNet. The different principles involved in the proposed methodology are distinguished into 5 phases. Figure 1 shows the diagram of the brief outline of the methods applied.

Phase 1: Image Acquisition.
Phase 2: Database Generation.
Phase 3: Data Augmentation of Images.
Phase 4: Transfer Learning and Fine Tuning.
Phase 5: Assessment of the Results.

## *2.2 Microscopic Image Acquisition*

The histopathological whole slide images were collected from RogNidaan Technologies Pvt. Limited, Guwahati. Whole slide images were outlined using H&E staining

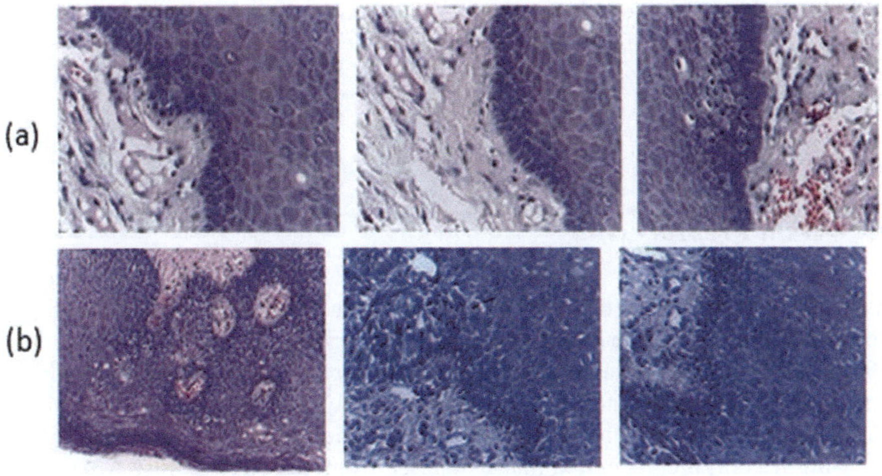


**Fig. 2** Some biopsy images from the image dataset. (**a**) Benign cells and (**b**) abnormal cells

on tumor sections from various oral cancer locations, like the left and right buccal mucosa, esophagus, larynx, lower and upper lip, vocal cords, floor of the mouth, posterior pharyngeal wall, tongue floor, and lower alveolus (Chiranjit Dutta et al. 2023). It was later provided to perform the various tasks. The H&E stained biopsy slides consisted of 633 images of size 2048 × 1536 pixels each, focal length of about 35 mm and depth of about 23 bit depth. The whole histopathological slide images consisted of 400× magnification (objective lens: 40x; eyepiece: 10x) magnifications, captured using Leica ICC 50 HD digital microscope connected to a high-configured software and computer (Renuka et al. 2024). Figure 2 presents some of the biopsy image samples collected.

## 2.3 *Database Generation*

Expert pathologists at RogNidaan Technologies Pvt. Limited, Guwahati, manually outlined the regions of interest (RoI) on the biopsy slide images to ensure accurate training and validation of the deep learning models. The labeling process included various sections of oral epithelial tissue such as the adipose tissue, muscle and connective tissues. The outlined regions were then processed further through segmentation techniques for removal (Jiang et al. 2023). To facilitate proper analysis, image patches of 1536 × 2048 pixels were extracted from the outlined regions of interest (ROIs) in each whole slide image, resulting in a total of 633 patches. A Sequential Movement Technique was used to select image patches from left to right, ensuring that any patches overlapping with background glass were excluded from the selection.

**Table 2** Data Augmentation parameters for a randomly chosen image

| Augmentation parameters | Settings |
|---|---|
| Shuffle | True |
| Horizontal flip | True |
| Rotation | 0.5 |

## *2.4 Data Preprocessing*

Each of the images in the dataset was represented in pixel values of range 0 to 255. To normalize the data, the data was divided by every pixel value by 255 (Zhu et al. 2020). The images (benign and abnormal cells) were split into two datasets. Of the total images, eighty percent of the total image patches were assigned for the training dataset, while the remaining twenty percent were designated for validation dataset by using tf.keras.preprocessing.image_dataset_from_directory utility. Afterwards, the images were plotted, labeled and displayed in batches from the training dataset. The images were loaded using buffered prefetching from disk without having I/O blockage.

## *2.5 Data Augmentation*

Data augmentation techniques are valuable for deep learning models since they help expand the training dataset, reducing overfitting and enhancing the model's ability to generalize (Hamida et al. 2021). To mitigate overfitting, two augmentation methods were randomly applied to the training images: rotation at finer angles and horizontal flipping (Cruz-Roa et al. 2013). Only these two augmentation techniques were used since the images were invariant to the other techniques. Table 2 shows the Data Augmentation parameters for a randomly chosen image. The OSCC images were augmented carefully in such a way that there was no loss of visual information in the Region of Interest (RoI). Figure 3 shows some augmented Images.

## *2.6 Transfer Learning and Fine Tuning*

To improve the performance of the pre-trained model utilized in this study, fine-tuning was implemented. MobileNet, featuring multiple layers, was chosen as the pre-trained network derived from the ImageNet database to extract deep features (Varalakshmi et al. 2024). The images of OSCC were classified into two categories: Benign and Abnormal. To differentiate between benign and cancerous cells, the model's original softmax layer was substituted with a new fully connected softmax layer featuring two nodes. The model was trained effectively using the "Transfer

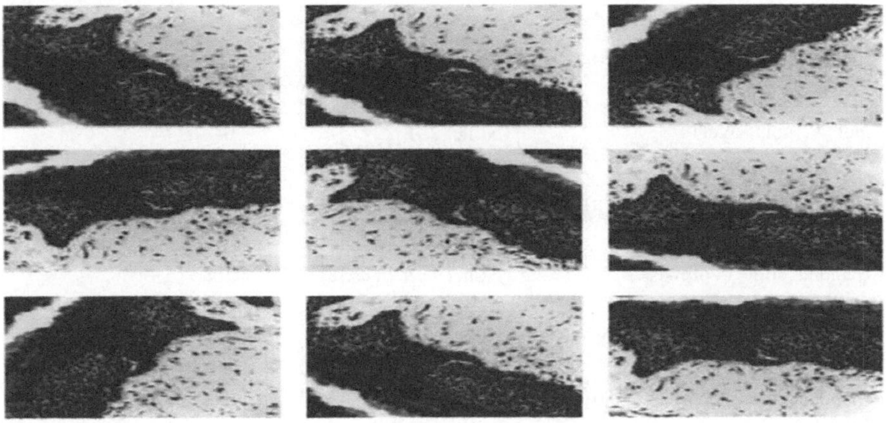

**Fig. 3** Some augmented images using rotation and flipping techniques

Learning" approach, which utilizes a pre-trained model from datasets like ImageNet, while retaining its weights for the newly added classes (Hari Mohan Rai and Joon Yoo 2023).

## *2.7 Base Model from Pre-trained ConvNets*

Following data pre-processing and augmentation, the pre-trained image classification model was further made to proceed for the training phase. The base model (originally trained on the ImageNet dataset) was built using Google's MobileNet V2 architecture (Mehran Ahmad et al. 2023). ImageNet is a comprehensive research dataset used for training models, covering a diverse array of categories such as jackfruits and syringes (Kassani et al. 2019). Firstly, a layer needs to be picked up by ImageNet which will be used for feature extraction. The bottom layer (often referred to as the "top" layer in diagrams that depict machine learning models vertically) is generally not very useful for this purpose. Instead, it is a standard practice to utilize the layer just before the flatten operation, known as the "bottleneck layer," as it retains more generalized features as compared to the top layer (Hu et al. 2018). A MobileNet V2 model, initialized with ImageNet-trained weights was instantiated with the argument include_top=False, which omits the top classification layers, thereby making it ideal for feature extraction (Muhammed et al. 2024). Table 3 gives the summary of the base model.

**Table 3** Summary of the base model

| Model: "mobilenetv2_1.00_224" | | |
|---|---|---|
| Layer (type) | Output shape | No. of parameters |
| input_4 (InputLayer) | (None, 224, 224, 3) | 0 |
| Conv1 (Conv2D) | (None, 112, 112, 32) | 864 |
| bn_Conv1 (BatchNormalization) | (None, 112, 112, 32) | 128 |
| Conv1_relu (ReLU) | (None, 112, 112, 32) | 0 |
| expanded_conv_depthwise (Depthwise) | (None, 112, 112, 32) | 288 |
| expanded_conv_depthwise_BN (BatchNormalization) | (None, 112, 112, 32) | 128 |
| expanded_conv_depthwise_relu (ReLU) | (None, 112, 112, 32) | 0 |
| expanded_conv_project (Conv2D) | (None, 112, 112, 16) | 512 |
| expanded_conv_project_BN (BatchNormalization) | (None, 112, 112, 16) | 64 |
| block_1_expand (Conv2D) | (None, 112, 112, 96) | 1536 |
| block_1_expand_BN (BatchNormalization) | (None, 112, 112, 96) | 384 |
| block_1_expand_relu (ReLU) | (None, 112, 112, 96) | 0 |
| block_1_pad (ZeroPadding2D) | (None, 113, 113, 96) | 0 |
| block_16_expand_relu (ReLU) | (None, 7, 7, 960) | 0 |
| block_16_depthwise (Depthwise | (None, 7, 7, 960) | 8640 |
| block_16_depthwise_BN (BatchNormalization) | (None, 7, 7, 960) | 3840 |
| block_16_depthwise_relu (ReLU) | (None, 7, 7, 960) | 0 |
| block_16_project (Conv2D) | (None, 7, 7, 320) | 307,200 |
| Conv_1 (Conv2D) | (None, 7, 7, 1280) | 409,600 |
| Conv_1_bn (BatchNormalization) | (None, 7, 7, 1280) | 5120 |
| out_relu (ReLU) | (None, 7, 7, 1280) | 0 |
| **Total parameters**: 2,257,984 | | |
| **Trainable parameters**: 0 | | |
| **Non-trainable parameters**: 2,257,984 | | |

## *2.8 Feature Extraction*

Feature extractor was the convolutional base that was created to be freezed. An additional classifier was incorporated into the model, and the top-level classifier was subsequently trained. Freezing the convolutional base model was necessary before it can be compiled and trained (Kumar and Das 2022). The setting layer.trainable=False was applied to ensure the weights remained unchanged during training. Given that

**Table 4** Table for the base model summary after freezing

| Model: "model_1" | | |
|---|---|---|
| Layer (type) | Output shape | No. of parameters |
| input_5 (InputLayer) | (None, 224, 224, 3) | 0 |
| sequential_1 (Sequential) | (None, 224, 224, 3) | 0 |
| tf.math.trudiv_1 (TFOpLambd) | (None, 224, 224, 3) | 0 |
| tf.math.subtract_1 (TFOpLambd) | (None, 224, 224, 3) | 0 |
| mobilenetv2_1.00_224 (Functi) | (None, 7, 7, 1280) | 2,257,984 |
| global_average_pooling2d_1 | (None, 1280) | 0 |
| dropout_1 (Dropout) | (None, 1280) | 0 |
| dense_1 (Dense) | (None, 1) | 1281 |
| **Total parameters**: 2,259,265 | | |
| **Trainable parameters**: 1281 | | |
| **Non-trainable parameters**: 2,257,984 | | |

MobileNet V2 comprises many layers, the trainable flag for the whole model was set to False to freeze all layers (Kritsasith Warin and Siriwan Suebnukarn 2024).

## 2.9 *Freezing the Convolutional Base*

In the MobileNet architecture, 2.5 million parameters were frozen, while 1,200 trainable parameters were retained in the dense layer (Pullaiah Pinnika and Venkata Rao 2024) as shown in Table 4.These parameters were split into two tf.Variable objects: weights and biases. The learning curve for training and validation accuracy/loss was generated using the MobileNet V2 base model and it was used as a fixed feature extractor, as shown in the results section.

### 2.9.1 Unfreezing the Topmost Layers of the Model

Initially, the top layer was unfrozen, while the bottom layers were kept untrainable. The model was subsequently recompiled to complete the training process. MobileNet was used as the structural framework for network design with 28 convolutional layers, including both depthwise convolution and pointwise convolutions, with a kernel size of $3 \times 3$. The model takes image patches of size $224 \times 224$ as input (Rahman et al. 2017, 2020a, b). To normalize the positive features without biasing them strategically so that it does not make it through to the next convolution layer, batch normalization was to be incorporated only after the ReLU activation function. It helps to normalize the feature distribution, particularly when certain features were truncated due to nonlinearity, like that introduced by ReLU. Pooling layers

**Table 5** Table for the base model after unfreezing the top layers

| Model: "model_1" | | |
|---|---|---|
| Layer (type) | Output shape | No. of parameters |
| input_5 (InputLayer) | (None, 224, 224, 3) | 0 |
| sequential_1 (Sequential) | (None, 224, 224, 3) | 0 |
| tf.math.trudiv_1 (TFOpLambd) | (None, 224, 224, 3) | 0 |
| tf.math.subtract_1 (TFOpLambd) | (None, 224, 224, 3) | 0 |
| mobilenetv2_1.00_224 (Functi) | (None, 7, 7, 1280) | 2,257,984 |
| global_average_pooling2d_1 | (None, 1280) | 0 |
| dropout_1 (Dropout) | (None, 1280) | 0 |
| dense_1 (Dense) | (None, 1) | 1281 |
| **Total parameters**: 2,259,265 | | |
| **Trainable parameters**: 1,862,721 | | |
| **Non-trainable parameters**: 396,544 | | |

contribute in reducing network parameters which helps in decreasing the training time while tackling overfitting issues (Fu 2020). To decrease the image dimensions, max pooling layers are incorporated, which select the maximum value from the matrix representing the image at the pixel level. A flattened and dense layer is a component of a fully connected neural network that links each input node to the output node in the subsequent layer (Kumar et al. 2022). To calculate the probability of an image belonging to one of two classes, a softmax layer was added as shown in Table 5.

### 2.9.2 Platform Used

The CNN model was built using the Python, along with the deep learning libraries such as TensorFlow and Keras. Implementation of the codes was done in an interactive environment called Google Colab. Some of the coding was done from the in-built function on the Colab and available source codes.

## 3 Dataset

The biopsy datasets for OSCC that were retrieved from RogNidaan Technologies Pvt. Limited, Guwahati included Benign Cells, MDSCC, and WDSCC. For simplicity, these categories were consolidated into a binary classification of Benign Cells and Abnormal Cells, with Abnormal Cells encompassing both MDSCC and WDSCC for categorization purposes. Consequently, the suggested model applies this dataset to evaluate and predict oral cancer. Of the total 633 biopsy image datasets, 508

**Table 6** OSCC biopsy dataset

| Classes | No. of images |
|---|---|
| Benign cells | 159 |
| Abnormal cells (MDSCC + WDSCC) | 379 + 95 = 474 |

images were allocated for training while the rest 125 were reserved for validation. Table 6 gives the details of the dataset classes, Table 7 gives image details and Fig. 4 represents the sample of image datasets.

Table 7 is the total number of biopsy slide images with H&E staining.

**Table 7** Details of the biopsy slide images

| Images | Details |
|---|---|
| Total slides | 633 |
| Total slides with benign cells | 159 |
| Total slides with abnormal cells (MDSCC + WDSCC) | 474 (379 + 95) |
| Image format | JPEG |
| Magnification | 400× magnification (objective lens: 40x; eyepiece: 10x) |
| Microscope used | Leica ICC 50 HD microscope |
| Size | 2048 × 1536 pixel |
| Focal length and depth | 35 mm; 23 bit depth |
| Horizontal and vertical resolutions | 96 dpi |

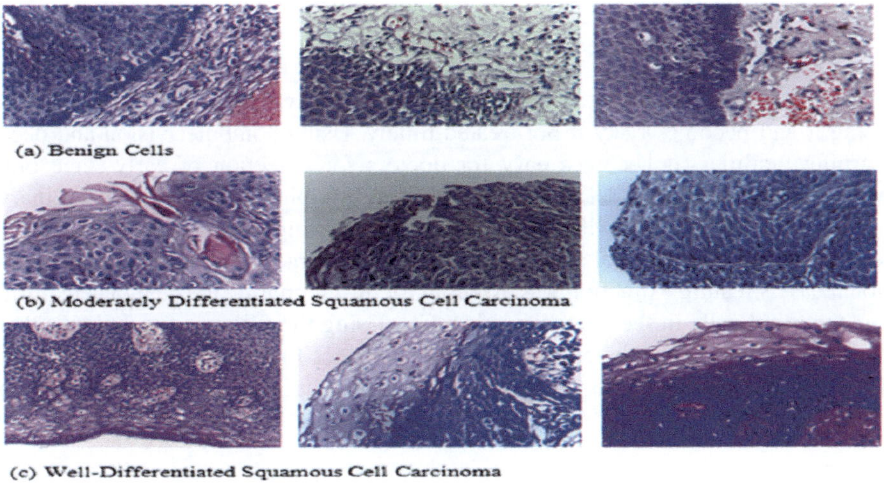


**Fig. 4** Benign cells, MDSCC and WDSCC biopsy image dataset

## 4 Results and Discussion

We have used two assessment measures to make a comparative analysis of the models considered in this study—

(i) Loss Graph and
(ii) Accuracy Graph

Using the proposed MobileNet Architecture, the learning rate was kept at 0.0001. Binary Crossentropy has been used as a loss function. ReLU activation function has been employed in the hidden layers. RMSprop has been used as an optimizer. After training different models, the best accuracy thus achieved after 30 epochs was 95.8%. Using Softmax as an activation function and Adam as an optimizer, the training accuracy achieved after 20 epochs was 96%. Figures 5 and 6 shows the Learning curves for analysis of the training and validation datasets. Tables 8 and 9 shows Performance of CNN under different parameters.

| Kernel size | Loss function | Activation function | Optimizers | Epoch | Classification accuracy | Learning rate |
|---|---|---|---|---|---|---|
| $3 \times 3$ | Binary crossentropy | ReLU | RMSprop | 30 | 95.8 | 0.0001 |
| $3 \times 3$ | Binary crossentropy | Softmax | Adam | 20 | 96 | 0.0001 |

The loss and accuracy of the graph has been shown below.

## 5 Conclusion

Early detection of oral cancer is very important as it may save several people undergoing it and becomes risky if not treated timely. Using computer vision and deep learning facilities, it becomes easy for doctors for detection of early-stage oral carcinoma. Here, the model is constructed using MobileNet to perform the binary classification of OSCC. RMSprop optimizer gave the highest accuracy with proper learning curves. The model achieved the highest accuracy of 95.8%. The convolutional base's feature extraction was essential for differentiating the benign cells from the abnormal cells in the diagnosis of OSCC. Using the following techniques that have been implemented here, one can classify images for real-time clinical testing. Thus, this technique will be of great help in the medical field to detect cancer at an early stage.

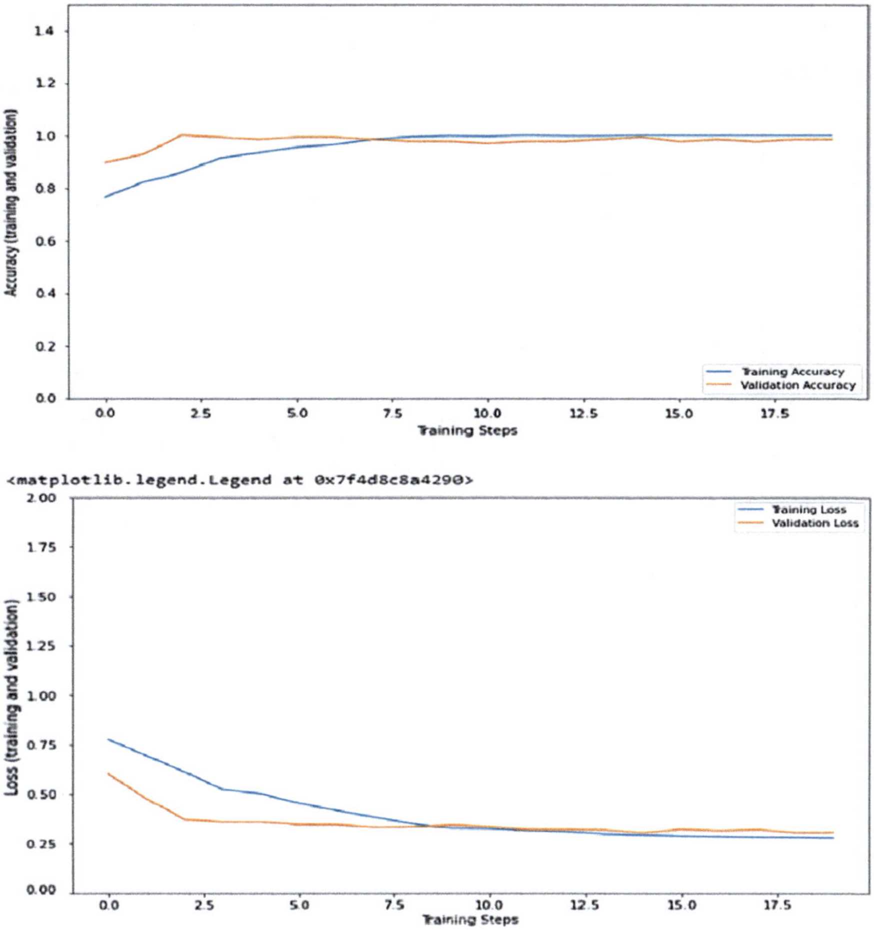


**Fig. 5** Learning curves for analysis of the training and validation datasets. Classification accuracy—95.8%

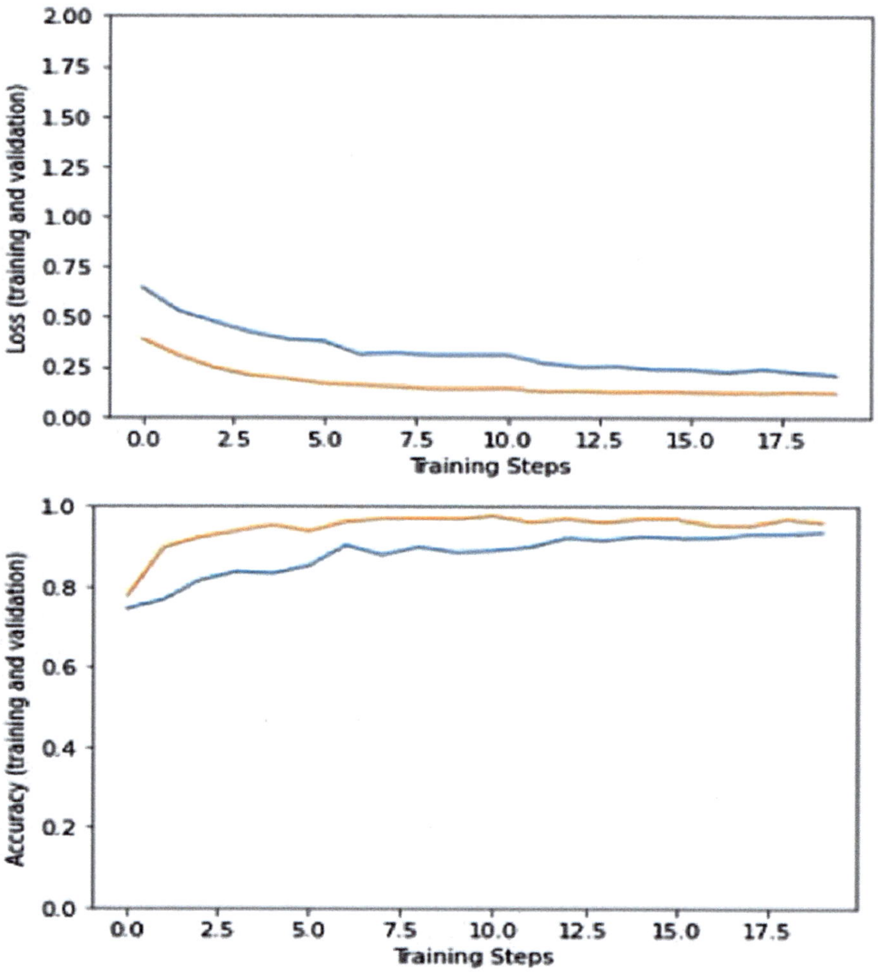


**Fig. 6** Learning curves for analysis of the training and validation datasets. Classification accuracy—96%

**Table 8** Performance of CNN under different parameters (Adam as the optimizer)

| Kernel size | Loss function | Activation function | Optimizers | Epoch | Classification accuracy (%) | Learning rate |
|---|---|---|---|---|---|---|
| 3 × 3 | Binary crossentropy | Softmax | Adam | 10 | 52 | 0.0003 |
| 3 × 3 | Binary crossentropy | Softmax | Adam | 150 | 65 | 0.0001 |
| 3 × 3 | Binary crossentropy | Softmax | Adam | 100 | 65 | 0.0001 |
| 3 × 3 | Binary crossentropy | Softmax | Adam | 150 | 70 | 0.0001 |
| 3 × 3 | Binary crossentropy | Softmax | Adam | 100 | 70 | 0.0001 |
| 3 × 3 | Binary crossentropy | Softmax | Adam | 10 | 82.5 | 0.0001 |
| 3 × 3 | Binary crossentropy | Softmax | Adam | 20 | 95 | 0.0001 |
| 3 × 3 | Binary crossentropy | Softmax | Adam | 20 | 96 | 0.0001 |

**Table 9** Performance of CNN under different parameters (RMSprop as the optimizer)

| Kernel size | Loss function | Activation function | Optimizers | Epoch | Classification accuracy (%) | Learning rate |
|---|---|---|---|---|---|---|
| 3 × 3 | Binary crossentropy | ReLU | RMSprop | 10 | 52 | 0.0001 |
| 3 × 3 | Binary crossentropy | ReLU | RMSprop | 40 | 52.5 | 0.00001 |
| 3 × 3 | Binary crossentropy | ReLU | RMSprop | 20 | 72.7 | 0.0002 |
| 3 × 3 | Binary crossentropy | ReLU | RMSprop | 15 | 75.4 | 0.000001 |
| 3 × 3 | Binary crossentropy | ReLU | RMSprop | 10 | 76 | 0.0001 |
| 3 × 3 | Binary crossentropy | ReLU | RMSprop | 20 | 87.1 | 0.000001 |
| 3 × 3 | Binary crossentropy | ReLU | RMSprop | 40 | 87.5 | 0.01 |
| 3 × 3 | Binary crossentropy | ReLU | RMSprop | 30 | 87.9 | 0.0001 |
| 3 × 3 | Binary crossentropy | ReLU | RMSprop | 40 | 87.9 | 0.0001 |
| 3 × 3 | Binary crossentropy | ReLU | RMSprop | 30 | 94.06 | 0.0001 |
| 3 × 3 | Binary crossentropy | ReLU | RMSprop | 30 | 95.8 | 0.0001 |

## References

Adwan A. Alanazi, Manal M. Khayyat, Mashael M. Khayyat, Bushra M. Elamin Elnaim and Sayed Abdel-Khalek (2022). Intelligent Deep Learning Enabled Oral Squamous Cell Carcinoma Detection and Classification Using Biomedical Images.

Asthana, S., Patil, R. S., & Labani, S. (2016). Tobacco-related cancers in India: A review of incidence reported from population-based cancer registries. Indian Journal of Medical and Paediatric Oncology, 37(3), 152–157.

Balasundaram Ananthakrishnan, Ayesha Shaik, Soham Kumar, S. O. Narendran, Khushi Mattu and Muthu Subash Kavitha (2023). Automated Detection and Classification of Oral Squamous Cell Carcinoma Using Deep Neural Networks.

Basavegowda, H. S., & Dagnew, G (2020). Deep Learning approach for microarray cancer data classification. CAAI Transactions on Intelligence Technology, 5(1), 22–33.

Bhadra, S and Kumar, C.J. (2022), An insight into diagnosis of depression using machine learning techniques: a systematic review", Current Medical Research and Opinion, IF: 2.705 38:5, 749–771, https://doi.org/10.1080/03007995.2022.2038487.

Bhadra,S and Kumar,C.J. (2023) Enhancing the efficacy of depression detection systems using optimal feature selection from EHR, Computer Methods in Biomechanics and Biomedical Engineering, IF: 1.933. https://doi.org/10.1080/10255842.2023.2181660.

Centre, Disease, Cancer and Programme. (2013). National centre for disease informatics and research Indian council of medical research consolidated report of hospital based cancer registries an assessment of the burden and care of cancer Patients (September).

Choudhury, A., Biswas, A., Singh, T. P., & Ghosh, S. K. (Eds.). (2022). Smart Agriculture Automation Using Advanced Technologies: Data Analytics and Machine Learning, Cloud Architecture, Automation and IoT. Springer Nature.

Chiranjit Dutta, Prasad Sandhya, Kandasamy Vidhya, Ramanathan Rajalakshmi, Devasahayam Ramya, Kotakonda Madhubabu (2023). Effectiveness of deep learning in early-stage oral cancer detections and classification using histogram of oriented gradients.

Cruz-Roa, A. A., ArevaloOvalle, J. E., Madabhushi, A., & González Osorio, F. A. (2013).A deep learning architecture for image representation, visual interpretability and automated basal-cell carcinoma cancer detection. In Medical Image Computing and Computer-Assisted Intervention–MICCAI 2013: 16th International Conference, Nagoya, Japan, September 22-26, 2013, Proceedings, Part II 16 (pp. 403–410). Springer Berlin Heidelberg.

Durairaj Varalakshmi, Mayakrishnan Tharakeswari, Thirunavukarasou Anand, Konda Mani Saravanan (2024). Transforming oral cancer care: The promise of deep learning in diagnosis.

Gaye Keser, Filiz Namdar Pekiner, Ibrahim Sevki Bayrakdar, Ozer Celik, Kaan Orhan (2024). A deep learning approach to detection of oral cancer lesions from intra oral patient images: A preliminary retrospective study.

Hamida, A. B., Devanne, M., Weber, J., Truntzer, C., Derangère, V., Ghiringhelli, F., &Wemmert, C. (2021). Deep learning for colon cancer histopathological images analysis. Computers in Biology and Medicine, 136, 104730.

Hari Mohan Rai and JoonYoo (2023). A comprehensive analysis of recent advancements in cancer detection using machine learning and deep learning models for improved diagnostics.

Hu, Z., Tang, J., Wang, Z., Zhang, K., Zhang, L., & Sun, Q. (2018). Deep learning for image-based cancer detection and diagnosis– A survey. Pattern Recognition, 83, 134–149.

Iyer, S.G, Pradhan, S.A, Pai, P.S, &Patil, S. (2004).Surgical treatment outcomes of localized squamous carcinoma of buccal mucosa. Head and Neck, 26(10), 897–902.

Kassani, S.H, Wesolowski, M.J & Schneider, K.A (2019). Classification of histopathological biopsy images using ensemble of deep learning networks.

Kritsasith Warin and Siriwan Suebnukarn (2024).Deep learning in oral cancer- a systematic review.

Kumar, M., Singh, T. P., Choudhury, T., & Gupta, S. C. (2019). ICT-the smart education system in India. In 2019 International conference on contemporary computing and informatics (IC3I) (pp. 279–282). IEEE.

Kumar,C.J, Das,P.R., Hazarika,A. (2022) "Autism spectrum disorder diagnosis and machine learning: a review", International Journal of Medical Engineering and Informatics, 14(6), pp. 512–527.

Kumar, C.J. and Das, P.R. (2022) The diagnosis of ASD using multiple machine learning techniques, International Journal of Developmental Disabilities, IF:2.170, https://doi.org/10.1080/20473869.2021.1933730,Vol-68,no-6 Pp 973–983.

Marc Aubreville, Christian Knipfer, Nicolai Oetter, Christian Jaremenko, Erik Rodner, Joachim Denzler, Christopher Bohr, Helmut Neumann, Florian Stelzle & Andreas Maier (2017). Automatic Classification of Cancerous Tissue in Laser Endomicroscopy Images of the Oral Cavity using Deep Learning.

Mehran Ahmad, Muhammad Abeer Irfan, Umar Sadique, Ihtishamul Haq, Atif Jan, Muhammad Irfan Khattak, Yazeed Yasin Ghadi and Hanan Aljuaid (2023). Multi-Method Analysis of Histopathological Image for Early Diagnosis of Oral Squamous Cell Carcinoma Using Deep Learning and Hybrid Techniques.

Muhammed Yaseer P, Arul Xavier V M, Shyni S S (2024). Oral Cancer Using Deep Learning and Auto-Fluorescence Image Analysis.

Navarun Das, ElimaHussain, Lipi B. Mahanta (2020). Automated classification of cells into multiple classes in epithelial tissue of oral squamous cell carcinoma using transfer learning and convolutional neural network.

Pullaiah Pinnika, K. Venkata Rao (2024). Analysis of Oral Cancer Detection based Segmentation and Classification using Deep Learning Algorithms.

Qiuyun Fu, Yehansen Chen, Zhihang Li, Qianyan Jing, Chuanyu Hu, Han Liu, JiahaoBao, Yuming Hong, Ting Shi, Kaixiong Li, HaixiaoZou, Yong Song, Hengkun Wang, Xiqian Wang, Yufan Wang, Jianying Liu, Hui Liu, Sulin Chen, Ruibin Chen, Man Zhang, Jingjing Zhao, Junbo Xiang, Bing Liu, Jun Jia, Hanjiang Wu, Yifang Zhao, Lin Wan, Xuepeng Xiong (2020). A deep learning algorithm for detection of oral cavity squamous cell carcinoma from photographic images: A retrospective study.

Renuka Chavva, Jeba Priya S, Mathu (2024). Oral Cancer Detection Using Deep Learning.

Roshan Alex Welikala, Paolo Remagnino, Jian Han Lim, Chee Sang Chan, Senthilmani Rajendran, Thomas George Kallarakkal & Rosnah Binti Zain (2020). Automated Detection and Classification of Oral Lesions Using Deep Learning for Early Detection of Oral Cancer.

Sankaranarayanan, R., Ramdas, K., Thomas, G., Muwonge, R., Thara, S., Mathew, B., et al. (2005). Effect of screening on oral cancer mortality in kerala, India: A cluster-randomized controlled trial.

Tabassum Yesmin Rahman, C. Chakraborty, Lipi B. Mahanta, Anup K Das, Jagannath D. Sarma (2017). Textural pattern classification for oral squamous cell carcinoma.

Tabassum Yesmin Rahman, Lipi B. Mahanta, Anup K Das, Jagannath D. Sarma (2020a). Histopathological imaging database for oral cancer analysis.

Tabassum Yesmin Rahman, Lipi B. Mahanta, Hiten Choudhury, Anup K Das, Jagannath D. Sarma (2020b). Study of morphological and textual features for classification of oral squamous cell carcinoma by traditional machine learning techniques.

Xiaoyan Jiang, Zuojin Hu, Shulhua Wang and Yudong Zhang (2023). Deep Learning for Medical Image-Based Cancer Diagnosis.

Zhu, W., Xie, L., Han, J., & Guo, X. (2020). The application of deep learning in cancer prognosis prediction. Cancers, 12(3), 603.

# Deep Learning-Driven CNN Models for Enhanced Brain Tumor Classification

Kirti Aggarwal, Kunal Kartikeya, and Vikalp Srivastava

**Abstract** A crucial aspect of medical diagnosis is the classification of brain tumors. Identification of tumor type is very much needed for the correct and effective treatment. In this research work a deep learning approach for the classification of brain tumor based on MRI images using Convolutional Neural Networks (CNNs) is introduced. The CNN model developed in this research work extract features from the preprocessed brain MRI images and categorize them based on the tumor identified. The CNN model was also compared against the conventional machine learning classification models such as such as Recurrent Neural Network (RNN), K-Nearest Neighbor (KNN), Random Forest, Long Short-Term Memory (LSTM), Support Vector Machine (SVM), and Artificial Neural Network (ANN) based on the experiments performed. The outcome shows that the CNN model outperformed the other algorithms by achieving the accuracy of 96.98%. This book chapter highlights the potential of CNNs to classify the brain tumor accurately and provide support in diagnosis decisions.

**Keywords** Magnetic resonance imaging (MRI) · Convolutional neural networks (CNNs) · Brain tumor classification

## 1 Introduction

Human brain is responsible for the wide range of important functions of human body, from regulating basic physiological processes to controlling complex cognitive activities (Carter 2019). However, the different brain functions can be affected by a range of health conditions and brain tumors are among the most severe. A brain tumor can present significant risks to the person's health (Rasheed et al. 2021). The tumor cells as shown in Fig. 1, consume the nutrients needed by the healthy brain

K. Aggarwal (✉) · K. Kartikeya · V. Srivastava
Department of Computer Science and Engineering and Information Technology, Jaypee Institute of Information Technology, Noida, India
e-mail: aggarwalkirti25@gmail.com



T. P. Singh et al. (eds.), *Revolutionizing Healthcare: Impact of Artificial Intelligence on Diagnosis, Treatment, and Patient Care*, Studies in Computational Intelligence 1182,
https://doi.org/10.1007/978-3-031-80813-5_10

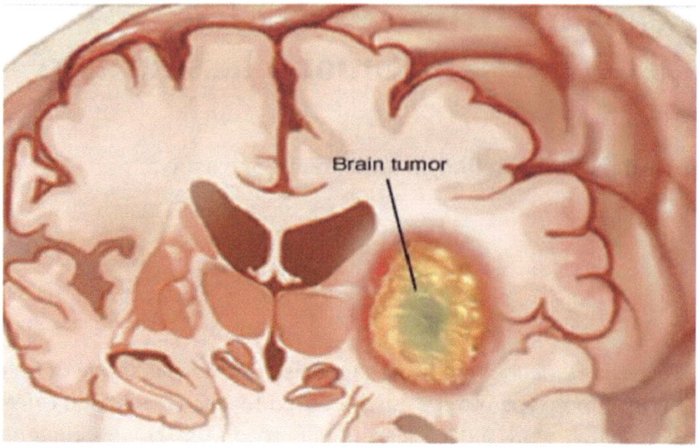


**Fig. 1** Brain tumor

cells, depleting healthy cells of the nutrients and causing deterioration in the brain performance. It can also result in life-threating consequences (Kumar et al. 2022). Brain tumor represents a major challenge in the field of healthcare because of its ability to cause severe neurological damage and dysfunction. This serious condition highlights that it is important to make early and accurate diagnosis, so that effective treatment can be started timely (Bakshi et al. 2024).

Magnetic Resonance Imaging (MRI) is an effective approach to identify the brain tumors in human body. Various types of MRI scans as shown in Fig. 2, like T1-weighted, T2-Weighted, and FLAIR (Fluid Attenuated Inversion Recovery), that can be utilized to track tumor induced changes (Young 2007). These traditional methods also have certain drawbacks. They are time taking and prone to human error, resulting in delayed or inaccurate diagnosis (Bhadra et al. 2024). The complex nature of human brain and brain tumors with the varying appearance of different imaging techniques further makes the diagnosis more difficult. So, an alternative approach was needed to detect and classify brain tumor detection. An automated system was required in the field of healthcare to make the diagnosis process accurate and correct. In this research work we seeks to overcome the drawbacks of these traditional methods by developing an advanced system that uses convolution Neural Networks (CNNs) and advanced image processing technique that can help healthcare professionals to correctly classify the brain tumor and provide accurate diagnosis. The system analyzes the MRI Images to identify and categorize the brain tumors with high accuracy and precision.

The approach proposed in this research work has multiple stages. It starts with the preprocessing of input to increase the standard of MRI images, followed by the image Segmentation process. Image Segmentation isolates the tumor region from the other part of image. After that feature extraction process is utilized to identify the main properties of the tumor, these characteristics are then used to classify the

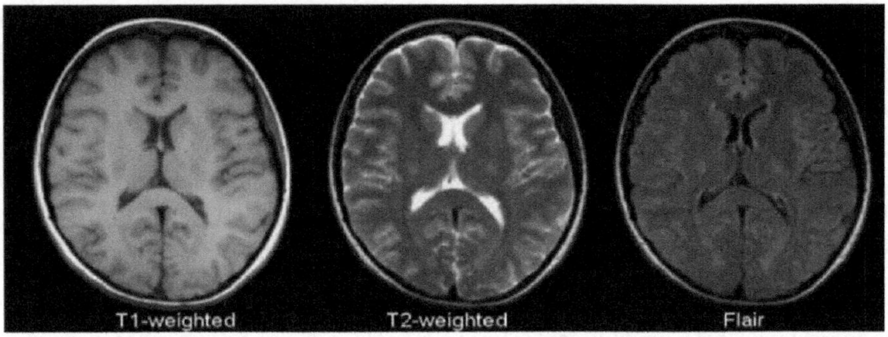


**Fig. 2** Brain MRI

tumor and to determine its type and the severity (Agarwal et al. 2024). This system with all the automatic process tries to minimize the human intervention, that further minimizes the risk of errors in diagnosis and it also improves the speed of the process to present the result in timely manner. The importance of this work moves beyond the technical challenges of identification of brain tumor cells. These cells whether benign or malignant, have a great effect on the life of the individual and sometimes leading to physical impairments. The development of a system that is reliable and provide efficient results timely has great consequences in providing personalized treatment. CNNs in medical imaging has many more applications in the healthcare field that represents the advancement of artificial intelligence to revolutionize the complex medical diagnosis and treatment.

## 2 Literature Survey

Image segmentation techniques utilized in medical imaging has been evolved substantially. It involves multiple advancements and methods to enhance the accuracy and efficacy of tumor detection. This literature survey is presented to summarize the key developments and recent research in this field. Kaur et al. (2012) explores different image segmentation techniques and gives a comparison between edge detection and thresholding. In this study they have evaluated different clustered approaches of Image segmentation. The authors have also performed the assessment of metrics used for performance evaluation like sensitivity, accuracy and specificity. This reveals the success of image segmentation techniques on different image types. However, this work was confined to traditional methods and did not consider the critical factors, such as lighting conditions and image resolution, that are crucial for accurate analysis.

Wu et al. (2008) proposed a tumor detection and segmentation approach based on texture analysis and using the Seeded Region Growing (SRG) method. This study addressed the challenge of non-sharp tumor tissue edges affecting segmentation accuracy. Implemented in MATLAB, their technique showed effective texture evaluation and segmentation. Nevertheless, the necessity of a large training dataset for improved accuracy remains a challenge, given the difficulty in obtaining extensive medical data. Mahmoud and Mohamed (Hussein et al. 2012) introduced an Artificial Neural Network (ANN) model for brain tumor identification. Their computerized recognition system, employing the Elman network, improved both accuracy and processing time. The use of the sigmoid function in the network enhanced tumor segmentation accuracy. Despite these advancements, their study focused solely on ANNs, limiting exploration of other potentially complementary techniques. Subashini and Sahoo (2014) proposed a brain MR image segmentation technique using ANNs, incorporating Pulse-Coupled Neural Networks (PCNN) and noise removal strategies. Their results showed that Convolutional Neural Networks (CNNs) got superior precision in classification of the image compared to Random Forests (RF) and K-Nearest Neighbors (KNN), against CNN achieving an accuracy of 98%. However, the paper did not address the practical implications or clinical applications of the proposed method.

Recent advancements in image segmentation have continued to build upon these foundational studies. Cheng et al. (2020) introduced an approach for brain tumor segmentation based on deep learning technique using a 3D U-Net architecture. Their method demonstrated considerable progress in segmentation precision and robustness compared to traditional methods. The study emphasized integrating contextual information and multi-scale features for accurate tumor segmentation. Li et al. (2022) examined hybrid deep learning models combining CNNs and Transformers for medical image segmentation. This approach leveraged the strengths of both models, achieving superior performance in segmenting complex anatomical structures in MRI images. The work emphasizes the potential of combining different deep learning architectures to improve segmentation results.

Another significant contribution is by Apostolopoulos et al. (2023), who developed a novel attention-based CNN model for brain tumor segmentation. This model incorporated spatial and channel-wise attention mechanisms to focus on relevant features, achieving improved segmentation performance over traditional CNN architectures. Their approach demonstrated enhanced ability to detect and delineate tumors in complex MRI images. Additionally, Zhou et al. (2023) highlighted a self-supervised learning approach for brain tumor segmentation, which leverages large-scale unlabeled data to pre-train models before fine-tuning on labeled datasets. This approach showed promising results in reducing the dependency on labeled data and improving segmentation accuracy, especially in scenarios with limited annotated samples (Hussein et al. 2012).

## 3 Neural Network

Neural Networks (NN), especially Convolutional Neural Networks (CNN), have emerged as powerful tools for segmentation and classification tasks, especially in medical imaging. The basic operation of NN involves layers of neurons that analyze and interpret input data to recognize patterns and predict outputs (Konde and Thakur

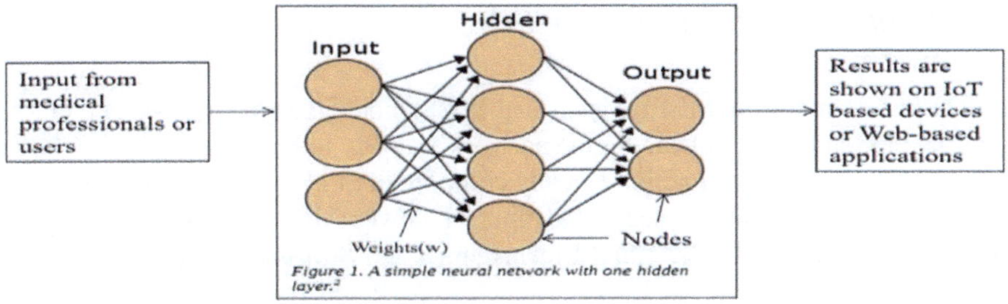


**Fig. 3** Simple neural network

2021). In medical imaging, this process is critical for tasks such as tumor detection, where the network must accurately classify regions of interest in MRI images. Figure 3 illustrates a simple Neural Network, where each neuron in one layer is connected to neurons in the next layer through weighted channels. This connectivity allows the network to process complex data and make predictions. The training process involves forward propagation, where data is passed through the network, and backpropagation, where errors are used to adjust the weights and improve accuracy.

## 3.1 Basic Neural Network Operation

A Neural Network is composed of neuron layers, with each neuron linked to neurons in subsequent layers through weighted channels. Every neuron calculates the weighted sum of the inputs and implement an activation function (Oh et al. 2021). For a neuron $j$ in layer $l$, the output $a_j^l$ of Forward Propagation in NN can be described as:

$$z_j^l = \sum_i w_{ij}^l a_i^{l-1} + b_j^l \quad (1)$$

$$a_j^l = \sigma\left(z_j^l\right) \quad (2)$$

where $w_{ij}^l$ represents the weight connecting neuron $i$ in layer $l-1$ to neuron $j$ in layer $l$ and $b_j^l$ is the bias term, and $\sigma$ denotes the activation function. Errors are spread backward via the network to adjust weights and biases. The weight update rule using gradient descent is:

$$w_{ij}^l \leftarrow w_{ij}^l - \eta \frac{\delta L}{\delta w_{ij}^l} \quad (3)$$

where $\eta$ is the learning rate, and $L$ represents the loss function.

### 3.2 Multi-layer Perceptron (MLP)

Multi-Layer Perceptron (MLP) (Masih et al. 2021) model is a type of NN that includes multiple hidden layers, enhancing the network's ability to learn from data as shown in Fig. 4. The activation function is a key component of NN, determining whether a neuron should be activated based on the input it receives. Various activation functions, such as Sigmoid, ReLU (Rectified Linear Unit), and Softmax, are used depending on the classification task. ReLU, for instance, is particularly effective in introducing non-linearity into the network, making it well-suited for image classification tasks (Al Bataineh et al. 2022). Sigmoid Function can be defined as:

$$\sigma(\mathrm{x}) = \frac{1}{1+e^{-x}} \tag{4}$$

where $x$ represents the input to the function, which is the weighted sum of inputs to a neuron, plus a bias term and $x$ can be calculated as:

$$\mathrm{x} = \sum_{i} \mathrm{w_i a_i} + \mathrm{b} \tag{5}$$

ReLU activation function can be defined as:

$$\mathrm{f(x)} = \max(0, \mathrm{x}) \tag{6}$$

For $x > 0$, the output is simply $x$. This means that if the input is positive, it is passed through unchanged. For $x \leq 0$, the output is 0. This means that if the input is zero or negative, it is set to zero. ReLU is particularly effective in handling non-linearity and is widely used in deep networks (Mathivanan et al. 2024).

Softmax activation function converts raw output scores into probabilities, used for multi-class classification tasks. It can be defined as:

$$\mathrm{Softmax(x_i)} = \frac{\mathrm{e_x}}{\sum_j e^{x_j}} \tag{7}$$

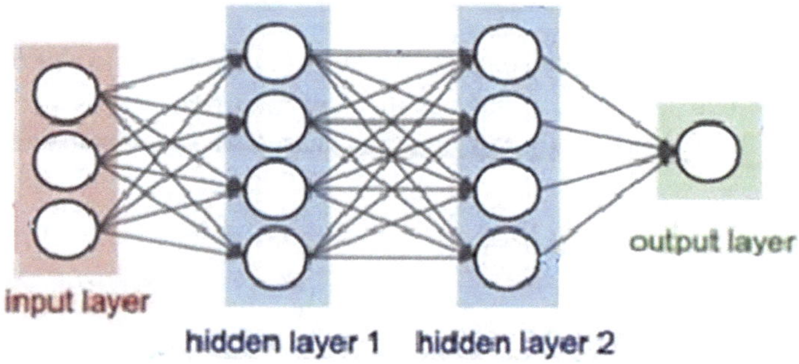


**Fig. 4** Multi-layer perceptron model

### 3.3 Convolutional Neural Networks (CNNs)

CNNs are specialized neural networks that process image data in a way that closely resembles human visual perception (Lindsay 2021) as shown in Fig. 5. They automatically learn mid-level and high-level representations from input data, making them highly effective for tasks like brain tumor segmentation (Tie and Peng 2024). The CNN architecture's ability to handle RGB layers of an image and analyze them one at a time allows it to identify and classify important features, such as tumor regions in MRI scans (Liu et al. 2021). The architecture of CNNs involves several key layers:

1. **Convolution Layer**: The convolution operation is defined by:

$$(X*K)(i,j) = \sum_m \sum_n X(i+m, j+n)K(m,n) \quad (8)$$

   where $X$ is the input image, $K$ is the kernel (filter), and $i, j$ are the coordinates of the convolution output.
2. **Pooling Layer**: Pooling reduces the spatial dimensions of feature maps. For instance, max pooling with a $2 \times 2$ window is:

$$\text{MaxPool}(X)i, j = \max\left(X_{2i:2i+2, 2j:2j+2}\right) \quad (9)$$

3. **Fully Connected Layer**: After feature extraction and pooling, the final layer is fully connected and performs classification:

$$y = W \cdot \text{flatten}(X) + b \quad (10)$$

   where $W$ represents the weights and $b$ is the bias term. CNNs are particularly relevant in the context of brain tumor detection as it provides robust solutions for complex image classification and segmentation tasks in medical imaging by leveraging these mathematical operations and architectures.

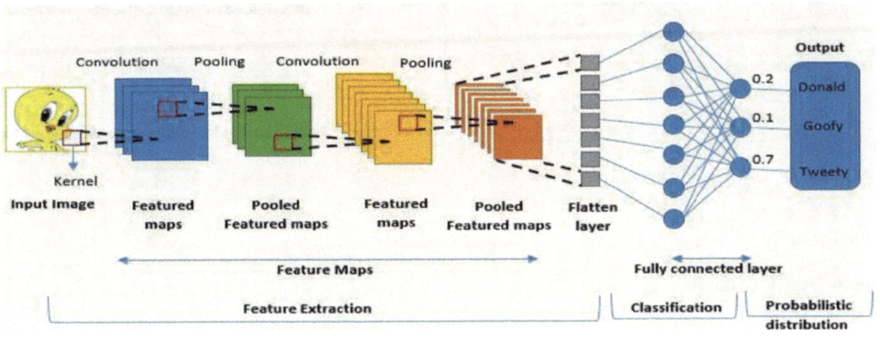


**Fig. 5** Convolution Neural Network (Sanghavi 2023)

# 4 Methdology

The methodology for developing a system to classify brain tumors utilizing the Convolutional Neural Networks (CNNs) involves several key steps as shown in Fig. 6, including data acquisition, pre-processing, CNN model development, training, and evaluation. Each step is crucial for ensuring the effectiveness and accuracy of the system.

## *4.1 Data Acquisition*

The first phase involves data acquisition, where brain tumor images are gathered from multiple sources such as hospitals, research institutions, or publicly available medical image databases. These images can be acquired from various imaging approaches, like Magnetic Resonance Imaging (MRI) and Computed Tomography (CT). The dataset should be varied and extensive, encompassing different tumor types and imaging conditions to guarantee robustness of the model and its capacity to simplify across diverse cases. For this research work dataset titled "Brain Tumor Classification (MRI)" available on the Kaggle is used (Brain Tumor Classification (MRI), n.d.).

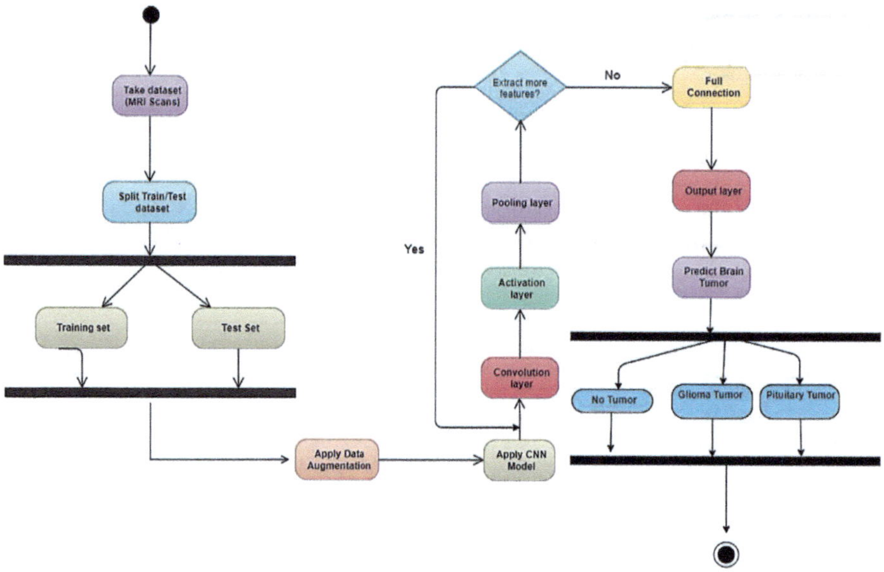


**Fig. 6** Activity chart

## 4.2 Data Pre-processing

After acquiring the images, they undergo pre-processing to prepare them for analysis. This process involves several methods to improve the standard of the input data. Noise reduction is achieved through filtering methods, while contrast enhancement improves the visibility of important features in the images. Additionally, normalization of intensity values ensures consistency across the dataset, making the data suitable for input into the CNN model. Proper pre-processing is crucial for maximizing the effectiveness of the CNN and improving the accuracy of tumor classification.

## 4.3 CNN Model Development

During the development phase of the CNN model, an appropriate architecture is crafted to efficiently learn from the pre-processed images. The CNN structure generally includes multiple convolutional layers, as illustrated in Fig. 7, which apply filters to drive characteristics from the images. These layers are succeeded by ReLU (Rectified Linear Unit) activation functions, which add non-linearity to the model. Pooling layers are used to decrease the dimensions of the feature maps, while fully connected layers consolidate the extracted features to execute classification. The architecture is customized according to the dataset's size and complexity to enhance the model's efficiency. In our mode we have used 6 convolution layers (from EfficientNetB1), 1 pooling layer (Global Average Pooling in the custom part, others are embedded within EfficientNetB1) and 1 fully connected (Dense) layer for classification. The Base Model, EfficientNetB1, is loaded with pre-trained weights (weights = "imagenet") and used without the top classification layer (include_top = False). This means the convolutional layers from EfficientNetB1 are used as the backbone. EfficientNetB1 has a complex architecture with multiple convolution layers, pooling layers, and residual connections. GlobalAveragePooling2D() layer reduces the dimensions after the base model's output, replacing the need for flattening. Dropout(0.5) layer is added for regularization, to avoid overfitting by randomly disabling 50% of neurons during training. Fully Connected Layer, Dense(4, activation = "softmax"), a final dense layer with 4 neurons for multi-class classification, corresponding to the 4 brain tumor types.

## 4.4 CNN Model Training

During the training phase the pre-processed images are inputted into the CNN model. During this stage, the model learns to categorize the images by fine-tuning its parameters to minimize the errors. The training procedure involves using a labeled portion of the dataset and refining the model's weights through backpropagation and a loss

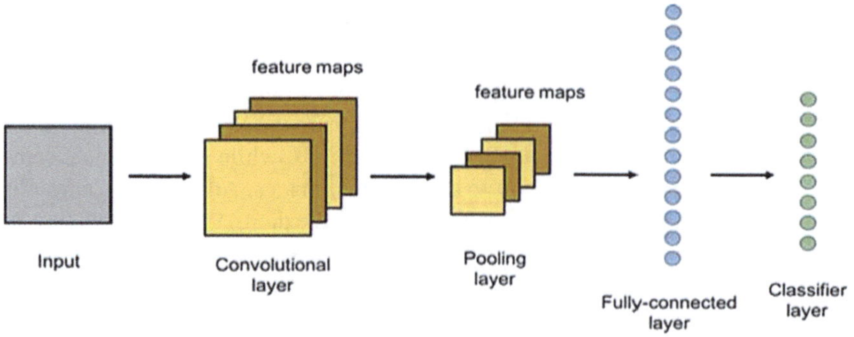


**Fig. 7** Layers in a CNN architecture, including convolutional, ReLU, pooling, and fully connected layers (Zaniolo and Marques 2020)

function like Cross-Entropy Loss. An optimizer like Adam or SGD, is employed to adjust the weights and improve model performance. The Training continues over multiple epoch runs that tunes hyperparameters to achieve the best results.

### *4.5 Model Evaluation*

After the training is complete, the CNN model is tested on a distinct dataset to evaluate its performance. Evaluation metrics such as accuracy, precision, recall, and F1 score are determined to measure how the model classifies the brain tumors. Cross-validation techniques are also used to check the model's reliability and generalizability. The results from this evaluation phase are crucial for deciding whether the model is ready for deployment.

## 5 Experimentation and Results

In this section, the experimentation process and outcome by applying CNNs for brain tumor classification are presented. The experimentation process involves different steps that includes data acquisition, preprocessing, model development, training, and evaluation. The outcomes are evaluated based on key performance metrics such as accuracy, loss, precision, and recall.

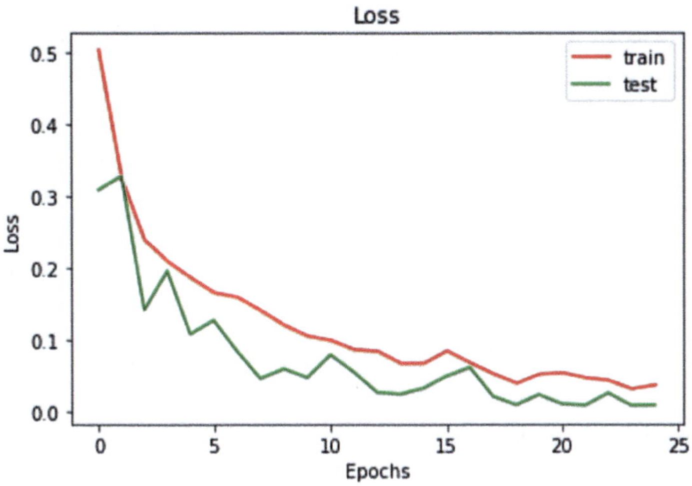


**Fig. 8** Loss versus epoch

## 5.1 Results Analysis

The efficacy of the suggested CNN model was assessed using essential metrics. Which includes accuracy, loss, precision, recall, and F1 score. The outcomes from the training and validation phases are discussed in this section.

The graph shown in Fig. 8 depicts the behavior of loss function during epoch runs. The value of loss decreased steadily as the model learned and indicates effective training. The validation loss also decreased over the epoch duration suggesting that the model did not overfit and could generalize effectively to unseen data.

Figure 9 depicts the accuracy of the suggested model during training. The training accuracy increases consistently with each epoch and reaching a high value of approximately 96% at the end of the 25th Epoch. The validation accuracy also increases showing the capacity of the model for the classification of brain tumor images.

## 5.2 Comparative Analysis of Classifier Performance

To further assess the effectiveness of the proposed CNN model, it was evaluated with the performance of other well-known machine learning classifiers, such as KNN, Random Forest, SVM, LSTM, RNN, and ANN. The comparison shown in Table 1 is done using different metrics such as accuracy, recall, specificity, and precision.

***Accuracy***: The CNN model proposed in this work achieves the highest accuracy among all the considered classifiers and reaches the accuracy of 96.98%. This

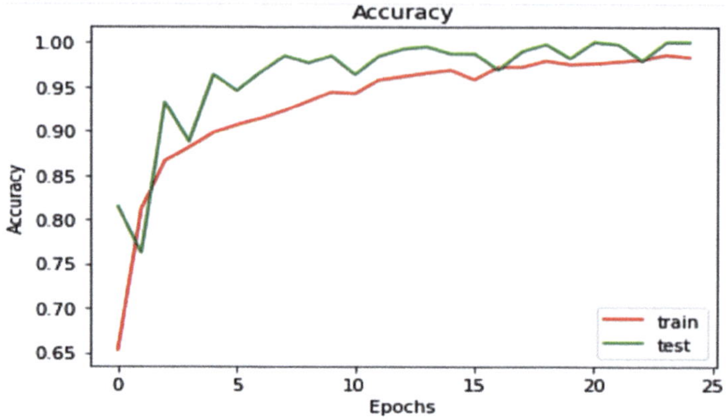


**Fig. 9** Accuracy versus epochs

**Table 1** Comparison across different classifiers

| Classifier | Accuracy (%) | Recall | Specificity | Precision |
|---|---|---|---|---|
| K-nearest neighbour | 89.39 | 0.949 | 0.428 | 0.933 |
| Random forest | 89.39 | 0.983 | 0.167 | 0.903 |
| SVM | 92.42 | 0.983 | 0.428 | 0.935 |
| LSTM | 74.39 | 0.789 | 0.717 | – |
| RNN | 88.54 | 0.903 | 0.881 | – |
| ANN | 82.56 | 0.665 | 0.891 | – |
| Proposed CNN model | 96.98 | 0.987 | 0.932 | 0.967 |

outcome shows a significant improvement over other models, like SVM has an accuracy of 92.42%. KNN and Random Forest classifiers both achieves the accuracy of 89.39%. This highlights the CNN model's capability to effectively capture complex patterns in brain MRI images, making it superior to the traditional machine learning models in terms of accuracy.

***Recall***: Recall is utilized to assess the capability of the machine learning model to accurately determine positive instances (tumors). The suggested CNN model surpasses the other models in terms of recall as well, achieving a recall value of 0.987. It signifies that the CNN model was the most effective at detecting tumors. It surpasses the recall of both the Random Forest and SVM classifiers as both models achieves recall value of 0.983. The high recall rate of the CNN model underscores its potential in minimizing false negatives that is crucial in medical diagnosis.

***Specificity***: Specificity is used to measure the capability of the machine learning model to accurately determine negative instances (non-tumors). The proposed CNN model outperforms other considered models with a specificity of 0.932. This specificity rate was higher than that of the KNN and SVM models both of which had a specificity of 0.428. The result shows that the CNN model is more reliable in correctly identifying non-tumor cases that further reduce the likelihood of false positives.

***Precision***: Precision is used to measure the accuracy of positive predictions. The proposed CNN model achieves a precision value of 0.967. The precision of the CNN model is higher than that of the SVM (0.935) and KNN (0.933) classifiers. This high precision indicates that the CNN model not only detected tumors effectively but also made accurate positive predictions to reduce the number of false alarms.

## *5.3 Findings*

Our model primarily focuses on binary classification of the MRI images in two regions tumor and non-tumor. This is a fundamental task in brain tumor detection. Our model demonstrated robust performance in this area that shows strong potential for clinical application. The CNN model exhibited sensitivity to variations in imaging protocols for example, differences in scanners and imaging parameters. This sensitivity is a known issue in CNN-based medical image analysis since, these variations can lead to reduced model accuracy. Addressing this requires additional data normalization or domain adaptation techniques. To increase the model's efficiency transfer learning was used. By refining a pre-trained CNN model, we were able to significantly improve classification accuracy. Especially, in scenarios where the available labeled data was limited. Transfer learning has been proven to be a beneficial technique in overcoming the small data problem of the medical image analysis.

The findings from our experiments align with recent research trends in the field of brain tumor classification using CNNs. Below are the key observations:

1. High Accuracy: The CNN model utilized in this research achieves high accuracy in classifying brain tumors. The accuracy of the used CNN model ranges between 92 and 96%. This finding is consistent with the recent studies where CNN models have shown accuracies ranging between 80 and 95%.
2. Comparison with Traditional algorithms: The efficiency of the CNN model was compared with the traditional machine learning approaches, including SVMs, KNN, and Random Forest. The comparison evaluation was done regarding accuracy, recall, specificity, and precision. The CNN consistently outperformed other classifiers, demonstrating its superiority in handling the complexities of brain tumor classification tasks.

## 6 Conclusion

This research work demonstrates the utilization of Convolutional Neural Networks (CNNs) for classifying brain tumors using MRI images. The suggested CNN model demonstrates enhanced performance compared to conventional machine learning classification techniques across all mentioned metrics as it achieves high accuracy, recall, specificity, and precision. These findings highlight the efficiency of CNNs in handling the complex patterns that ae found in medical imaging data. It makes it a valuable tool in clinical settings. The CNN model's ability to balance sensitivity (recall) and specificity, combined with high precision, indicates its potential for the reduction of both the false negatives and the false positives, that are very important in medical diagnostics. This work also specifies the limitations of traditional models like SVM and Random Forest when compared with the deep learning models, particularly, when it involves complex image datasets. The results indicate that the CNNs hold significant promise for advancing brain tumor diagnosis that offers a powerful, accurate, and reliable solution for experts in medical field. while the CNN model showed excellent performance some areas still needs improvement. The specificity is high, but it still indicates that the model could be further refined to reduce false positives. Additionally, the study's reliance on a dataset that might not completely present the diversity of real-world medical imaging conditions, suggests a need for broader validation.

## 7 Future Work

Future research concentrates on multiple critical aspects to improve the performance and effectiveness of the CNN model proposed in this work. One important step is to expand the dataset with a large and more diverse set of samples. Dataset should include images from various imaging protocols and conditions that can enhance the robustness and generalizability of the model. Additionally, advanced data augmentation approaches can be explored to further reduce the problem of small data size that will further reduce the risk of overfitting. Another focus could be to further integrate the CNN model into clinical workflows. This can be done by developing a user-friendly GUI that will allow medical experts to upload MRI images and receive outcome of classification in real time. Hybrid models that combine CNNs with other deep learning architectures, like RNNs or Transformers can also be used in the future to enhance the ability of model to record temporal and contextual information that will lead to more accurate classifications.

## References


Agarwal, M., Rani, G., Kumar, A., K, P. K., Manikandan, R., & Gandomi, A. H. (2024). Deep learning for enhanced brain Tumor Detection and classification. *Results in Engineering, 22*, 102117. https://doi.org/10.1016/j.rineng.2024.102117

Al Bataineh, A., Kaur, D., & Jalali, S. M. J. (2022). Multi-Layer Perceptron Training Optimization Using Nature Inspired Computing. *IEEE Access, 10*, 36963–36977. https://doi.org/10.1109/ACCESS.2022.3164669.

Apostolopoulos, I. D., Aznaouridis, S., & Tzani, M. (2023). An Attention-Based Deep Convolutional Neural Network for Brain Tumor and Disorder Classification and Grading in Magnetic Resonance Imaging. *Information 2023, Vol. 14, Page 174, 14*(3), 174. https://doi.org/10.3390/INFO14030174.

Bakshi, I., Manish Pamnani, M., & Singh Rawat, H. (2024). Unravelling the Complexities of Inoperable Glioblastoma Multiforme: A Comprehensive Review of Current Strategies and Future Directions. *Article in International Journal For Multidisciplinary Research.* https://doi.org/10.36948/ijfmr.2024.v06i02.19244.

Brain Tumor Classification (MRI). (n.d.). Retrieved August 21, 2024, f0072om https://www.kaggle.com/datasets/sartajbhuvaji/brain-tumor-classification-mri.

Bhadra, S.; Kumar, C.J.; Bhattacharyya, D.K. (2024). Multiview EEG signal analysis for diagnosis of schizophrenia: an optimized deep learning approach. *Multimedia Tools and Applications.* https://doi.org/10.1007/s11042-024-20205-y.

Carter, R. (2019). *The brain book: An illustrated guide to its structure, functions, and disorders.* https://books.google.co.in/books?hl=en&lr=&id=icXhDwAAQBAJ&oi=fnd&pg=PT114&dq=The+Brain+Book:+An+Illustrated+Guide+to+its+Structure,+Functions,+and+Disorders&ots=BNsgzxoLa9&sig=tKQacSvcztTqEZ3ydQYTvlUBzCE.

Cheng, X., Jiang, Z., Sun, Q., & Zhang, J. (2020). Memory-efficient cascade 3d u-net for brain tumor segmentation. *Lecture Notes in Computer Science (Including Subseries Lecture Notes in Artificial Intelligence and Lecture Notes in Bioinformatics), 11992 LNCS*, 242–253. https://doi.org/10.1007/978-3-030-46640-4_23/TABLES/3.

Hussein, E. M., Mahmoud, D., & Mahmoud, A. (2012). Brain Tumor Detection Using Artificial Neural Networks. *Journal of Science and Technology, 13*(2). www.sustech.edu.

Kaur, J., Agrawal, S., & Vig, R. (2012). A Comparative Analysis of Thresholding and Edge Detection Segmentation Techniques. *International Journal of Computer Applications, 39*(15), 975–8887.

Konde, A., & Thakur, A. (2021). *Fundamentals of Neural Networks. 9.* https://doi.org/10.22214/ijraset.2021.37362.

Kumar, C.J.; Das, P.R.; Hazarika, A. (2022). Autism spectrum disorder diagnosis and machine learning: a review. *International Journal of Medical Engineering and Informatics*, Inderscience, 14(6).

Li, Z., Li, D., Xu, C., Wang, W., Hong, Q., Li, Q., & Tian, J. (2022). TFCNs: A CNN-Transformer Hybrid Network for Medical Image Segmentation. *Lecture Notes in Computer Science (Including Subseries Lecture Notes in Artificial Intelligence and Lecture Notes in Bioinformatics), 13532 LNCS*, 781–792. https://doi.org/10.1007/978-3-031-15937-4_65/TABLES/5.

Lindsay, G. W. (2021). Convolutional Neural Networks as a Model of the Visual System: Past, Present, and Future. *Journal of Cognitive Neuroscience, 33*(10), 2017–2031. https://doi.org/10.1162/JOCN_A_01544.

Liu, Y., Pu, H., & Sun, D. W. (2021). Efficient extraction of deep image features using convolutional neural network (CNN) for applications in detecting and analysing complex food matrices. *Trends in Food Science & Technology, 113*, 193–204. https://doi.org/10.1016/J.TIFS.2021.04.042.

Masih, N., Naz, H., & Ahuja, S. (2021). Multilayer perceptron based deep neural network for early detection of coronary heart disease. *Health and Technology, 11*(1), 127–138. https://doi.org/10.1007/S12553-020-00509-3/TABLES/5.

Mathivanan, S. K., Sonaimuthu, S., Murugesan, S., Rajadurai, H., Shivahare, B. D., & Shah, M. A. (2024). Employing deep learning and transfer learning for accurate brain tumor detection. *Scientific Reports*, *14*(1), 1–15. https://doi.org/10.1038/s41598-024-57970-7.

Monica Subashini, M., & Sahoo, S. K. (2014). Pulse coupled neural networks and its applications. *Expert Systems with Applications*, *41*(8), 3965–3974. https://doi.org/10.1016/J.ESWA.2013.12.027.

Oh, S., Shi, Y., del Valle, J., Salev, P., Lu, Y., Huang, Z., Kalcheim, Y., Schuller, I. K., & Kuzum, D. (2021). Energy-efficient Mott activation neuron for full-hardware implementation of neural networks. *Nature Nanotechnology 2021 16:6*, *16*(6), 680–687. https://doi.org/10.1038/s41565-021-00874-8

Rasheed, S., Rehman, K., & Akash, M. S. H. (2021). An insight into the risk factors of brain tumors and their therapeutic interventions. *Biomedicine & Pharmacotherapy*, *143*, 112119. https://doi.org/10.1016/J.BIOPHA.2021.112119.

Sanghavi, R. (2023). *Exploring GANs and CNNs: Unleashing Creativity and Analyzing Visual World Introduction: In the… | by Rajveer Sanghavi | Medium.* https://medium.com/@rajveersanghavi167/exploring-gans-and-cnns-unleashing-creativity-and-analyzing-visual-world-introduction-in-the-9da56b84518d.

Tie, J., & Peng, H. (2024). Multi-Level Parallel Network for Brain Tumor Segmentation. *CMES - Computer Modeling in Engineering and Sciences*, *139*(1), 741. https://doi.org/10.32604/CMES.2023.043353.

Wu, J., Poehlman, S., Noseworthy, M. D., & Kamath, M. V. (2008). Texture feature based automated seeded region growing in Abdominal MRI segmentation. *BioMedical Engineering and Informatics: New Development and the Future - Proceedings of the 1st International Conference on BioMedical Engineering and Informatics, BMEI 2008*, *2*, 263–267. https://doi.org/10.1109/BMEI.2008.352.

Young, G. S. (2007). Advanced MRI of Adult Brain Tumors. *Neurologic Clinics*, *25*(4), 947–973. https://doi.org/10.1016/J.NCL.2007.07.010.

Zaniolo, L., & Marques, O. (2020). On the use of variable stride in convolutional neural networks. *Multimedia Tools and Applications*, *79*(19–20), 13581–13598. https://doi.org/10.1007/s11042-019-08385-4.

Zhou, H. Y., Lu, C., Chen, C., Yang, S., & Yu, Y. (2023). A Unified Visual Information Preservation Framework for Self-supervised Pre-Training in Medical Image Analysis. *IEEE Transactions on Pattern Analysis and Machine Intelligence*, *45*(7), 8020–8035. https://doi.org/10.1109/TPAMI.2023.3234002.

# Improving Remote Patient Monitoring and Care Using Machine Learning

**Sangeeta Borkakoty, Atowar Ul Islam, and Kanak Chandra Bora**

**Abstract** This study studies the potential of the Random Forest (RF) algorithm in Remote Patient Monitoring (RPM) and its impact on improving patient care standards. This research attempts to significantly enhance the effectiveness of telehealth-enabled RPM systems by focusing on the use of advanced machine learning (ML) techniques. Through meticulous analysis and interpretation of a huge amounts of patient data, the RF algorithm can provide vital insights into the health status of the patients. This would empower the healthcare providers to take timely decisions and steps. Moreover, the use of ML techniques enables the implementation of predictive analysis, advanced risk assessment, and customized patient management approaches. This in turn would help in optimizing healthcare delivery and enhance patient satisfaction levels. This paper examines both the advantages and challenges in integrating the Random Forest algorithm with telehealth RPM systems. The RF algorithm has numerous benefits, including the capacity to handle complex and large datasets, robustness to noise and outliers, and the capacity to provide high accuracy results. The algorithm is also well-suited for analyzing diverse healthcare data and providing actionable insights, as it can efficiently handle nonlinear relationships among the variables. How-ever, implementing the RF algorithm in telehealth RPM systems comes with some challenges as well. There may arise issues related to data privacy and security, model interpretability, and scalability. Apart from this, to ensure the reliability and accuracy of any ML model in healthcare settings, careful validation and monitoring is necessary. Despite these challenges, the integration of the RF algorithm and telehealth RPM systems holds great promise in efficient healthcare delivery. This innovative approach can immensely impact the future of healthcare delivery by providing timely and personalized healthcare, enhancing patient outcomes and satisfaction and in-creasing the healthcare efficiency.

**Keywords** Telehealth · Remote patient monitoring · Machine learning · Random forest

---

S. Borkakoty (✉) · A. U. Islam · K. C. Bora
Department of Computer Science, University of Science and Technology, Baridua, Meghalaya, India
e-mail: s.borkakoty06@gmail.com

 
T. P. Singh et al. (eds.), *Revolutionizing Healthcare: Impact of Artificial Intelligence on Diagnosis, Treatment, and Patient Care*, Studies in Computational Intelligence 1182,
https://doi.org/10.1007/978-3-031-80813-5_11

# 1 Introduction

In recent years, advancements in technology have transformed the landscape of healthcare delivery, starting an era of telehealth and remote patient monitoring (RPM). Telehealth refers to a broad range of healthcare services provided remotely through the use of technology. This includes methods like video conferencing, mobile applications, wearable devices, and remote monitoring tools. On the other hand, Remote Patient Monitoring (RPM) is a specific type of telehealth technique that deals with the continuous monitoring of the health data and vital signs of patients remotely (Bhadra and Kumar 2022).

The need for telehealth and RPM arises from a number of challenges that are faced in traditional healthcare delivery. This includes managing chronic disease, addressing the needs of the aging population, and meeting the ever-growing demand for healthcare services (Bhadra et al. 2024). Telehealth technologies eliminates geographical barriers, which allows patients especially from rural or remote areas to access healthcare services remotely. This reduces travel requirements and the costs associated with it. RPM provides management and active monitoring of health conditions, which enables patients to take an active role in their health management, reducing the likelihood of complications. Moreover, for the global aging population, telehealth and RPM can provide support to the aging individuals in monitoring their health. Telehealth technologies also allow timely identification of health issues through regular monitoring. This enables faster interventions and preventive measure in reducing risks.

Telehealth and RPM encompass a range of technologies that are designed to facilitate remote healthcare delivery and monitoring. This include teleconsultation platforms for virtual consultations between the patients and the healthcare providers, remote monitoring devices such as wearable sensors for collecting and transmitting patient data, mobile applications that help the patients to track their health, and advanced analytics tools and algorithm, that analyze patient data and provide personalized healthcare recommendations.

Machine learning (ML) has immense potential to transform telehealth and RPM. ML algorithms can analyze huge datasets to identify patterns, relationships, and trends (Kumar and Das 2021). This helps in predictive analytics and prompt decision-making. In context of telehealth and RPM, ML offers facilities such as predictive analytics for risk reduction, personalized healthcare recommendations, anomaly detection and early warning systems, optimization of remote monitoring devices, and medical decision support systems. These applications of ML can help in improving the outcomes for the patients, optimizing the utilization of resources, and providing healthcare remotely in distant locations (Jeddi and Bohr 2020).

Machine learning (ML) offers healthcare providers the ability to identify ideal models for predicting the necessity of remote consultations, which in turn facilitate prompt interventions and optimized resource allocation (Cascella et al. 2022). RPM provides various advantages including improved healthcare care quality, enhanced health literacy, patient satisfaction, increased healthcare accessibility, and enhanced patient engagement (Dhinakaran et al. 2022). However, despite some trial studies, the lack of complete trust from medical professionals in error correction methods within technology persists (Malasinghe et al. 2019). This necessitate further research to examine the acceptance among both healthcare providers and patients thinking (Fong and Usman 2020). Inconsistency in telehealth access and usage, that were highlighted during the COVID-19 pandemic (Shaik et al. 2023), have given rise to the rapidly growing field of digital determinants of health, which covers factors such as digital literacy, access to AI/technology, and community infrastructures like WiFi/broadband internet access (Andrikopoulou 2023).

This study investigates the potential of the Random Forest algorithm in Remote Patient Monitoring (RPM) and its impact on enhancing patient care standards (Bhadra et al. 2024). Through dedicated use of advanced ML techniques, particularly emphasizing the efficacy of Random Forest, the research aims to improve telehealth-enabled RPM systems. Integrating ML algorithms aims to enhance patient outcomes and care efficiency. The inclusion of Random Forest within telehealth-enabled RPM systems signifies a significant healthcare delivery advancement, offering critical insights into patient health status and empowering real-time decision-making.

## 2 Data Collection and Preprocessing

Data collection and preprocessing are important steps in evaluating a machine learning (ML). In data collection, it is crucial to gather relevant and representative data that adequately reflects the real-world scenarios the model is going to encounter (Sharma et al. 2024). This involves identifying sources, collecting data, and ensuring quality of the data. This includes handling issues like missing values, outliers, and data imbalance.

Preprocessing is a sub-step which involves preparing the collected data for ML model training. This includes cleansing, transformation, and encoding of the data. Preprocessing may also involve handling of missing values, feature scaling, encoding of categorical variables, and splitting the dataset into training set, validation set, and test set. Proper data collection and preprocessing is crucial for the construction of ML models that generalize effectively to new data and generate dependable predictions. These steps ascertain that the model learns relevant patterns from the data that has been provided and is robust to variations and noise.

### *2.1 Data Collection*

The dataset (Moon et al. 2019) under consideration has been taken from a cross-sectional retrospective study conducted on adult patients admitted to two distinct emergency departments over a span of one year.

The dataset includes 1267 systematically selected records that captures a range of patient demographics, medical histories, and clinical assessments in a total of 24 variables. Among these, some important features include chief complaints, which provide insights into the patient's primary health concerns upon admission, vital signs recorded at the beginning of their emergency department visit, and the clinical outcomes, which detail the course of treatment and eventual disposition of each patient.

Additionally, the KTAS Scores were calculated by registered nurses which were then verified and modified by three KTAS experts, who assessed and assigned suitable KTAS values to reflect the acuity level of each patient's condition.

KTAS (Korean Triage and Acuity Scale) (Park and Lim 2017) is a system used for triaging patients in emergency departments in South Korea. Triaging refers to conducting a preliminary assessment in order to determine the urgency of the patient's need for treatment and the nature of treatment required. The scale categorizes patients into five levels based on the severity of their condition. The criteria for each level include physiological parameters, chief complaints, and the need for urgent interventions. While the specific details of the KTAS criteria may vary, the general framework typically includes the scores as described in Table 1.

This is a comprehensive dataset with a wide array of clinical information, which makes it suitable for analysis and modeling aimed at understanding and predicting emergency department outcomes. The dataset features are described in Table 2.

**Table 1** KTAS scores and descriptions

| KTAS score | Color code | Description | Examples |
|---|---|---|---|
| 1 | Red | Requires immediate medical attention | Cardiac arrest, severe trauma, respiratory distress, and severe burns |
| 2 | Orange | Requires urgent medical attention | Chest pain suggestive of a heart attack, stroke symptoms, and significant trauma without immediate life-threatening injuries |
| 3 | Yellow | Requires semi-urgent medical attention | Moderate respiratory distress, uncomplicated fractures, and minor trauma |
| 4 | Green | Requires non-urgent medical attention | Minor injuries, mild pain, and minor illnesses |
| 5 | Blue | Requires minimal medical attention | Minor cuts, bruises, and minor illnesses without significant symptoms |

**Table 2** Dataset features

| Feature | Description |
|---|---|
| Group | Type of ED (1 = local, 2 = regional) |
| Sex | Biological gender of patient (1 = female, 2 = male) |
| Age | Age of patient |
| Patients number per hour | Patient Identifier in current hour |
| Arrival mode | Type of transportation to the hospital ([1 = walking, 2 = public ambulance, 3 = private vehicle, 4 = private ambulance, 5, 6, 7 = other) |
| Injury | Whether patient is injured or not (1 = no, 2 = yes) |
| Chief_ complain | Paient's primary complaint |
| Mental | Mental state of the patient (1 = alert, 2 = verbal response, 3 = painful response, 4 = unresponsive) |
| Pain | Whether patient is in pain or not (1 = yes, 0 = no) |
| NRS_pain | Nurse's assessment of pain for patient |
| SBP | Systolic blood pressure |
| DBP | Diastolic blood pressure |
| HR | Heart rate |
| RR | Respiratory rate |
| BT | Body temperature |
| Saturation | Oxygen saturation levels |
| KTAS_RN | KTAS score as per registered nurse (1, 2, 3 = emergency, 4, 5 = non-emergency) |
| Diagnosis in ED | Diagnosis or medical condition identified in Emergency Department |
| Disposition | Outcome or plan for a patient's care (1 = dischange, 2 = admission to ward, 3 = admission to ICU, 4 = discharge, 5 = transfer, 6 = death, 7 = surgery) |
| KTAS_ expert | KTAS score as per expert (1, 2, 3 = emergency, 4, 5 = non-emergency) |
| Error_group | Mismatch scale between KTAS_RN and KTAS_expert in a scale of 0–9 |
| Length of stay_min | Minimum duration of hospital stay required for patient |
| KTAS duration_ min | KTAS duration (time patient was in emergency state) |
| Mistriage | Mismatch between the KTAS score of the emergency nurses and the true KTAS score of the triage experts |

The primary objective of the prediction task is to develop a model that is capable of accurately predicting the emergency level of patients admitted to the emergency department, leveraging the KTAS values as predictive features. This entails utilizing machine learning techniques to classify patients into distinct emergency level categories based on their assessed acuity levels as indicated by the KTAS values.

## 2.2 Data Preprocessing

The dataset comprises a wide range of variables, including both categorical and numerical data types. This heterogeneous nature necessitates tailored preprocessing techniques to ensure optimal model performance.

As a preliminary step, any gibberish values existing in the dataset are identified and then replaced with NaN (Not a Number) placeholders. This process of standardization streamlines subsequent data processing steps by ensuring uniformity in data representation.

The bar plot depicted in Fig. 1 displays the proportion of values that are missing for each of the variable. This visualization provides a quick over-view of missing data, making it easier to pin down the columns that has substantial values missing.

The missing numerical values in the dataset are assigned values based on the mode of the existing numerical values, which refers to the value that occurs most frequently within each of the respective numerical variable. By replacing the missing values with the mode, the data integrity is preserved, which minimizes the impact of missing data on subsequent analysis.

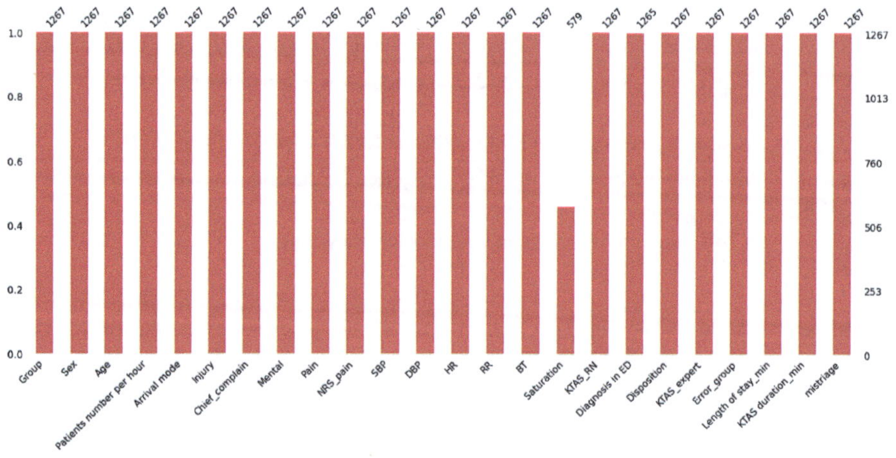


**Fig. 1** Proportion of values that are missing for each of the variable

To gain a deeper comprehension of the distributional characteristics of the variables, histograms and density plots are generated. These graphical representations offer visual insights into the frequency and density distributions of variable values. This helps in identifying data skewness, central tendencies, and potential outliers. Figures 2 and 3 show the density and histograms of the variables respectively.

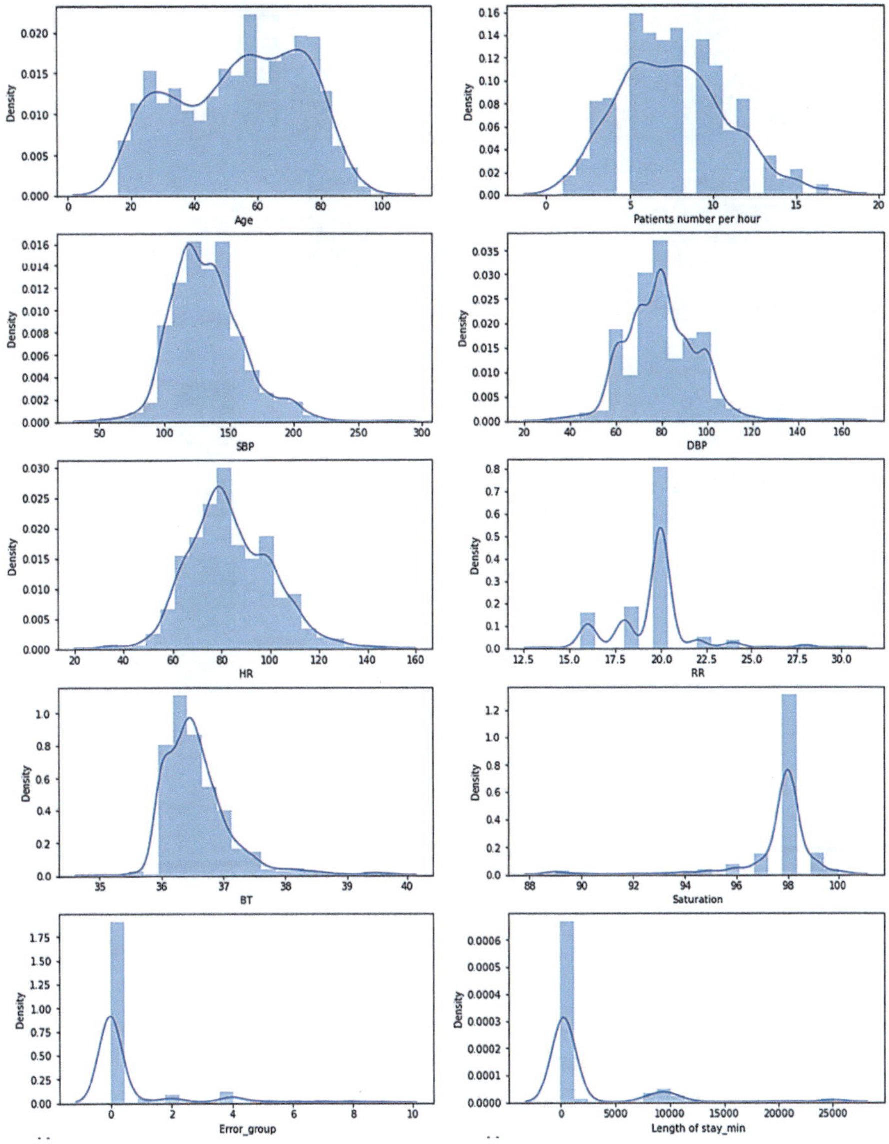


**Fig. 2** Density of the variables

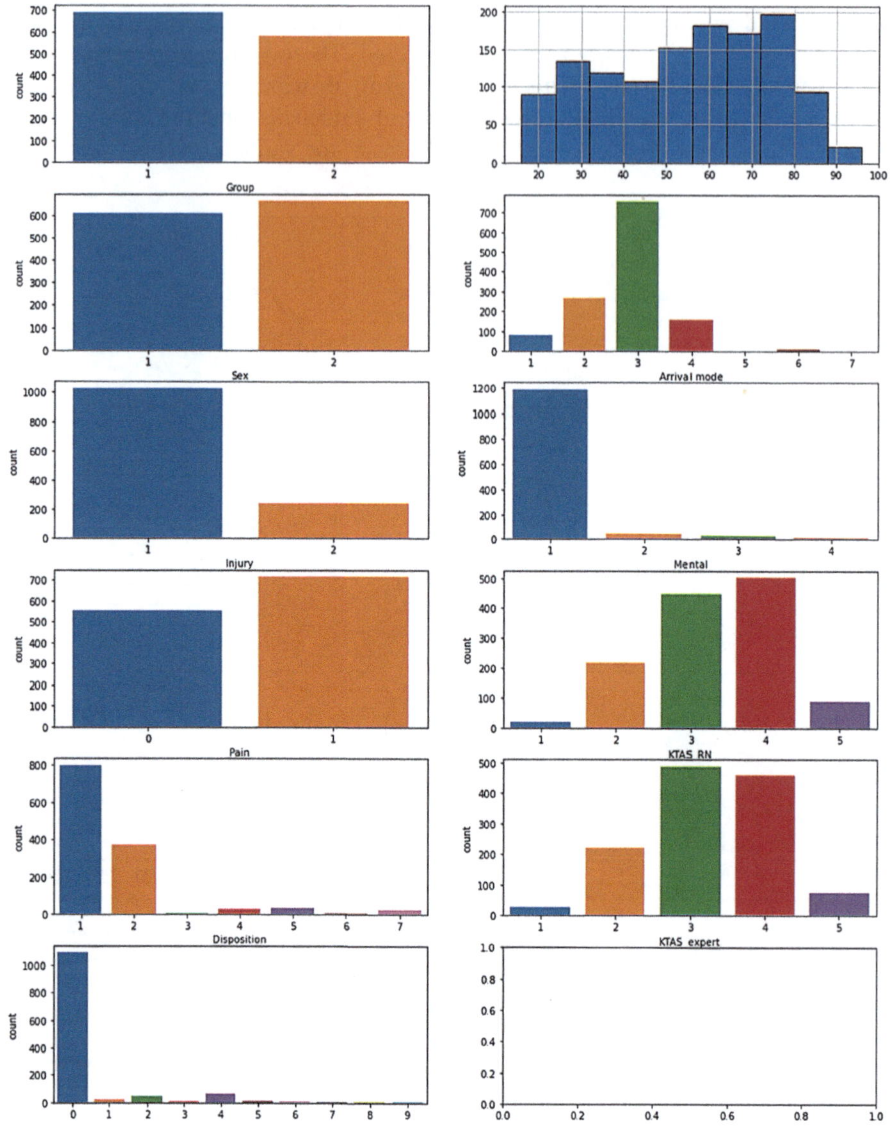


**Fig. 3** Histogram of the variables

Figure 4 shows the correlation matrix, which is used to examine the relationships between the variables in the dataset. This matrix helps in identifying potential patterns and dependencies, which in turn is useful for suitable feature selection and model building.

Figure 4 shows the correlation matrix, which is used to examine the relationships between the variables in the dataset. This matrix helps in identifying potential patterns

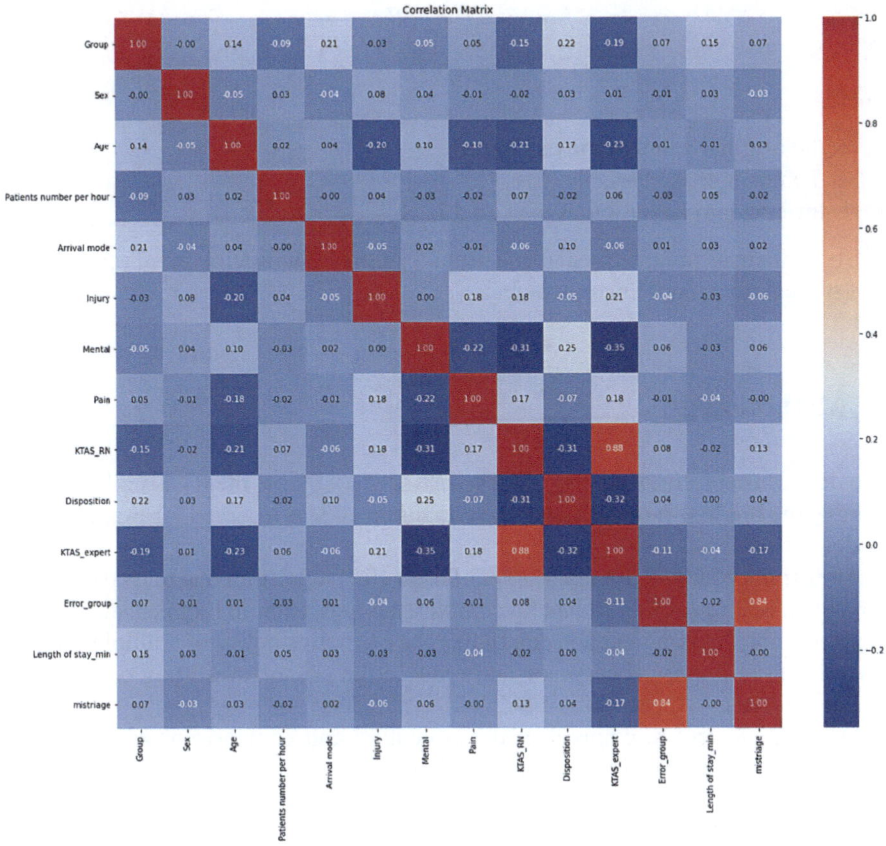


**Fig. 4** Correlation matrix

and dependencies, which in turn is useful for suitable feature selection and model building.

## 2.3 Feature Scaling

Feature scaling aims to modify the scale or range of features of a dataset to ensure more effective model training and improved performance. Here, Z-score normalization is used, which is a method that rescales the numerical values of the dataset in a manner such that their mean is 0 and standard deviation is 1. This process standardizes the distribution of the numerical data, making it easier to compare variables with different scales and ensuring that they contribute equally to the model training.

The formula for calculating the Z-score of a data point x in a variable X is as shown in Eq. (1):

$$Z = \frac{x - \mu}{\sigma} \tag{1}$$

where,

$x$ is the original value of the data point
$\mu$ is the mean of the variable X
$\sigma$ is the standard deviation of the variable X

This enhances the quality and consistency of data, facilitating more accurate and robust predictive modeling outcomes.

## 3 Methodology

The dataset is split into training set and testing set. 80% of the data is allocated for training the predictive models, and the remaining 20% is reserved for evaluating the model performance.

The primary focus is on the Random Forest (RF) algorithm due to its robustness, versatility, and the ability to handle complex datasets with high dimensionality. RF is an ensemble learning method combining the predictive power of multiple decision trees. It offers improved accuracy and reduces the risk of overfitting compared to individual trees.

However, to ensure comprehensive evaluation and comparison, other machine learning algorithms are also explored, including:

- **K-Nearest Neighbor (KNN)**: KNN (Laaksonen and Oja 1996) is a simple but functional algorithm for classifying data points based on the majority class of their nearest neighbors in the given feature space.
- **Support Vector Machine (SVM)**: SVM (Hearst et al. 1998) is a powerful algorithm. It can be used for both classification and regression tasks. It finds the optimal hyperplane that separates the different classes in the feature space, maximizing the margin between the classes. SVM can effectively handle high-dimensional data as well as non-linear relationships by making use of kernel functions.
- **Gaussian Naive Bayes (NB)**: Gaussian Naive Bayes (Reddy et al. 2022) is a probabilistic algorithm that is based on the Bayes' theorem. It specifically assumes that the numerical features follow a Gaussian (normal) distribution. It is efficient computationally and often used as a baseline model in classification tasks.
- **Decision Tree Classifier**: Due to their simplicity and comprehensibility, decision trees (Podgorelec et al. 2002) are a popular choice for classification tasks. Here the feature space in divided into regions based on feature values. The algorithm makes predictions by traversing the tree from the root to a leaf node. Decision trees (Quinlan, 1990) can be used for both numerical and categorical data and are resistant to outliers.
- **LightGBM Classifier (LGBM)**: LightGBM (Ke et al. 2017) is a gradient boosting framework. It uses decision trees as the base learners. It is highly efficient and

scalable, making it suitable for large-scale datasets. LGBM employs a leaf-wise growth strategy and can handle categorical features directly, leading to faster training times and improved predictive performance.

By comparing the performance of Random Forest with the other algorithms, the model that is most suited for this specific prediction task can be identified. This comprehensive evaluation will allow us to assess each algorithm's strengths and limitations in handling the dataset's characteristics and help in selecting the best-performing model for deployment.

## 4 Results and Discussion

This study investigates the prospects of the Random Forest algorithm in Remote Patient Monitoring (RPM) and its impact on enhancing patient care standards. Through dedicated use of advanced ML techniques, particularly emphasizing the efficacy of Random Forest, the research aims to improve telehealth-enabled RPM systems. Integrating ML algorithms aims to enhance patient outcomes and care efficiency. The inclusion of Random Forest within telehealth-enabled RPM systems signifies a significant healthcare delivery advancement, offering critical insights into patient health status and empowering real-time decision-making.

The performance of the models were evaluated using different popular metrics—accuracy, recall, precision, F-score, and Root mean squared error (RMSE).

**Accuracy**: Accuracy is the proportion of the correctly classified instances out of the total number of instances. It is calculated as:

$$Accuracy = \frac{Number\ of\ correct\ predictions}{Total\ number\ of\ predictions} \tag{2}$$

A higher accuracy means that out of the total predictions, the model is making more correct predictions.

**Recall**: Recall (also called sensitivity or true positive rate) is the proportion of actual positive instances, which have been correctly identified by the model. It is calculated as:

$$Recall = \frac{Number\ of\ true\ positive\ predictions}{Total\ number\ of\ actual\ positive\ predictions} \tag{3}$$

Recall value is particularly important when there is high cost associated with false negatives (missed detections), such as in medical diagnosis. A higher recall value stipulates that the model is correctly identifying a larger proportion of actual positive instances, and minimizing false negatives.

**Precision**: Precision is the proportion of true positive predictions compared to all the positive predictions made by the model. It is calculated as:

$$Precision = \frac{Number\ of\ true\ positive\ predictions}{Total\ number\ of\ positive\ predictions} \tag{4}$$

Here, the total number of positive predictions consists of both true positives and false positives. Precision indicates how many of the predicted positive instances are in reality actually positive. A higher precision stipulates that the model will make fewer false positive predictions, leading to higher quality positive predictions.

**F-score**: F-score (or F1 score) is the harmonic mean of precision and recall. The formula is:

$$F - Score = 2 \times \frac{precision \times recall}{precision + recall} \tag{5}$$

It is a score that balances both precision and recall. If the F-score is higher, it indicates a better balance between precision and recall.

**Root Mean Squared Error (RMSE)**: RMSE is the average measure of the errors between actual values and predicted values. It is calculated as:

$$RMSE = \sqrt{\frac{\sum_{i=1}^{n}(y_i - \hat{y}_i)^2}{n}} \tag{6}$$

where,

$n$ is the number of observations
$y_i$ is the actual value of the target variable for observation i

*$\hat{y}_i$ is the predicted value of the target variable for observation i*

A lower RMSE value indicate better model performance, as they suggest that the model's predictions are closer to the actual values on average.

The results of the experimental setup is given in Table 3.

The performance of the algorithms is visually depicted in Fig. 5.

**Table 3** Testing results

| Algorithm | Accuracy (%) | Precision (%) | Recall (%) | F1 Score (%) | RMSE (%) |
|---|---|---|---|---|---|
| KNN | 92.911243 | 93.019969 | 80.818996 | 86.404579 | 0.386791 |
| SVM | 94.882133 | 91.477022 | 89.834297 | 90.644348 | 0.381666 |
| DTC | 96.061432 | 97.303698 | 93.404939 | 95.268858 | 0.166009 |
| RF | 97.640012 | 97.496162 | 97.259071 | 97.377599 | 0.166009 |
| LGBM | 96.854314 | 99.368281 | 91.571429 | 95.370065 | 0.198419 |

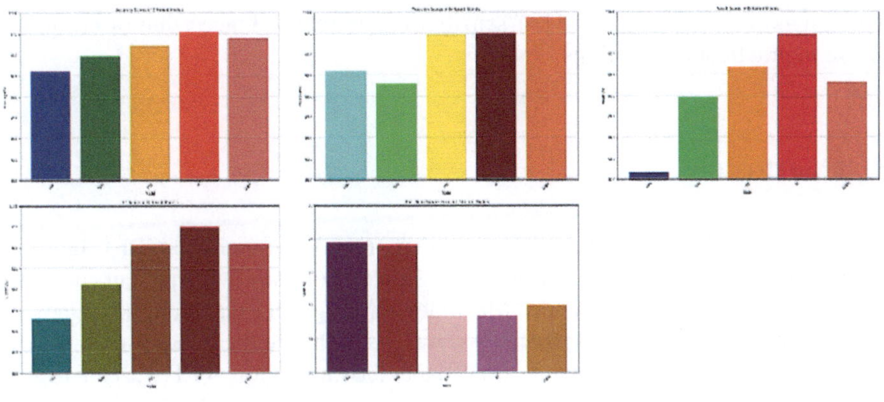

**Fig. 5** Performance of the different models

From the results, we see that Random Forest (RF) has the highest accuracy of 97.64%, followed closely by the Decision Tree Classifier (DTC) with an accuracy of 96.06%. RF and DTC have the lowest RMSE values of approximately 0.166%, indicating that their predictions are closest to the actual values on average. Random Forest (RF) has high precision, recall, and F1-score, indicating a good balance between the precision and recall values. It correctly identifies and classifies a high proportion of positive instances.

Although, LightGBM (LGBM) has the highest precision of 99.37%, indicating a low rate of false positives, but its recall is comparatively lower at 91.57%. This suggests that while LGBM makes precise positive predictions, it may miss some actual positive instances. In a spam email detection system, precision might be prioritized to avoid incorrectly classifying legitimate emails as spam. However, in a medical diagnosis system, recall and accuracy needs to be prioritized to ensure that all instances of a disease are detected, even if it means accepting some false positives.

Thus, Random Forest (RF) and Decision Tree Classifier (DTC) appear to be the most promising models, as they demonstrate high accuracy, precision, recall, F1 Score, and low RMSE.

However, while both RF and DTC are tree-based algorithms, RF outperforms DTC in several aspects, making it a preferable choice. RF's ensemble approach reduces overfitting by building multiple decision trees on random subsets of the training data and features. This results in in more robust predictions. In comparison, DTC is more prone to overfitting, especially on complex datasets. RF generally achieves higher accuracy by aggregating the predictions from multiple trees, which reduces the variance and bias in predictions. Its ensemble nature also leads to improved generalization, which captures a wider range of patterns and reduces overfitting of the training data. Moreover, RF provides feature importance estimation, which helps in understanding the significance of each feature in predictions. RF is also more stable and robust, and is less affected by outliers and data imbalances compared to a single decision tree. Finally, RF requires less hyperparameter tuning and optimization. This makes it more user-friendly and less sensitive to parameter choices, thus offering ease of use in practical applications.

## 5 Conclusion

In conclusion, this study explored the potential of the Random Forest (RF) algorithm in Remote Patient Monitoring (RPM) and its impact on improving patient care standards. Through the use of advanced Machine Learning (ML) techniques, particularly highlighting the effectiveness of RF, we aimed to enhance the effectiveness of telehealth RPM systems. Our research showed that the integration of RF with telehealth-enabled RPM systems represents a significant enhancement in healthcare delivery. The RF algorithm can provide critical insights into the health status of patients. This enables the healthcare providers to make informed decisions and perform interventions in real-time. Moreover, the use of ML techniques enables predictive analytics, advanced risk assessment methodologies, and customized patient management approaches. This optimized healthcare delivery and increases patient satisfaction.

While Random Forest offers numerous benefits such as its capacity to handle large and complex datasets, robustness to noise and outliers, and the ability to provide interpretable results, its deployment in telehealth RPM systems presents certain challenges as well, including issues related to data privacy and security, the need for suitable data preprocessing and feature engineering, model interpretability, and scalability. To address these challenges, proper validation, continuous monitoring, and collaboration between data experts and healthcare providers is necessary.

Despite these challenges, the amalgamation of Random Forest and telehealth RPM systems holds immense potential for transforming healthcare delivery. Through continues research, collaboration, and innovation, the integration of ML techniques into telehealth RPM systems can pave the path for a more patient-centered, data-driven, and efficient healthcare ecosystem.

## References

Andrikopoulou, E. (2023). The rise of AI in telehealth. In *Emerging practices in telehealth* (pp. 183–207). Academic Press.

Bhadra, S.; Kumar, C. J. (2022). An insight into diagnosis of depression using machine learning techniques: a systematic review. *Current Medical Research and Opinion*, 38(5), 749–771. https://doi.org/10.1080/03007995.2022.2038487.

Bhadra, S.; Kumar, C. J.; Bhattacharyya, D. K. (2024). Multiview EEG signal analysis for diagnosis of schizophrenia: an optimized deep learning approach. *Multimedia Tools and Applications*. https://doi.org/10.1007/s11042-024-20205-y.

Cascella, M., Coluccia, S., Monaco, F., Schiavo, D., Nocerino, D., Grizzuti, M., ... & Cuomo, A. (2022). Different machine learning approaches for implementing telehealth-based cancer pain management strategies. *Journal of Clinical Medicine*, 11(18), 5484.

Dhinakaran, M., Phasinam, K., Alanya-Beltran, J., Srivastava, K., Babu, D. V., & Singh, S. K. (2022). A System of Remote Patients' Monitoring and Alerting Using the Machine Learning Technique. *Journal of Food Quality*.

Fong, A. C. M., & Usman, M. (2020, December). Machine Thinking for Telehealth–Toward the Next Generation of Healthcare. In *2020 IEEE 2nd International Workshop on System Biology and Biomedical Systems (SBBS)* (pp. 1–3). IEEE.

Hearst, M. A., Dumais, S. T., Osuna, E., Platt, J., & Scholkopf, B. (1998). Support vector machines. *IEEE Intelligent Systems and Their Applications*, 13(4), 18–28.

Jeddi, Z., & Bohr, A. (2020). Remote patient monitoring using artificial intelligence. In *Artificial intelligence in healthcare* (pp. 203–234). Academic Press.

Ke, G., Meng, Q., Finley, T., Wang, T., Chen, W., Ma, W., ... & Liu, T. Y. (2017). Lightgbm: A highly efficient gradient boosting decision tree. *Advances in Neural Information Processing Systems*, 30.

Kumar, C. J., Das, P. R. (2021). The diagnosis of ASD using multiple machine learning techniques. *International Journal of Developmental Disabilities*, 68(6), 973–983. https://doi.org/10.1080/20473869.2021.1933730.

Laaksonen, J., & Oja, E. (1996, June). Classification with learning k-nearest neighbors. In *Proceedings of International Conference on Neural Networks (ICNN'96)* (Vol. 3, pp. 1480–1483). IEEE.

Malasinghe, L. P., Ramzan, N., & Dahal, K. (2019). Remote patient monitoring: a comprehensive study. *Journal of Ambient Intelligence and Humanized Computing*, 10, 57–76.

Moon, S. H., Shim, J. L., Park, K. S., & Park, C. S. (2019). Triage accuracy and causes of mistriage using the Korean Triage and Acuity Scale. *PloS One*, 14(9), e0216972.

Park, J., & Lim, T. (2017). Korean triage and acuity scale (KTAS). *Journal of The Korean Society of Emergency Medicine*, 28(6), 547–551.

Podgorelec, V., Kokol, P., Stiglic, B., & Rozman, I. (2002). Decision trees: an overview and their use in medicine. *Journal of Medical Systems*, 26, 445–463.

Quinlan, J. R. (1990). Decision trees and decision-making. *IEEE Transactions on Systems, Man, and Cybernetics*, 20(2), 339–346.

Reddy, E. M. K., Gurrala, A., Hasitha, V. B., & Kumar, K. V. R. (2022). Introduction to Naive Bayes and a review on its subtypes with applications. *Bayesian Reasoning and Gaussian Processes for Machine Learning Applications*, 1–14.

Shaik, T., Tao, X., Higgins, N., Li, L., Gururajan, R., Zhou, X., & Acharya, U. R. (2023). Remote patient monitoring using artificial intelligence: Current state, applications, and challenges. *Wiley Interdisciplinary Reviews: Data Mining and Knowledge Discovery*, 13(2), e1485.

Sharma, M.; Kumar, C. J.; Bhattacharyya, D. K. (2024). Machine/deep learning techniques for disease and nutrient deficiency disorder diagnosis in rice crops: A systematic review. *Biosystems Engineering*, 244, 77–92.

# Detecting Depression Using Textual Data from Social Media

**Bapynshngainlang Nongkynrih, Hemashree Bordoloi, Gypsy Nandi, and Samar Jyoti Saikia**

**Abstract** Depression is an important psychological problem that affects an extensive group of individuals all around the world, including an assortment of folks today. Depression is inclined to become worse during a host of conditions, including social isolation, poverty, the hurried and high-pressure characteristics of present-day life, and a variety of other stressors. In this work, a comparative comparison was conducted of four machine learning models for classifying mental illness: Support Vector Machine, Naive Bayes, Logistic Regression, along with Decision Tree. We trained and tested these models using a dataset of mental health-related text. Accuracy, precision, recall, and F1 score were used to assess them. Evaluation of our tests demonstrated that SVM had the best and most constant accuracy and F1 score, outperforming both Logistic Regression, Decision Tree, and Naïve Bayes models. Advantages and disadvantages of each model are illustrated in this paper, along with recommendations on ways to enhance its functioning. As per our results, SVM is a promising approach for spotting mental illness in text data overall.

**Keywords** Mental health · Machine learning · Classification · SVM · Naive Bayes · Decision tree · Comparative analysis

B. Nongkynrih · G. Nandi
Department of Computer Applications, Assam Don Bosco University, Guwahati, Assam, India
e-mail: bapynshngain.nongkynrih@gmail.com

G. Nandi
e-mail: gypsy.nandi@dbuniversity.ac.in

H. Bordoloi (✉) · S. J. Saikia
Department of Electronics and Communication Engineering, Assam Don Bosco University, Guwahati, Assam, India
e-mail: hemashree.bordoloi@dbuniversity.ac.in

S. J. Saikia
e-mail: samar.saikia@dbuniversity.ac.in



T. P. Singh et al. (eds.), *Revolutionizing Healthcare: Impact of Artificial Intelligence on Diagnosis, Treatment, and Patient Care*, Studies in Computational Intelligence 1182,
https://doi.org/10.1007/978-3-031-80813-5_12

# 1 Introduction

Mental illnesses are having an impact on millions of people every year. Of late, the process of creating automated ways to diagnose and identify mental illness has heightened. The outcome has been the creation of a variety of machine learning algorithms that assess text input and determine whether they exhibit signs of mental illness, using natural language processing (NLP) methods. This research shows a comparative analysis of four different machine learning models: Naive Bayes, Decision Trees, SVM, and Logistic Regression are, correspondingly. The models were developed and appraised using a group of texts and tweets about depression, which had previously been classified. This investigation seeks to establish the most effective machine learning model, from the group of four chosen, based on performance criteria such as accuracy, precision, and the F1 score, and to assess the factors that influence each model's ability to reach its desired outcome. The Support Vector Machine technique is frequently selected in machine learning for purposes of classification. It discovers the optimal hyperplane which splits the data into two groups and has revealed beneficial effects in former studies (Chung and Teo 2022). There is regular benefit to activities of classification from Naive Bayes. Essentially, in principle, Bayesian theory in its normal form supposes that different events' likelihoods are not connected to each other, provided the label classes. Prior studies indicated that Naive Bayes successfully performs text categorization and is applied in the discovery of mental illness. Decision Trees, a format of machine learning algorithm, can be employed in types of categorization tasks (Bhadra and Kumar 2022). The Decision Trees method classifies data into diverse classes based on the attributes that most strongly determine those distinctions. The technique proceeds by creating a tree framework that represents each feature as a node and the varying values of the feature as branches. Recent studies applied decision trees to detect mental health disorders. In this paper, the dataset that was used in the research is described, along with the feature extraction methods that were used to convert the data into a format that is suitable for machine learning models (Das et al. 2024). The text data preprocessing stages are also detailed. Moreover, it offers a good examination of the cross-validation techniques that were used to validate the performance of each model and the hyperparameters that were used (Tadesse et al. 2019).

Ultimately, the observed results from the study—including the metrics derived from each model—were presented. The elements contributing to each model's performance, alongside directions for improving those results, are explored. According to the analysis, machine learning methods show their value in revealing mental illness from text-based data, creating more capable and accurate systems for detecting mental health conditions and their corresponding treatments.

## 2 Related Works

A number of studies have analysed using machine learning techniques for anticipating mental health challenges and identifying vital posts on social media (Bhadra and Kumar 2023). A thorough review of machine learning techniques applied to mental health prognoses, concentrating mostly on anxiety and depression, was done by Chung and Teo (Chung and Teo 2022). The researchers detailed a variety of methods for gathering data, extracting features, and choosing models, focusing on the difficulties associated with using machine learning for mental health anticipations. The study shows that a support vector machine reached an accuracy of 74%.

Tadesse and Colleagues (2019) used natural language processing and machine learning to find depression-related messages on the platform Reddit. They looked over a dataset of more than 14,000 messages, applying a variety of feature extraction strategies as well as machine learning methods for classifying messages as either related to depression or unrelated. Using Linguistic Inquiry Word Count (LIWC), Latent Dirichlet Allocation (LDA), and bigrams together, they assessed the field related to the early discovery of mental health disorders. They achieved an accuracy level of 80% with their Support Vector Machine classifier.

Blasco-Fontecilla and peers (Blasco-Fontecilla et al. 2012) have delved into patterns of suicide across many countries throughout a three-decade span. The aim of their study was to examine these patterns and their influencing factors using mixed effect models, furnishing insights suitable for mental health policy and intervention methods.

Alee and Hasan (Alee et al. 2018) examined psychological tension and mental health issues in the politically unstable and conflict disturbed areas of North East India. Their data was gathered through a technique that integrated multiple methodologies for the study of mental health disorders, and they assessed the consequences of this for mental health policy as well as program design in the region.

Using information from the World Health Organization and other sources, Sudhakar (2021) performed an analysis of suicide patterns around the world. By analysing a diversity of topics, including diverse mental health conditions, this inquiry concluded the need for both preventive and intervention strategies for suicide.

## 3 Methodology

### *3.1 Data Collection*

Three datasets containing textual data (Havigerová et al. 2019) related to mental health is collected from Kaggle, two of them consisted of tweets taken from Twitter and labelled as depressive or non-depressive, while the other dataset was named

Twitter US Airline Sentiment and it consisted of labels such as 'depressive', 'non-depressive' and 'neutral'. The other dataset consists of data taken from subreddits r/SuicideWatch and r/Depression and each of them labelled as suicidal or depressive.

## 3.2 *Data Pre-processing*

Various preprocessing methods are implemented on the collected textual data to adapt and improve data quality for machine learning models (Amanat et al. 2022). If there were any, missing values have been taken out from the dataset. The use of text cleaning techniques that eliminate special characters and punctuation raises the standard of text quality (Nijhawan et al. 2020). Tokenization is commonly defined as the process of taking text apart into tokens, which signifies every important word within the entire text. The objective is to reduce noise and improve model performance by doing away with stopwords, which are routinely used words (they, them, how, both, can), since they do not contribute to important information. To achieve a better insight, tokens were transformed into their rudimentary words. Once completed, TF-IDF assesses the text material, giving each term a rank of importance within the document in contrast to the larger set of documents. Count Vectorizer is often employed in text analyses to tally term frequency, but TF-IDF Vectorizer continues to carry out this task with higher levels of precision (Figs. 1, 2 and 3; Tables 1 and 2).

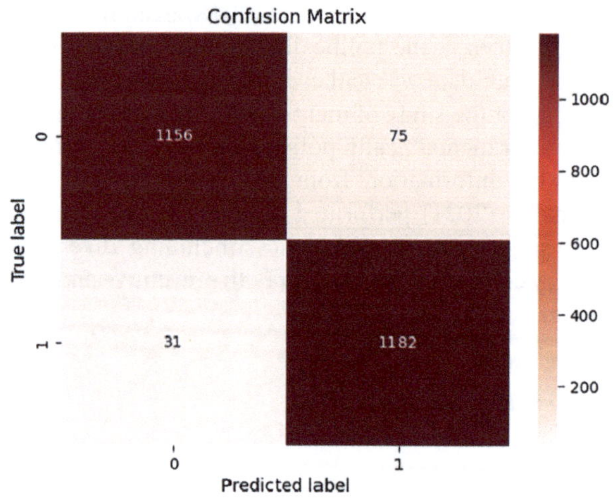


**Fig. 1** The confusion matrix from SVM for the combined dataset with 1199 true positives and 1198 for true negatives while giving very low values for false positives and false negatives

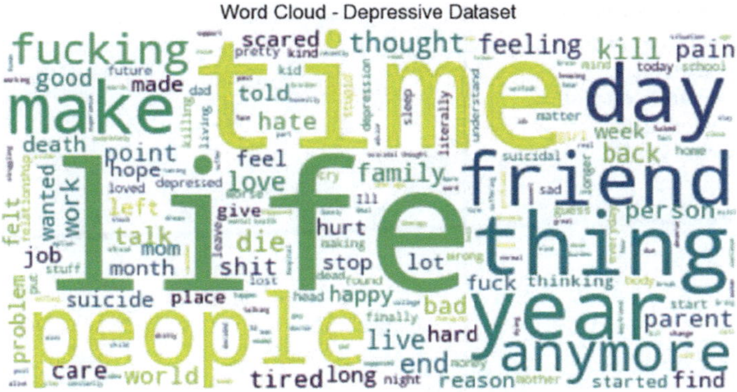


**Fig. 2** Word cloud of the keywords that are found in the 'depressive' labelled data

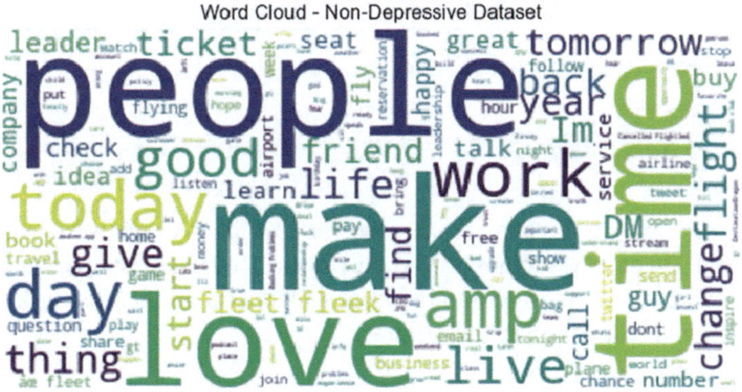


**Fig. 3** Word cloud of the keywords that are found in the 'not-depressive' labelled data

**Table 1** The various metrics for each model using count vectorizer

| Model | Train accuracy (%) | Test accuracy (%) |
|---|---|---|
| Support vector machine | 99.50 | 95.66 |
| Naïve Bayes | 96.84 | 92.35 |
| Decision tree | 99.90 | 92.14 |
| **Logistic regression** | 99.36 | 96.03 |

**Table 2** The TF-IDF vectorization of the texts shows improved metrics for the models, with decreased gaps between training and testing accuracies, suggesting better dataset fit

| Model | Train accuracy (%) | Test accuracy (%) |
|---|---|---|
| Support vector machine | 98.84 | 96.52 |
| Naïve Bayes | 96.98 | 92.59 |
| Decision tree | 99.90 | 92.92 |
| **Logistic regression** | 99.71 | 96.56 |

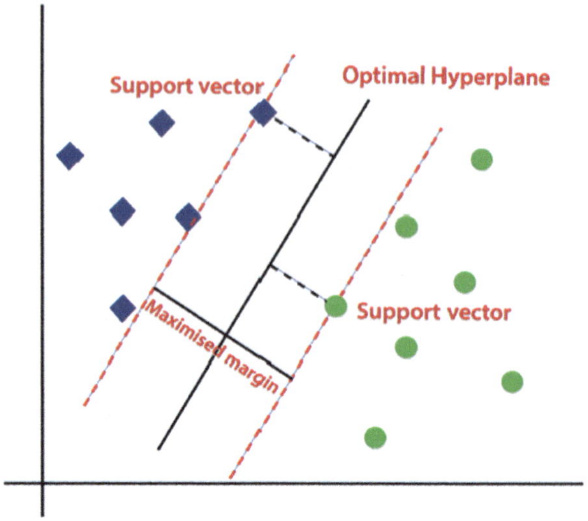


**Fig. 4** Diagram for SVM with the optimal hyperplane being the one with the highest distance from each of the support vectors from both the classes

## 3.3 Model Selection

Four machine learning models are selected for comparative analysis: Naive Bayes (NB), Support Vector Machines (SVM), Logistic Regression (LR), and Decision Tree (DT). These models are commonly used for text classification tasks and have shown promising results in various studies.

## 3.4 Training and Evaluation

In our case, a division of the training and testing sets was done to 80% and 20% respectively. This means that 80% of the dataset was used to train the model and the rest 20% was used for model testing. For each model, hyperparametric values such as regularization in logistic regression or parameter C and kernel in SVMs, criterion and random state in case of decision trees, are searched for using the grid search approach. Each model is assessed using relevant evaluation metrics to measure how effectively they are able to classify mental illness based on text data (Figs. 4 and 5).

## 3.5 Model Comparison

To identify the best model in detecting mental illness, the evaluation metrics of each model are compared. The model with the highest general metrics is chosen (Babu and Kanaga 2021). The comparison analysis illustrates the advantages and

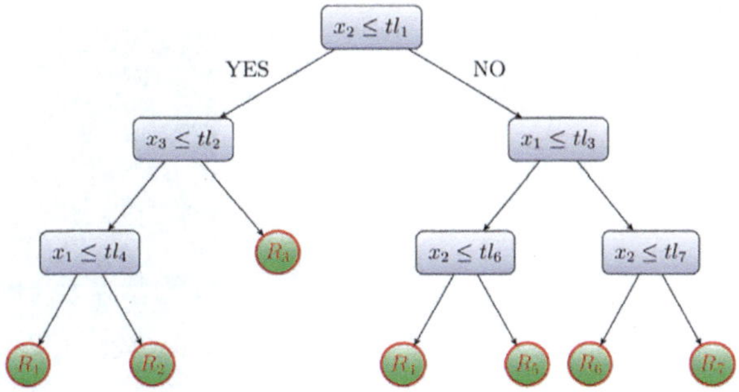


**Fig. 5** Basic structure of a decision tree with decision nodes and leaf nodes

disadvantages of each model, as well as its effectiveness in detecting mental illnesses through messages.

### *3.6 Model Integration and Deployment*

Following the analysis of the performance of the four machine learning models built, the best performing model is dumped into a pickle file for easy storage and retrieval. The model is then serving through a Flask application. The Flask application loads the model from the saved pickle file and makes predictions based on the user requests.

The Flutter application accesses the Flask API through one of its endpoints. As a user types information in the Flutter app, the app pushes this information to the Flask API. The API uses the saved model to process the input, predicts the output, and sends it back to the Flutter application.

## 4 Experimental Evaluations

The details of the experiments conducted for comparing the four machine learning models are reported in this section. Some of the evaluation metrics used to determine the effectiveness of the model include accuracy, precision, recall and F1-score. The experiments were first carried out on the Twitter depressive tweets, then the models were trained on both Twitter and the reddit dataset together.

### *4.1 Dataset Description*

In addition to the above, several datasets on Kaggle that featured similar content and classification from Reddit—in this case, sub-reddits 'r/Depression' and 'r/SuicideWatch'—and a Twitter one which was the Twitter US Airline Sentiment

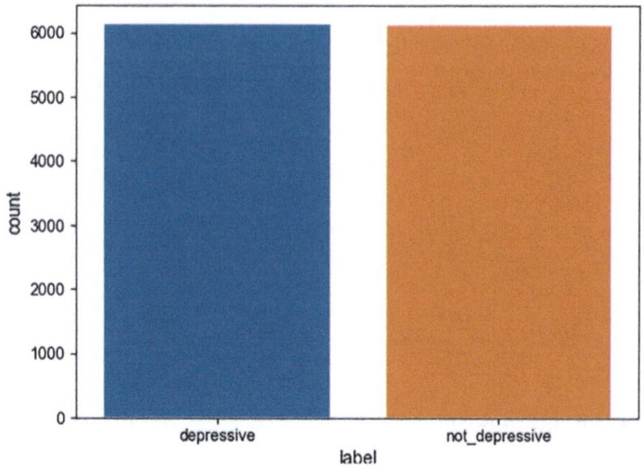


**Fig. 6** Class distribution is well balanced

**Table 3** The machine learning models' performance metrics

| Model | Accuracy (%) | Precision (%) | Recall (%) | F1-Sco re (%) | Time (s) |
|---|---|---|---|---|---|
| SVM | 96.52 | 96.59 | 96.52 | 96.52 | 34.12 |
| NB | 92.59 | 93.40 | 92.59 | 92.56 | 0.01 |
| DT | 92.92 | 92.92 | 92.92 | 92.92 | 1.78 |
| **LR** | 96.56 | 96.56 | 98.12 | 98.12 | 18.77 |

dataset were employed (Shah et al. 2020). The data was cleaned to eliminate noise and provide uniformity in the information presented. An 80:20 ratio was observed in splitting the dataset into a training set and a testing set. The models were supervised and their parameters adjusted based on the training set while the performance was evaluated on the testing set (Fig. 6).

## *4.2 Model Performance*

See Table 3.

### 4.2.1 Support Vector Machine (SVM)

When it comes to SVM model for classifying tweets that are depressed, it displayed remarkable performance with an accuracy of 96.52%. The precision, recall, and f1 score were also high implying good classification performance. The SVM model with kernel linear with $C = 1$ was chosen as the best model through grid search and cross validation. The SVM Model was trained and tested in a time period of

34.12 s. The SVM Model also displays the best balance between the training and testing dataset which shows less chance for overfitting or underfitting.

### 4.2.2 Naïve Bayes

The Naïve Bayes model performed with an accuracy of 92.59% which was a little performed lower than the SVM model but still showed remarkable classification ability. The results were also good which suggests good classification results. The best performance Naive Bayes was taken from the grid search applying alpha of 0.1. The execution time for training and testing the Naive Bayes model was almost negligible which was only 0.01 s.

### 4.2.3 Decision Tree

The Decision tree model achieved 92.92% accuracy. Similarly, the performance of this model was much lower than the SVM and Naive Bayes models, nevertheless it also gave satisfactory classification results. All the metrics were the same, which meant that the classification performance was well fitted. The best performing Decision Tree model was taken from grid search using Gini index and random state as 42. It took 1.78 s to train and test the Decision Tree model.

### 4.2.4 Logistic Regression

In classifying depressed tweets, the Logistic Regression model attained an accuracy rate of 96.56%, exhibiting exemplary results in practice. The precision, recall and F1-score were also high, which shows balanced classification performance was achieved. Based on the grid search results, the best-performing Logistic Regression model was chosen by way of 'liblinear', C value of 1. The total amount of time spent training and testing the Logistic Regression model was 18.77 s (Fig. 7).

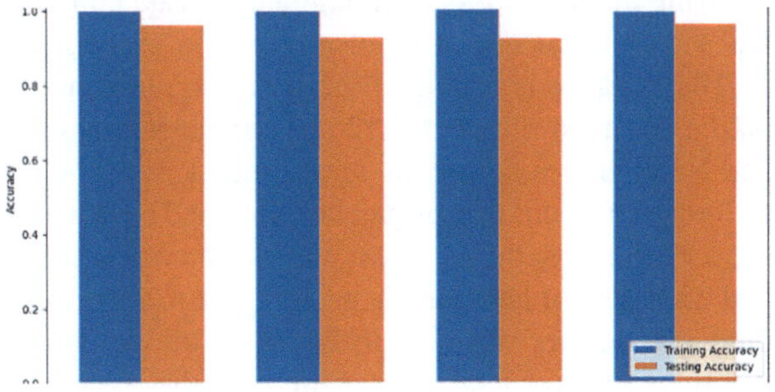


**Fig. 7** A comparison between the training and testing accuracy

We can see that they are almost balanced in case of SVM and Logistic Regression, which indicates that they do not show any signs of overfitting or underfitting. Overfitting happens when a machine learning model memorizes all training data rather than understanding the underlying pattern, causing the model to perform poorly when fed new data.

## 5 Discussion

According to the results of the experiments, all four machine learning models performed well in classifying the depressed tweets. The highest level of performance was recorded in the SVM and Logistic Regression models which achieved accuracy rate of 96.52% and 96.56% respectively, while the Naive Bayes model followed closely with an accuracy of 92.59%. The accuracy level of the Decision Tree model was 92.92%.

The SVM model achieved excellent results with balancing precision, recall and F1 score. Naive Bayes achieved compromise performance relative to effectiveness and performance, a reason why it is applicable in such scenarios where time is of the essence. The Decision Tree model gave fairly good results but was slightly lesser in accuracy to the SVM and Naive Bayes models. The time taken to achieve a high accuracy in the SVM model was similar to that of the Logistic Regression model, but execution speed was still extremely long.

## 6 Conclusion

In summary, we have assessed the effectiveness of four machine learning models designed to infer mental health disorders from language data. Several performance metrics were used to assess the models and the time taken to train each model was recorded. It was shown in this study that SVM and Logistic Regression ceiling metrics were quite close, scoring above 96%. These models had high accuracy when inferring mental illness from language based content. Conversely, Naive Bayes and Decision Tree models were also effective, garnering more than 92% scores.

The variance in training time was also quite great among the models, with SVM taking the most time of 34.12 s, Logistic Regression the second longest at 18.77 s and Naive Bayes taking the least time of just 0.01 s and Decision Tree 1.78 s. The difference in training time can be explained by other factors such as how complex the model is, how big the dataset is and its features, what hyperparameters need adjusting, and what type of hardware is present. We took longer to complete execution because more parameters were set to be tuned as we used GridSearchCV for tuning and it trains the model on every single of the parameters in order to seek 'the best' parameters which can be applied to the model. These 'best' parameters are those which ensure that the model gives us the highest metric values possible.

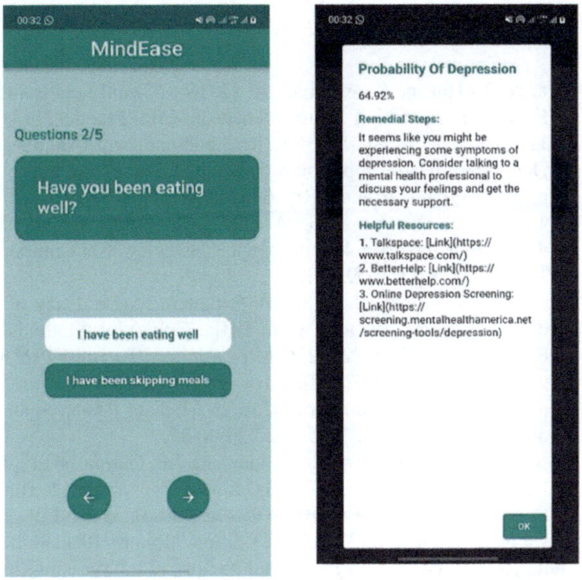


**Fig. 8** Sample interface of the flutter application

According to the performance metrics and training time presented in the machine learning models' performance metrics table, the Support Vector Machine (SVM) model at an accuracy of 96.52% with balanced precision, recall, the F1 score is found to be the best model for identifying whether a text data is about a mental illness in the study comparison. The same cannot be said for the Logistic Regression model, as it produced similar results to that of SVM in terms of accuracy and precision although not the best. Thus, implementing training with the Logistic Regression may be a better alternative if one wishes to cut on cost as far as model training turnaround time is concerned while still achieving high performance level. But here considering the slightly better balance between the training and testing accuracy of the SVM, we considered that as the best performing model.

This research can be used to devise an application that would identify a mental disorder from written text. The appropriate model of the current study, SVM, is incorporated in a Flutter application to offer users real-time assessment of their mental health and assistance. The model is also trained, saved as a pickle file for use at a later date and consists of a Flask application that acts as a bridge for the Flutter app and model (Fig. 8).

As a whole, this work sheds light on the use of machine learning models aimed at the assessment of clinical texts with regards to mental disorders, including their efficacies, limitations, and other operational aspects of diverse models. It opens doors for more research and development in the area of mental health analysis using machine learning.

## References


Alee, N. T., Hasan, A., & Aijaz, A. (2018). Mental health and psychological problems in North East India. Journal of Neurosciences in Rural Practice, 9(3), 369–375.

Amanat, A., Rizwan, M., Javed, A. R., Abdelhaq, M., Alsaqour, R., Pandya, S., & Uddin, M. (2022). Deep learning for depression detection from textual data. Electronics, 11(5), 676. https://doi.org/10.3390/electronics11050676.

Babu, N. V., & Kanaga, E. G. M. (2021). Sentiment analysis in social media data for depression detection using artificial intelligence: A review. SN Computer Science, 3(1). https://doi.org/10.1007/s42979-021-00958-1.

Bhadra, S., & Kumar, C. J. (2023). Enhancing the efficacy of depression detection system using optimal feature selection from EHR. Computer Methods in Biomechanics and Biomedical Engineering, 27(2), 222–236. https://doi.org/10.1080/10255842.2023.2181660.

Bhadra, S.; Kumar, C. J. (2022). An insight into diagnosis of depression using machine learning techniques: A systematic review. Current Medical Research and Opinion, 38(5), 749–771. https://doi.org/10.1080/03007995.2022.2038487.

Blasco-Fontecilla, H., Perez-Rodriguez, M. M., Garcia-Nieto, R., Fernandez-Navarro, P., Galfalvy, H., De Leon, J., & Baca-Garcia, E. (2012). Worldwide impact of economic cycles on suicide trends over 3 decades: Differences according to level of development. A mixed effect model study. BMJ Open, 2(3), e000785. https://doi.org/10.1136/bmjopen-2011-000785.

Chung, J., & Teo, J. (2022). Mental health prediction using machine learning: Taxonomy, applications, and challenges. Applied Computational Intelligence and Soft Computing, 2022, 1–19. https://doi.org/10.1155/2022/9970363.

Das, P. R., Talukdar, R. R., & Kumar, C. J. (2024). Exploring the interplay of abstinence self-efficacy, locus of control, and perceived social support in substance use disorder recovery. Current Medical Research and Opinion, 40(9), 1625–1635. https://doi.org/10.1080/03007995.2024.2390046.

Havigerová, J. M., Haviger, J., Kučera, D., & Hoffmannová, P. (2019). Text-based detection of the risk of depression. Frontiers in Psychology, 10. https://doi.org/10.3389/fpsyg.2019.00513.

Nijhawan, T., Rana, N. P., Dwivedi, Y. K., & Nerur, S. (2020). Stress detection using natural language processing and machine learning over social interactions. Information Systems Frontiers, 22(2), 379-394.

Shah, F. M., Nisar, M. A., Khan, W. A., & Lee, S. (2020). Early depression detection from social network using deep learning techniques. Healthcare Informatics Research, 26(1), 61-69.

Sudhakar, S. (2021). Global suicide analysis. International Journal of Mental Health and Psychiatry, 8(3), 1-10.

Tadesse, M. M., Lin, H., Xu, B., & Yang, L. (2019). Detection of depression-related posts in Reddit social media forum. IEEE Access, 7, 44883–44893. https://doi.org/10.1109/access.2019.2909180.

# Machine Learning in Addiction Research: Advancements, Challenges, and Future Directions

**Rita Rani Talukdar and Priti Rekha Das**

**Abstract** Addiction is a chronic condition marked by the involvement of brain both anatomically and physiologically. Research in this domain has put forwarded numerous insights on this fact. However, inclusion of machine leaning models in addiction research is still in its early stage and there is a lot of scope especially in the domain of diagnosis of addiction. This chapter will discuss about various types of addiction, Advancements, Challenges, and Future Directions of use of machine learning in addiction researches.

**Keywords** Addiction research · Central nervous system · ML/DL models

## 1 Introduction

Addiction is a complex disorder of the brain which is menifested as repetative behaviors like use of harmful substances or engagement into activities like game, gambling etc. despite the harmful consequences. Addiction alters the reward pathways of the brain which ultimately results in insatiable craving for the substances or activities. The pervasive nature of the disorder encompasses individuals beyond age, gender, and socioeconomic status. The effect of addiction is widespread and it includes alterations in almost all dimensions of health. Addiction is chronic by nature and it develops gradually within the individual hampering the physical and mental wellbeing in a successive manner.

Addiction has been used as an umbrella term for indicating different compulsive acts like using ellicit drugs or engaging in various activities. One of the most prevelent types of addiction is harmfull substance addiction. The United Nations' World Drug Report 2023 claimed that during 2021 in every 17 people between the age group

R. R. Talukdar · P. R. Das (✉)
Department of Psychology, Gauhati University, Guwahati, Assam 781014, India
e-mail: pritirekhadas76@gmail.com

R. R. Talukdar
e-mail: ritapsychology@gauhati.ac.in



T. P. Singh et al. (eds.), *Revolutionizing Healthcare: Impact of Artificial Intelligence on Diagnosis, Treatment, and Patient Care*, Studies in Computational Intelligence 1182,
https://doi.org/10.1007/978-3-031-80813-5_13

15–64, one individual is using drugs across the globe (UNODC 2023). Cannabis was found to be the most prevalent drug of choice with 219 million users worldwide which is 4.3% of the world adult population (UNODC 2023). However, regarding the other types of addiction, a large scale systematic review has reported 26.99% pooled prevalance of smartphone addiction globally. The review also reported prevalance for social media addiction was 17.42%, for Internet addiction 14.22%, for cybersex addiction 8.23%, and for game addiction it was 6.04% (Meng et al. 2022).

A machine learning (ML) algorithm is a highly adaptable AI method, particularly effective for processing large datasets under challenging conditions. ML can also be utilized to analyze substance or non substance addiction. These disorders manifest as abnormal thoughts, emotions, and behaviors, leading to significant stress and confusion in daily life. They can include intellectual difficulties, mood swings, anxiety, depression, and schizophrenia, and are monitored and treated through specific mechanisms (Mauri et al. 2017; Chhetri et al. 2023). Healthcare professionals assess patients using questionnaires and interviews (structured or semi-structured) to evaluate physical, behavioral, and emotional characteristics (Lynskey and Strang 2013). Disorders may also be detected through imaging techniques such as EEG and MRI, with medical informatics at the forefront of integrating modern technologies. The datasets generated are complex, unstructured, and variable. The processing of such complex data involves collection, cleaning, transformation, analysis, and storage, all of which are vital for technological solutions. ML models identify trends in existing data to predict future outcomes, and while ML's role in addiction research is still emerging, it holds significant promise. During training, various parameters and attributes are selected, and the algorithm is optimized and tested on different subsets of the dataset. Validation of the algorithm and model is performed using case study samples, with performance measured through accuracy, sensitivity, and specificity. These metrics help in analyzing large datasets and devising diagnostic solutions, early predictions, and personalized treatments (Rehman et al. 2022; Das et al. 2024). In diagnostics, challenges include ensuring model efficiency, monitoring strategies, and accuracy. Early intervention and recovery support systems are essential for addressing substance use disorders, with techniques such as Decision Trees (DT), Random Forests (RF), Support Vector Machines (SVM), and Deep Learning (DL) models effectively identifying and extracting patterns from patient information (Yahyaoui et al. 2019, Kumar and Das 2021).

## 2 Types of Addiction

In behavioural sciences, it is believed that any object having the capacity to stimulate an individual, can be considered as addictive. When the use of addictive entities becomes an obligation and the individual compulsively engaged in that particular behaviour, it is called as addiction. According to various researches, on the basis of symptoms, addiction can be classified into two broad categories namely Substance addiction and Behavioral addiction. Substance addiction (or drug addiction) is a disorder of the brain characterized by a frequent desire to use drug as well as taking it compulsively despite of its harmful consequences. However, behavioural or non-substance addiction involves recurrent compulsive activities like gambling, mobile phone addiction, internet addiction, game addiction etc.

In Substance addiction the drug seeking behaviour is associated with strong desire to use and loss of control over the same. It can be rooted from genetic as well as various psychosocial origins. Historically substance addiction was considered more as a moral failure than a fully developed disorder of the brain. A number of recent neuroscientific studies in this context has now evidenced for the involvement of complex brain structures and pathways in substance addiction and hence establishing it as a neuropsychiatric disorder. These advancements in the understanding of the disorder have also aided the periodical changes in the concept of it over time with a lot more changes in the terminologies associated. According to the recent revision of Diagnostic and Statistical Manual of Mental Health Disorders (DSM) that is DSM 5 has defined the term Substance Use Disorder on the basis of pathological behavioral patterns associated with the ingestion of psychoactive substances (Zou et al. 2017). According to DSM-5 and International Classification of Diseases-10 (ICD-10), a number of psychoactive substances may give rise to addiction. Table 1 depicts various substances that may give rise to different types of addiction.

Behavioral or Non-substance addiction refers to a compulsive engagement in a specific behavior or activity like gambling, internet use, or gaming that may lead to significant impairment in the day today life activities (Zou et al. 2017). Like substance addiction behavioral addiction is also marked by an inability to control the repetitive behavior despite its harmful consequences. Certain behaviors other than psychoactive substance use, can also offer short-term rewards that can lead to persistent actions despite knowing the negative consequences, very similar to the core concept of addiction. Hence, the term "behavioral" addictions are also used analogically to substance addiction but with more focus to the behavioral domains (Grant et al. 2010). Table 2 enlisted various types of behavioral addictions and their characteristics.

## 3 Machine Learning and Addiction Research

Over the past decade, various biomedical research fields, including addiction, have effectively applied machine learning methods and techniques. Machine learning addresses numerous challenges, such as identifying individuals with SUD, evaluating treatment success, detecting patterns in brain images, understanding prognostic factors of SUD, and identifying genetic underpinnings of addiction. Despite these advantages, its application in addiction research remains limited (Morgado et al. 2022a). Machine learning can significantly enhance outcomes in health sciences by answering a broad set of research questions. However, its underutilization in addiction research may stem from a lack of understanding and the perception of machine learning methods as black boxes. Additionally, the terminology of machine learning

**Table 1** Psychoactive substances, their mechanism of action, commonly used forms and associated disorders

| Name of the class of the drug | Mechanism of action | Commonly used forms | Substance use disorder |
|---|---|---|---|
| Alcohol | Ethyl alcohol acts mainly by depressing the CNS through dissolving in the cell's lipid membrane, initially stimulating higher inhibitory centers but eventually depressing vital medullary centers, potentially causing cardiorespiratory failure and death | Alcoholic beverages | Alcohol use disorder |
| Nicotin | Nicotine binds to nicotinic acetylcholine receptors in the CNS, activating reward centers of the brain and leading to addiction. It affects memory, attention, and anxiety at moderate doses but causing convulsions at high doses | Tobacco, cigarette | Tobacco use disorder |
| Cannabis | Cannabis contains cannabinoids like cannabidiol and delta-9-THC, with THC's partial agonistic effects on cannabinoid-1 receptors in the brain inducing the "high" and impacting cognitive and behavioral functions | Marijuana, Hashish, Weed | Cannabis use disorder |
| Opioids | Opioid receptors, particularly mu-opioid receptors (MOR), are G protein-coupled receptors in the brain and other body areas that trigger euphoria, play roles in reward circuits and goal-directed behavior, and induce dopamine release, leading to drug-seeking behavior and changes in brain reward mechanisms with repeated opioid use | Heroin, morphin, prescribed analgesics like vicodin, oxycontin etc. | Opioid use disorder |
| Cocaine | Cocaine increases synaptic dopamine levels by blocking its reuptake transporter, enhancing reinforcing effects, while also affecting serotonin, norepinephrine, glutamate, GABA, endocannabinoid, and corticotropin-releasing hormone systems involved in reward, motivation, and memory | Cocaine | Cocaine use disorder |

(continued)

**Table 1** (continued)

| Name of the class of the drug | Mechanism of action | Commonly used forms | Substance use disorder |
|---|---|---|---|
| Hallucinogens | Hallucinogens mimic serotonin by binding to serotonin 2A receptors, altering their conformation, which changes the internal conditions and behavior of the neurons they affect | LSD, phencyclidine, psyilocybin, MDMA | Hallucinogen use disorder |
| Sedatives, tranquillizers, and hypnotics | It strengthens GABA's effects by attaching to GABA-A receptors and opening chloride channels in the neuronal membrane. The influx of chloride ions alters the neuron's electrical polarity, decreasing its excitability to nerve impulses | Barbiturates, benzodiazepines, chloral hydrate | Sedative, hypnotic or anxiolytic drug use disorder |
| Inhalants | Most abused inhalants induce pleasure by depressing the CNS, while nitrites, unlike other inhalants, work by dilating and relaxing blood vessels rather than providing anesthetic effects | Volatile solvents, aerosols, nitrites, gases | Inhalant use disorder |
| Amphetamines | It increases monoamine neurotransmitters like dopamine, norepinephrine, and serotonin at the synaptic levels | d-amphetamine | Amphetamine dependence |
| Miscellaneous | | Datura, caffeine, analgesics | |

*Notes* CNS—central nervous system, THC—tetra hydro cannabinol, GABA—gamma amino butyric acid, LSD—lysergic acid diethylamide, MDMA—3,4-methylenedioxymethamphetamine

differs significantly from that of applied statistics, despite overlapping concepts. Nonetheless, the maturity of the machine learning field means that many tools and techniques are now accessible to researchers, with free and open-source software implementations readily available, making it feasible to incorporate these powerful tools into addiction research (Morgado et al. 2022b).

Machine learning encompasses various computational and statistical tools designed to automate repetitive tasks typically performed by human experts. Far from being a magical solution, these tools extend traditional statistical approaches toward automation, bridging the gap between human-specified models and fully automated data analysis. Machine learning techniques learn from experience to enhance their performance in tasks like clustering, classification, and prediction by extracting actionable information from observed data to create models using algorithms (Morgado et al. 2022a; b; Bhadra et al. 2024).

In addiction research, machine learning has diverse applications, including predicting addction, distinguishing binge events from non-binge events, identifying

**Table 2** Various types of behavioral addictions and their characteristics

| Behavioral addiction | Key characteristics |
|---|---|
| Internet addiction | Internet addiction is characterized by a recurring, obsessive need to use the internet, resulting in an overwhelming urge to continue using it. This addiction generates both psychological and physical dependence, leading to increased time spent online despite the tensions and tolerance that develop (Yang 2008) |
| Mobile phone addiction | MPA is characterized by tolerance, withdrawal, continued use despite adverse consequences, neglecting important social activities, spending excessive time on the mobile phone, and unsuccessful attempts to reduce mobile phone use (Yen et al. 2009) |
| Pathological gambling | Pathological gambling is a disorder characterized by frequent, repeated episodes of gambling that take over a person's life, harming their social, occupational, material, and family responsibilities (Fauth-Bühler et al. 2017) |
| Food addiction | Chracterized by a chronic, relapsing condition resulting from the relationship of various complex factors that intensify cravings for specific foods to attain heightened pleasure, energy, or excitement, or to alleviate negative emotional or physical states (Imperatori et al. 2016) |

addiction-related genes, and analyzing tweets related to opioid misuse. Data for these tasks can come from sources such as electronic health records, surveys, genomic data, or social media, with performance measures tailored to each task (Morgado et al. 2022a; b; Bhadra et al. 2023). Machine learning is frequently used for making predictions, such as distinguishing individuals with specific behaviors (e.g., binge drinking) or predicting events (e.g., opioid overdoses). Supervised machine learning methods address these problems by identifying predictor variables and a target response. These methods can be divided into regression, where the response is a numerical value (e.g., a score or age of onset for SUD), and classification, where the response is categorical (e.g., occurrence of an opioid overdose).

Machine learning methods are generally classified into supervised and unsupervised categories. Supervised methods require a target outcome to guide the learning process, using independent variables to predict or explain this outcome. These methods can answer questions such as identifying at-risk populations, determining key variables associated with SUD, and distinguishing between patients with and without SUD (Kumar and Das 2021). While supervised methods rely on an underlying relationship between the outcome and the independent variables, unsupervised methods are primarily used to explore and uncover patterns within the data. These methods operate without a response variable, meaning there isn't a 'right answer' to guide the analysis. Unsupervised machine learning can group observations based on similarities into clusters and merge partially redundant features into new, more informative ones (Morgado et al. 2022a; b; Kumar and Das 2021). The complexity of the prediction model depends on the research question, available data, and the relationship between predictors and the outcome. Generalized linear models are often effective for small datasets with fewer predictors and are recommended as starting models. However, more complex tasks, such as those involving neuroimaging data with numerous features and dependencies, may require advanced techniques like artificial neural networks. Simpler models, such as linear models, operate under stronger assumptions, like assuming linear relationships, which may be too restrictive for capturing complex data patterns. More advanced methods relax these assumptions, creating more expressive models that can handle complex relationships. However, this increased expressivity comes at the cost of potential overfitting, where the model captures noise or anomalies specific to the training data, reducing its ability to generalize to new data (Sharma and Kumar 2022).

Balancing bias and variance is critical in model development to avoid overfitting (where the model fits training data too closely and fails to generalize) and underfitting (where the model is too simplistic and misses important data insights). Successful machine learning models achieve a balance, ensuring valuable and generalizable insights from the data (Bhadra and Kumar 2023). Overfitting is a significant issue, as models that fit too closely to training data fail to predict unseen data accurately. This can be compared to a student who memorizes answers without understanding underlying concepts, thus failing to apply knowledge to new questions. Additionally, simpler models are generally more interpretable than complex ones. For instance, the relationship between predictors and the response in neural networks can be opaque, making interpretability an essential consideration when selecting a method (Sharma et al. 2024).

In Tables 3 and 4 various supervised and unsupervised machine learning models, their application and use in addiction research is depicted.

## 4 Limitations and Future Scope

Recent advancements in ML/DL models have significantly enhanced one's understanding of the multi-dimensional patterns in medicine and behavior. These models are widely adopted due to their scalability and suitability for engineering and computational tasks. Consequently, it seems evident that data science has the potentiality to address drug-related disorders which may lead to more severe form of psychiatric illnesses. However, a number of limitations still exists in theseML/DL based approaches which may provide ample opportunity to the future researchers in this domain.

### *4.1 Lack of Reliable Datasets*

Large volumes of high-quality data are indispensable for effective functioning of ML/DL models. In medical and behavioral research, obtaining such data can be challenging due to privacy concerns, variability in data collection methods, and the rarity of certain conditions.

**Table 3** Supervised machine learning models, their application and use in addiction research

| Name of the models | Applications | Use in addiction research |
|---|---|---|
| Linear models for continuous outcomes | Multiple linear regressions are widely used due to its ease of interpretation and low computational demands. Penalized regression methods like lasso, ridge, and elastic net enhance linear regression by adding regularization to manage noise and improve generalization. For more complex relationships, polynomial regression, regression splines, and generalized additive models (GAMs) offer flexibility, with GAMs allowing for nonlinear associations between predictors and the target | Locke et al. (2015), Morozova et al. (2015), Linden-Carmichael et al. (2019), Chilcoat and Schütz et al. (1996) |
| Linear models for binary outcomes | Logistic regression is commonly used for predicting binary and categorical outcomes and often performs well against more complex models. Support vector machines (SVMs) are another efficient linear classifier | Kornfield et al. (2018)<br>Mete et al. (2016) |
| Bayesian approaches | The Naive Bayes algorithm, a straightforward and widely used classification method, assumes independence between predictive variables and selects the class with the highest conditional probability | Mumtaz et al. (2018), Abada and Bouramoul (2022) |
| K-nearest neighbors | The k-nearest neighbors (k-NN) algorithm, a non-parametric method for classification and regression, assigns values to new instances based on the closest training examples in the feature space. For classification, it uses voting, and for regression, averaging (possibly weighted by distance) | |
| Ensemble methods | Random forests<br>Boosting | These are the machine learning algorithms which combine multiple models<br>Squeglia et al. (2017), Lo-Ciganic et al. (2019), Segal et al. (2020), Wadekar (2020), Nasir et al. (2021) |
| Artificial neural networks | It analyzes the information from predictor variables by passing it through multiple sequential layers | Sakoglu et al. (2019), Che et al. (2018) |

**Table 4** Unsupervised machine learning methods their application and use in addiction research

| Name of the models | | Applications | Use in addiction research |
|---|---|---|---|
| Clustering techniques | K-means clustering | It is used to organize elements in a dataset in a way that more similar objects reside within the same group | Wilkinson et al. (2023) |
| Representation learning | | Generates changes in the feature space by transforming features into more meaningful entities for dimensionality reduction or embeddings | Wilkinson et al. (2023) |

### *4.2 Unavailability of Pre-trained DL Models*

Processing medical images is complex and time-consuming. However, existing models can be customized and integrated with diagnostic systems. Distributing and pre-processing medical images could result in reduced costs, decreased time requirements, and more efficient use of resources.

### *4.3 Small Amount of Research on Early-Onset Prediction*

Available data can predict the next stage of substance misuse, highlighting the need for on-site, localized, gender-specific, and geographic studies. Validating multidisciplinary knowledge at various stages of the decision process is critical.

### *4.4 Lack of More Longitudinal Studies*

Integrating image data with other types of data can add more efficiency to the analysis. In this aspect, a longitudinal design, despite the difficulty of labeling individuals, may provide important information.

### *4.5 Errors and Validation*

Drug use patterns vary from individual to individual, presenting challenges for researchers. Small datasets and case-based studies will always have variation and jitter, no matter how rigorously the validation process is executed. When scaled up, the ML/DL models often perform poorly, leading to misdiagnoses.

### *4.6 Specific Group-Based Studies*

A number of psychiatric disorders may be associated with substance abuse, so focusing on a specific substance is often the most effective approach. Substance abuse affects psyches across all social strata. By examining specific demographic groups and data, one can better understand the effect of addiction at societal level. The continuous improvement of ML applications can address research gaps.

### *4.7 Ethics and Interpretability*

In the quest for more advanced calculations and outcomes, the significance of ethics can sometimes be neglected. Machine learning predictions can be errored when used in studies of human substance abuse, potentially misclassifying individuals as drug users. As a result, clinical practices and tools may become disconnected from research areas, making it challenging to apply ML findings effectively.

In the future, machine learning technology could be used to address these challenges and bring about changes in addiction treatment. Although there are currently few studies on addiction studies, researchers need to develop strategies for longitudinal studies with multiple multivariate endpoints leading to addiction disorders. Clinical investigations remain the primary method for diagnosing substance use disorders. To develop practical solutions, researchers can leverage ML tools and other AI technologies. Using state-of-the-art methods for identifying, validating, and extracting features, along with empirical data from various sources, has shown differences in ML model performance. Sharing insights from other research areas such as COVID-19, retinopathy, and neurological and psychiatric conditions may also be beneficial. By integrating various ML/DL based advanced approaches, early screening for substance abuse can become more accurate and efficient, leading to better prevention and treatment strategies.

## 5 Conclusion

Drug-related disorders require a disease-based diagnostic approach. Brain activity is significantly affected by psychoactive substances, and behavioral changes associated with substance use are among the most striking findings. Without progress in research on substance abuse, its harms, treatment, and economic impact will remain unaddressed. Individuals seeking treatment for substance use disorders and those under suspicion are routinely monitored. Recently, ML algorithms have gained traction in addiction research, aiming to classify, identify, and predict addictions. Alongside technology-based interventions, planning, monitoring, and treatment management can reduce the burden of addiction. Disorders are associated with geographical variations, which can include or exclude substance use, but symptoms are often linked. Understanding the geographic, social, economic, and other variables that lead to addictive behaviors is crucial. Integrating multiple spheres and comprehensive clinical data into a model can further improve the capacity to predict vulnerabilities and risky behaviors. With the growing popularity of ML/DL applications, tremendous new opportunities have been unlocked for researchers and clinicians. These advancements are expected to further expand state-of-the-art research, from diagnosing SUD to providing effective treatment, support, and management.

## References

Abada, W., & Bouramoul, A. (2022). Using machine learning techniques to predict people at risk for drug addiction: A Bayesian-based model. In Proceedings of the 2022 4th International Conference on Pattern Analysis and Intelligent Systems (PAIS) (pp. 1–7). Oum El Bouaghi, Algeria.

Bhadra, S., Kumar, C.J. & Bhattacharyya, D.K. (2024). Multiview EEG signal analysis for diagnosis of schizophrenia: an optimized deep learning approach. *Multimed Tools Appl* https://doi.org/10.1007/s11042-024-20205-y.

Che, Z., Purushotham, S., Cho, K., et al. (2018). Recurrent neural networks for multivariate time series with missing values. Scientific Reports, 8, 6085. https://doi.org/10.1038/s41598-018-24271-9.

Bhadra, S., & Kumar, C. J. (2023). Enhancing the efficacy of depression detection systems using optimal feature selection from EHR. Computer Methods in Biomechanics and Biomedical Engineering, 27(2), 222–236. https://doi.org/10.1080/10255842.2023.2181660

Chhetri, B., Goyal, L. M., & Mittal, M. (2023). How machine learning is used to study addiction in digital healthcare: A systematic review. International Journal of Information Management Data Insights, 3(2), 100175. https://doi.org/10.1016/j.jjimei.2023.100175.

Chilcoat, H. D., & Schütz, C. G. (1996). Age-specific patterns of hallucinogen use in the US population: An analysis using generalized additive models. Drug and Alcohol Dependence, 43(3), 143–153.

Cresta Morgado, P., Carusso, M., Alonso Alemany, L., & Acion, L. (2022a). Practical foundations of machine learning for addiction research. Part I. Methods and techniques. The American Journal of Drug and Alcohol Abuse, 48(3), 260–271. https://doi.org/10.1080/00952990.2021.1995739.

Cresta Morgado, P., Carusso, M., Alonso Alemany, L., & Acion, L. (2022). Practical foundations of machine learning for addiction research. Part II. Workflow and use cases. The American Journal of Drug and Alcohol Abuse, 48(3), 272–283. https://doi.org/10.1080/00952990.2021.1966435.

Das, P. R., Talukdar, R. R., & Kumar, C. J. (2024). Exploring the interplay of abstinence self-efficacy, locus of control, and perceived social support in substance use disorder recovery. Current Medical Research and Opinion, 40(9), 1625–1635. https://doi.org/10.1080/03007995.2024.2390046.

Fauth-Bühler, M., Mann, K., & Potenza, M. N. (2017). Pathological gambling: A review of the neurobiological evidence relevant for its classification as an addictive disorder. Addiction Biology, 22, 885–897.

Grant, J. E., Potenza, M. N., Weinstein, A., & Gorelick, D. A. (2010). Introduction to behavioral addictions. The American Journal of Drug and Alcohol Abuse, 36(5), 233–241. https://doi.org/10.3109/00952990.2010.491884.

Imperatori, C., Fabbricatore, M., Vumbaca, V., Innamorati, M., Contardi, A., & Farina, B. (2016). Food addiction: Definition, measurement, and prevalence in healthy subjects and in patients with eating disorders. Rivista di Psichiatria, 51(2), 60–66.

Kornfield, R., Sarma, P., Shah, D., McTavish, F., Landucci, G., Pe-Romashko, K., & Gustafson, D. (2018). Detecting recovery problems just in time: Application of automated linguistic analysis and supervised machinelearning to an online substance abuse forum. Journal of Medical Internet Research, 20, e10136. https://doi.org/10.2196/10136 .

Kumar, C. J., & Das, P. R. (2021). The diagnosis of ASD using multiple machine learning techniques. International Journal of Developmental Disabilities, 68(6), 973–983.https://doi.org/10.1080/20473869.2021.1933730.

Kumar, C. J., Das, P. R. & Hazarika, A. (2021). Autism spectrum disorder diagnosis and machine learning: a review. International Journal of Medical Engineering and Informatics, 14(6), 512-527.

Linden-Carmichael, A. N., Stamates, A. L., & Lau-Barraco, C. (2019). Simultaneous use of alcohol and marijuana: Patterns and individual differences. Substance Use & Misuse, 54(13), 2156–2166.https://doi.org/10.1080/10826084.2019.1638407.

Locke, G. W., Shilkret, R., Everett, J. E., & Petry, N. M. (2015). Interpersonal guilt and substance use in college students. Substance Abuse, 36(1), 113–118.

Lo-Ciganic, W. H., Huang, J. L., Zhang, H. H., Weiss, J. C., Wu, Y., Kwoh, C. K., et al. (2019). Evaluation of machine-learning algorithms for predicting opioid overdose risk among Medicare beneficiaries with opioid prescriptions. JAMA Network Open, 2(3), e190968. https://doi.org/10.1001/jamanetworkopen.2019.0968.

Mauri, M. C., Di Pace, C., Reggiori, A., Paletta, S., & Colasanti, A. (2017). Primary psychosis with comorbid drug abuse and drug-induced psychosis: Diagnostic and clinical evolution at follow-up. Asian Journal of Psychiatry, 29, 117–122.

Meng, S.-Q., Cheng, J.-L., Li, Y.-Y., Yang, X.-Q., Zheng, J.-W., Chang, X.-W., et al. (2022). Global prevalence of digital addiction in general population: A systematic review and meta-analysis. Clinical Psychology Review, 92, 102128.

Mete, M., Sakoglu, U., Spence, J. S., Devous, M. D., Harris, T. S., & Adinoff, B. (2016). Successful classification of cocaine dependence using brain imaging: A generalizable machine learning approach. BMC Bioinformatics, 17(S13), 357.

Morozova, O., Levina, O., Uusküla, A., & Heimer, R. (2015). Comparison of subset selection methods in linear regression in the context of health-related quality of life and substance abuse in Russia. BMC Medical Research Methodology, 15(1), 71.

Mumtaz, W., Saad, M. N. b M., Kamel, N., Ali, S. S. A., & Malik, A. S. (2018). An EEG-based functional connectivity measure for automatic detection of alcohol use disorder. Artificial Intelligence in Medicine, 84, 79–89.

Nasir, J. A., Khan, O. S., & Varlamis, I. (2021). Fake news detection: A hybrid CNN-RNN based deep learning approach. International Journal of Information Management Data Insights, 1(1), Article 100007.

Rehman, A., Zhang, X., & Yang, Y. (2022). Leveraging machine learning for addiction studies: Diagnostic solutions and personalized treatment. Journal of Medical Informatics, 45(5), 601–615.

Sakoglu, U., Mete, M., Esquivel, J., Rubia, K., Briggs, R., & Adinoff, B. (2019). Classification of cocaine-dependent participants with dynamic functional connectivity from functional magnetic resonance imaging data. Journal of Neuroscience Research, 97(7), 790–803.

Segal, Z., Radinsky, K., Elad, G., Marom, G., Beladev, M., Lewis, M., et al. (2020). Development of a machine learning algorithm for early detection of opioid use disorder. Pharmacology Research & Perspectives, 8(6), e00669.

Sharma, M., & Kumar, C. J. (2022). Improving rice disease diagnosis using ensemble transfer learning techniques. International Journal on Artificial Intelligence Tools, 31(08), 2250040.

Sharma, M., Kumar, C. J., & Bhattacharyya, D. K. (2024). Machine/deep learning techniques for disease and nutrient deficiency disorder diagnosis in rice crops: A systematic review. Biosystems Engineering, 244, 77–92.

Shi-Qiu, M., Cheng, J.-L., Li, Y.-Y., Yang, X.-Q., Zheng, J.-W., Chang, X.-W., et al. (2022). Global prevalence of digital addiction in general population: A systematic review and meta-analysis. Clinical Psychology Review, 92, 102128.

Squeglia, L. M., Ball, T. M., Jacobus, J., Brumback, T., McKenna, B. S., Nguyen-Louie, T. T., et al. (2017). Neural predictors of initiating alcohol use during adolescence. American Journal of Psychiatry, 174(2), 172–185.

United Nations Office on Drugs and Crime. 2023. World drug report.

Wadekar, A. S. (2020). Understanding Opioid Use Disorder (OUD) using tree-based classifiers. Drug and Alcohol Dependence, 208, Article 107839.

Wilkinson, C. S., Luján, M. Á., Hales, C., Costa, K. M., Fiore, V. G., Knackstedt, L. A., & Kober, H. (2023). Listening to the data: Computational approaches to addiction and learning. Journal of Neuroscience, 43(45), 7547–7553. https://doi.org/10.1523/JNEUROSCI.1415-23.2023.

Yahyaoui, A., Rasheed, J., Jamil, A., & Yesiltepe, M. (2019). A decision support system for diabetes prediction using machine learning and deep learning techniques. UBMYK 2019, Article N/A.https://doi.org/10.1109/UBMYK48245.2019.8965556.

Yang, H. M. (2008). A research on the influence of internet addiction on college students. (Doctoral dissertation, Nanjing Normal University, Nanjing).

Yen, C. F., Tang, T. C., Yen, J. Y., Lin, H. C., Huang, C. F., Liu, S. C., & Ko, C. H. (2009). Symptoms of problematic cellular phone use, functional impairment and its association with depression among adolescents in Southern Taiwan. Journal of Adolescence, 32(4), 863–873.

Zou, Z., Wang, H., d'Oleire Uquillas, F., Wang, X., Ding, J., & Chen, H. (2017). Definition of substance and non-substance addiction. Advances in Experimental Medicine and Biology, 1010, 21–41. https://doi.org/10.1007/978-981-10-5562-1_2.

# MindEase: Systematic Review for Unravelling Mental Health with Machine Learning Approach in Adults

**Gypsy Nandi and Lal Omega Boro**

**Abstract** Machine learning (ML) has displayed satisfactory results in improving the treatment and diagnosis of mental health in adults. A systematic review has been carried out in this study to find out the effectiveness of ML in treating and diagnosing mental health conditions across different populations. Seven studies were identified for inclusion based on two-phase screening process. It has been observed that there are differences in ML effectiveness across various mental health conditions. The methods adopted in the selected studies highlight the complexity of ML applications in mental health conditions. An important discovery was made regarding data types and ML algorithms employed. Overall, this study highlights the capability of ML in treating mental health, while also stressing the need for more research and testing.

**Keywords** Machine learning · Mental health · Adults · Psychiatric disorders · Review

## 1 Introduction

Mental health conditions affect millions of lives globally (Jackson et al. 2007; Pagotto et al. 2015). Individuals, families, and societies face challenges in dealing with anxiety, depression, post-traumatic stress disorder (PTSD), and attention-deficit/hyperactivity disorder (ADHD). They pose significant challenges to providing accurate diagnosis, and treatment plans because of their complexity and differences in nature. Therefore, traditional approaches may fall short of expectations and standards in providing personalized care for the patient.

The popularity of ML applications in recent years (Mahesh 2020; Kotsiantis et al. 2006) has given us new ways to think and overcome the limitations of traditional approaches in dealing with mental health disorders. ML comprises a different set

G. Nandi (✉) · L. O. Boro
Assam Don Bosco University, Azara, India
e-mail: gypsy.nandi@gmail.com



T. P. Singh et al. (eds.), *Revolutionizing Healthcare: Impact of Artificial Intelligence on Diagnosis, Treatment, and Patient Care*, Studies in Computational Intelligence 1182,
https://doi.org/10.1007/978-3-031-80813-5_14

of flexible and adaptable algorithms that enable computers to learn and make decisions without human intervention. This makes them suitable for analyzing large and complex data in mental health disorder research.

By leveraging ML technologies, researchers and doctors can obtain important information about the root causes of mental health disorders. Furthermore, ML technologies can help identify biomarkers and develop diagnostic tools accordingly. Additionally, ML can also predict treatment outcomes, and customize the treatment plan according to the patient's needs (Miotto et al. 2018).

Despite gaining popularity of ML applications in mental health research, there remain several unanswered questions such as the generalizability of ML in different situations, the interpretability of the results, and the ethical issues of ML in real-life settings. Additionally, the differences in approaches in the existing studies call for a thorough and systematic evaluation of ML performance and applicability across different contexts.

The purpose of this systematic review is to investigate the effectiveness of ML in the diagnosis, and treatment plan for adult mental health disorders. A comprehensive review of several studies has been carried out to understand the impact of ML. In addition, we also look at significant patterns in the kinds of data used, ML algorithms, and assessment measures used in the literature to understand how ML is developing in the field of mental health research.

## 2 Research Method

This paper aims to systematically assess the current scenario of ML in diagnosing mental health conditions. This assessment encompasses examining the data origin, methods for extracting features, and the effectiveness of various ML. The paper is structured and designed based on the PRISMA flow diagram and protocol (Moher et al. 2009).

### *2.1 Eligibility Criteria*

Studies were selected based on the condition that they discussed the use of ML in addressing mental health issues. The selected studies were also required to be published in peer-reviewed journals and written in English. Studies that lacked original contributions to ML applications in mental health were excluded, and also those studies that are not focussing on mental health. Review articles were also excluded, or articles that were not published in English. Furthermore, duplicate articles, animal studies, and editorials were excluded to maintain the quality of the paper. Two independent reviewers assessed all studies and reached a consensus on the ones to include.

## 2.2 Information Sources and Search Strategy

Research question was formulated i.e., "In adults with mental health conditions, what is the effectiveness of ML approaches in diagnosis, prediction, and treatment outcomes?" using the PICO (Patient/Population/Problem, Intervention, Comparison, Outcome) scheme (Santos et al. 2007). Furthermore, search terms were constructed which combined the key concepts from the formulated research question to retrieve relevant articles. Search queries used: ("adults" OR "mental health" OR "mental illness" OR "psychiatric disorders" OR "mental health conditions") AND ("machine learning" OR "artificial intelligence" OR "predictive modeling") AND ("diagnosis" OR "prediction" OR "treatment" OR "outcome" OR "accuracy" OR "efficacy" OR "performance") in PubMed database on March 23rd, 2024. The search was limited between the years 2021–2023 due to the rapid growth of research.

## 2.3 Selection Process

Two reviewers independently evaluated the literature through a two-stage process according to predetermined criteria. Initially, titles and abstracts of the search results were examined and subsequently conducted a full-text assessment. Any disagreements were addressed through discussion.

## 2.4 Data Collection and Risk of Bias Assessment

Information was collected from the selected studies for qualitative assessment using pre-defined data collection template. This template included details like the research objective, mental health focus area, type of data, ML methods employed, and the outcomes. Two reviewers independently evaluated the quality of the selected literature using the QUADAS-2 tool (Whiting et al. 2011). Any differences in assessment were addressed through discussion.

# 3 Results

Trendy methods have been used to handle mental health disorders. Yet this study has revealed notable disparities in efficacy. The findings hint at the broad range of methodologies. The study also brought attention to their effect on the prognosis of treatment. Intriguing patterns also emerged concerning the categories of data that have been employed.

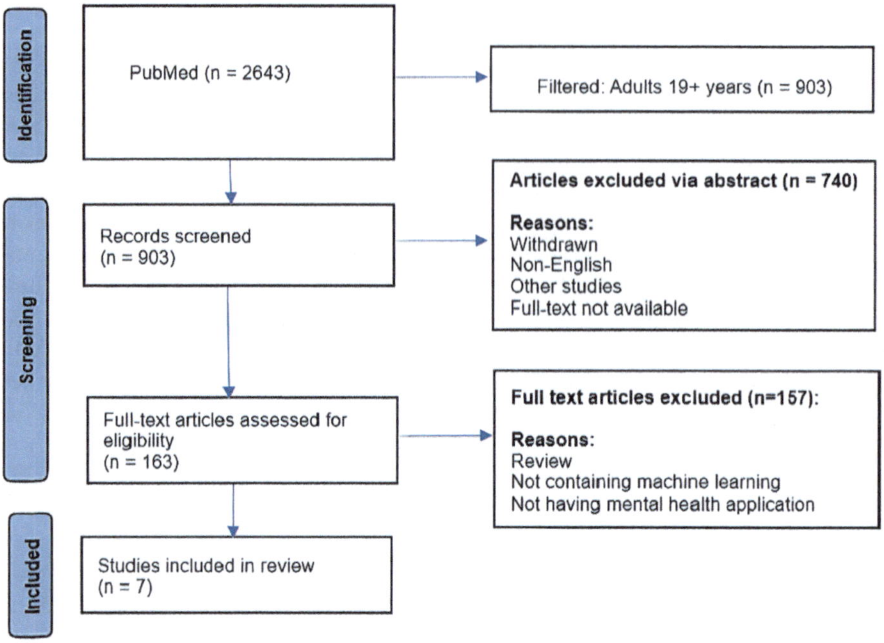


**Fig. 1** Flowchart showing 2-phase screening process

## 3.1 Study Selection

The results of 2-phase screening procedure are illustrated in Fig. 1. From the initial database search in PubMed, 2643 results were obtained, out of which 1740 were filtered out due to the criterion of including only adults aged 19 years and above. Subsequently, a total of 903 titles and abstracts underwent screening for eligibility. During the first screening phase, 740 articles were excluded based on the selection criteria detailed in Fig. 1. In the second screening phase, 157 articles were excluded. Ultimately, 7 studies met the criteria for inclusion in this systematic review.

## 3.2 Study Characteristic

The characteristics and results of the included studies are detailed in Table 1. Reviewed studies spanned a varied range of mental health issues including scope of populations and methodologies. This contributed significantly to a deeper understanding of utility of ML approaches.

Populations that were studied varied greatly. These included subclinical individuals with symptoms of anxiety and depression (Richter et al. 2021; Bhadra and Kumar, 2024). In addition, soldiers' pre-deployment to combat zones were also

**Table 1** Characteristic of the studies included in this systematic review

| Authors | Target | Dataset | Data size | Method | Performance |
|---|---|---|---|---|---|
| Richter et al. (2021) | Anxiety and depression | Clinical data questionnaires | 101 participants | Random forest | Anxiety/ depression/ mixed groups: sensitivity: 69.66%, specificity: 76.81%; and 80.50% success in anxiety classification; 66.46% in depression |
| Schultebrauck et al. (2021) | PTSD | Clinical data | 473 participants | Random forest and SVM | Random forest (AUC: 78%), SVM (AUC: 88%) |
| Gokten et al. (2021) | PTSD and depression | Clinical data | 482 children and adolescents | Random forest | Depression (AUC: 88.0%), PTSD (AUC: 76.0%) |
| Tachmazidis et al. (2020) | ADHD | Clinical data questionnaires | 69 patients | Knowledge-driven machine learning hybrid model | Accuracy: 95% |
| Mikolas et al. (2022) | ADHD | Clinical data | 299 participants | SVM | 30 features: (accuracy: 66.1%)<br>No demographic: (accuracy: 65.1%)<br>No missing data: (accuracy: 68.8%) |

(continued)

**Table 1** (continued)

| Authors | Target | Dataset | Data size | Method | Performance |
|---|---|---|---|---|---|
| Yin et al. (2022) | ADHD | Resting-state fMRI | 360 ADHD and TDC subjects | Extreme gradient boosting | ADHD vs. TDC: accuracy: 77% (CV), 74.46% (IT); ADHD severity prediction, $R^2$: 0.2794 (CV), 0.156 (IT) |
| Bhadra et al. (2023) | Depression | EHR data from the university of nice sophia-antipolis | 4,184 individuals | ANN, SVM, random forest, and XGBoost with feature selection algorithms like RFE, MI, PSO, GA, and FA | Accuracy: 84.18–88.46% |

studied Schultebraucks et al. (2021) along with children and adolescents who suffered sexual abuse (Gokten and Uyulan, 2021). Patients with ADHD were also studied Mikolas et al. (2022), Yin et al. (2022). Furthermore, adults under clinical diagnosis of ADHD were also probed (Tachmazidis et al., 2020). This diverse mixture of populations represents the broad potential of ML techniques. These techniques cut across various age groups and also cover various mental health conditions.

Each study utilized different methodologies to suit their research questions and datasets. The study done in (Richter et al. 2021), combined cognitive behavioral performance data and integrated it with ML algorithms to classify anxiety from depression. The study done in (Bhadra and Kumar, 2024), used optimal feature algorithm to diagnose depression at an early age. The study done in (Schultebrauck et al. 2021), used predictive modeling to predict PTSD symptoms in soldiers. The study done in (Gokten and Uyulan 2021), used forensic evaluations and psychiatric assessments to predict psychiatric disorders. The disorders were in victims of sexual abuse. The study done by (Mikolas et al. 2022) employed anonymized clinical records to target patients with ADHD. In addition (Tachmazidis et al. 2020) fused ML and knowledge-based models to diagnose ADHD in adults. The study done by (Yin et al. 2022) used a dataset that was collected from different locations to look into neural flexibility which was a biomarker for ADHD in children.

### 3.3 *Quality Assessment*

QUADAS-2 was used to assess any risk of bias contained in the selected studies as shown in Fig. 2. In the patient selection domain, the risk of bias was reportedly high in one study (Mikolas et al. 2022). This is because it's unclear how the participants were chosen for the study. Four studies (Schultebrauck et al., Gokten et al., Tachmazidis et al., Yin et al.) are found to have an unclear risk of bias since the precise patient selection procedures were not made explicit. In the index test domain, one study (Mikolas et al. 2022) is found to have a high risk of bias due to a lack of details on the ML model and feature selection process. Two studies (Gokten et al., Yin et al.) have unclear risk of bias. Gokten et al. did not provide specific information on random forest (RF) classifier model development, training, and validation. Yin et al. did not mention the details of the index test used to measure neural flexibility in children with ADHD. In the reference standard domain, one study (Mikolas et al. 2022) is classed as having a high risk of bias due to a lack of clarity on diagnostic criteria and application. Three studies (Schultebrauck et al., Gokten et al., Yin et al.) have unclear risk of bias. Schultebrauck et al., mentioned the use of provisional PTSD diagnosis based on self-report measures, but it is unclear if structured clinical interviews were used to verify PTSD diagnosis. Gokten et al., mentioned the use of DSM-V criteria as a reference standard, however, the specifics of how this standard was applied and its accuracy are not elaborated. Yin et al., provide no clear information regarding the reference standard used to diagnose ADHD severity. One study (Mikolas et al. 2022) is classed as high risk in the flow and timing domain due to insufficient information on the timing of assessments, and data collection, and four studies (Schultebrauck et al., Gokten et al., Tachmazidis et al., Yin et al.) have an unclear risk of bias since they did not provide details regarding the timing of assessments, and potential delays in the diagnostic process.

Three studies (Schultebrauck et al., Mikolas et al., and Bhadra et al.) were found to have a high risk in the patient selection domain in terms of applicability concerns. Schultebrauck et al., recruited a specific cohort of active-duty army personnel deployed to Afghanistan, which may limit the generalizability of the findings to a broader population. Mikolas et al., noted that the use of anonymized data from a clinical register for patient selection could introduce a bias. This is selection bias and this could limit the extensive applicability of findings. Bhadra et al., focuses on a specific demographic which could limits the generalizability of the findings.

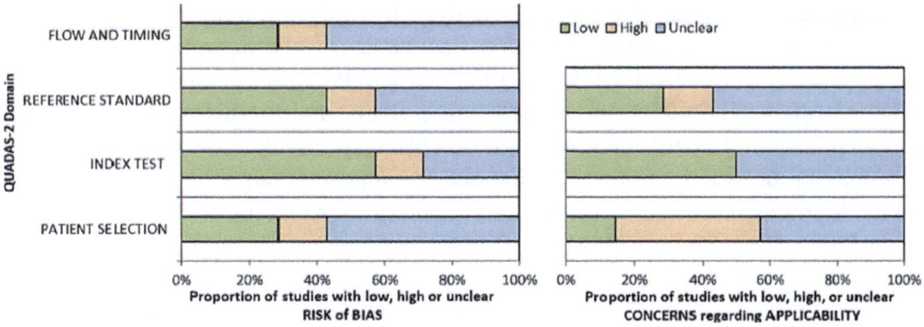


**Fig. 2** QUADAS-2

**Table 2** Summary of Risk of bias and applicability concerns for each domain of the included studies

| STUDY | RISK OF BIAS | | | | APPLICABILITY CONCERNS | | |
|---|---|---|---|---|---|---|---|
| | PATIENT SELECTION | INDEX TEST | REFERENCE STANDARD | FLOW AND TIMING | PATIENT SELECTION | INDEX TEST | REFERENCE STANDARD |
| (Richter et al. 2021). | ☺ | ☺ | ☺ | ☺ | ☺ | ☺ | ☺ |
| (Schultebrauck et al. 2021) | ? | ☺ | ? | ? | ☹ | ☺ | ☹ |
| (Gokten et al. 2021) | ? | ? | ? | ? | ? | ? | ? |
| (Tachmazidis et al. 2021) | ? | ☺ | ☺ | ? | ? | ☺ | ☺ |
| (Mikolas et al. 2022) | ☹ | ☹ | ☹ | ☹ | ☹ | ? | ? |
| (Yin et al. 2022) | ? | ? | ? | ? | ? | ? | ? |
| (Bhadra et al. 2023) | ☺ | ☺ | ☺ | ☺ | ☹ | ? | ? |

☺Low Risk ☹High Risk ? Unclear Risk

There is no risk of bias for the applicability concern of the index test domain. One study (Schultebrauck et al. 2021) is rated high in the applicability concern of the reference standard due to the lack of verification standard for PTSD diagnosis using standardized clinical assessments. Table 2 displays the separate assessment of the risk of bias and concerns regarding applicability for each individual.

## 3.4 Analysis and Discussion

Richter et al. aimed to introduce a novel diagnostic approach for distinguishing between anxiety and depression using cognitive behavioral performance data integrated with ML. They administered questionnaires to subclinical individuals categorized into four primary symptom groups: high anxiety, high depression, high anxiety, and depression, and controls with low symptoms of both. By employing a battery of six cognitive behavioral tasks, they assessed various biases in cognitive behaviors. Subsequently, an RF algorithm was employed to analyze the data, with the model accurately categorizing participants based on their aggregated cognitive performance. The specificity and sensitivity for distinguishing anxiety/depression/mixed groups from controls were 76.81% and 69.66% respectively, and for distinguishing anxiety from depression, they were 80.50% and 66.46% respectively.

Schultebrauck et al. applied ML methods to analyze a large, complex dataset of pre-deployment information for soldiers heading to Afghanistan in order to forecast PTSD. The dataset consisted of 473 army personnel from Kentucky. They utilized RF for predictive modeling, using clinical assessments, with SVM as a benchmark for comparison. The research indicated significant accuracy in predicting post-deployment symptom patterns and preliminary PTSD diagnosis based on pre-deployment data, with RF achieving a model performance of 78% AUC and SVM achieving 88%.

Gokten and Uyulan investigated children and adolescents who experienced sexual abuse. Through forensic evaluation and psychiatric assessment based on DSM-V diagnostic criteria, they developed a predictive model using the RF classifier. The model exhibited strong performance with AUCs of 88.0% for depression and 76.0% for PTSD.

Mikolas et al. presented a technique that uses de-identified clinical data analysis to identify people with ADHD among a variety of mental diseases. An SVM classifier applied to 30 distinct features yielded an accuracy rate of 66.1%. They also carried out additional classifications with accuracies of 65.1% and 68.8%, respectively, by excluding demographic characteristics and incomplete data.

Tachmazidis et al. employed a hybrid ML and knowledge-based model to diagnose ADHD in adults based on clinical data and questionnaires. They have tested their method in clinical settings and shown an amazing 95% accuracy rate.

Yin et al. looked at neural flexibility as a possible biomarker for differentiating between children with ADHD and typically developing children (TDC) using a multi-site ADHD dataset. They developed a regression model and a classification model to predict the severity of ADHD using the extreme gradient boosting (XGBoost) technique. The results suggested something significant. They hinted at neural flexibility changes in kids with ADHD. These changes could be a key to clinical utility that might help in identification and monitoring. The changes could also help in understanding treatment reactions. The model performance was quite accurate achieving 77% accuracy for differentiating ADHD from TDC. The model also had a 0.2794 R2 for predicting ADHD severity.

Bhadra et al. employed several classifiers along with feature selection methods to diagnose depression at an early age. Their findings reveal that choosing the right feature selection leads to model improvement. Their findings also become more relevance for resource-constrained clinical settings, and it is expected to lead to better depression diagnosis and mental health results.

## 4 Conclusion

To summarize the overall findings, it has been observed that ML is effective in diagnosing mental health issues. Additionally, it is adept in predicting outcomes of treatment in adults. It is noticed that several studies report successful detection of ADHD in adults and children. We also found similar success in detecting PTSD in soldiers prior to deployment. Furthermore, the detection of psychiatric illnesses in victims of sexual abuse has been successful and lastly, there is promising outcomes in discerning differences between anxiety and depression using ML.

Studies, such as Richter et al. (2021), Schultebrauck et al. (2021), Gokten et al. (2021), Tachmazidis et al. (2020), Mikolas et al. (2022), Yin et al. (2022), Bhadra et al.

(2023) evidence that ML algorithms can use clinical data to give accurate diagnoses. They can also forecast symptom trajectories and monitor treatment responses. These discoveries show that ML models can also integrate with traditional diagnostic and prognostic strategies to provide improving care to the patient.

On one hand, there are these studies that emphasize the promise of ML in mental healthcare. On the other hand, there is a need for more study and validation. The validation is required across all the different groups. It is important to recognize this need across a myriad of clinical settings. Additionally, there is concern regarding ethical issues on data privacy, transparency of the ML algorithm, and equitable access to the technologies. These need to be addressed to ensure that these technologies are responsibly put into practice. It's not just responsible, the implementation should also be equitable. In conclusion, all the collective evidence is rather compelling. It strongly suggests that ML has a lot of potential. It could help a lot in augmenting diagnosis. It could also help in prediction. It could even improve the outcome of treatment. This is the potential of ML in adults dealing with mental health issues.

## References

Bhadra, S., & Kumar, C. J. (2024). Enhancing the efficacy of depression detection system using optimal feature selection from EHR. Computer Methods in Biomechanics and Biomedical Engineering, 27(2), 222–236.

Gokten, E. S., & Uyulan, C. (2021). Prediction of the development of depression and post-traumatic stress disorder in sexually abused children using a random forest classifier. Journal of Affective Disorders, 279, 256–265.

Jackson, J. L., Passamonti, M., & Kroenke, K. (2007). Outcome and impact of mental disorders in primary care at 5 years. Psychosomatic medicine, 69(3), 270–276.

Kotsiantis, S. B., Zaharakis, I. D., & Pintelas, P. E. (2006). Machine learning: a review of classification and combining techniques. Artificial Intelligence Review, 26, 159–190.

Mahesh, B. (2020). Machine learning algorithms-a review. International Journal of Science and Research (IJSR). [Internet], 9(1), 381–386.

Mikolas, P., Vahid, A., Bernardoni, F., Süß, M., Martini, J., Beste, C., & Bluschke, A. (2022). Training a machine learning classifier to identify ADHD based on real-world clinical data from medical records. Scientific Reports, 12(1), 12934.

Miotto, R., Wang, F., Wang, S., Jiang, X., & Dudley, J. T. (2018). Deep learning for healthcare: review, opportunities and challenges. Briefings in bioinformatics, 19(6), 1236-1246.

Moher, D., Liberati, A., Tetzlaff, J., Altman, D. G., & PRISMA Group, T. (2009). Preferred reporting items for systematic reviews and meta-analyses: the PRISMA statement. Annals of internal medicine, 151(4), 264–269.

Pagotto, L. F., Mendlowicz, M. V., Coutinho, E. S. F., Figueira, I., Luz, M. P., Araujo, A. X., & Berger, W. (2015). The impact of posttraumatic symptoms and comorbid mental disorders on the health-related quality of life in treatment-seeking PTSD patients. Comprehensive psychiatry, 58, 68-73.

Richter, T., Fishbain, B., Fruchter, E., Richter-Levin, G., & Okon-Singer, H. (2021). Machine learning-based diagnosis support system for differentiating between clinical anxiety and depression disorders. Journal of Psychiatric Research, 141, 199–205.

Santos, C. M. D. C., Pimenta, C. A. D. M., & Nobre, M. R. C. (2007). The PICO strategy for the construction of the research question and search for evidence. Latin American Journal of Nursing, 15, 508–511.

Schultebraucks, K., Qian, M., Abu-Amara, D., Dean, K., Laska, E., Siegel, C., ... & Marmar, C. R. (2021). Pre-deployment risk factors for PTSD in active-duty personnel deployed to Afghanistan: a machine-learning approach for analyzing multivariate predictors. Molecular psychiatry, 26(9), 5011–5022.

Tachmazidis, I., Chen, T., Adamou, M., & Antoniou, G. (2020). A hybrid AI approach for supporting clinical diagnosis of attention deficit hyperactivity disorder (ADHD) in adults. Health Information Science and Systems, 9(1), 1.

Whiting, P. F., Rutjes, A. W., Westwood, M. E., Mallett, S., Deeks, J. J., Reitsma, J. B., ... & QUADAS-2 Group*. (2011). QUADAS-2: a revised tool for the quality assessment of diagnostic accuracy studies. Annals of internal medicine, 155(8), 529–536.

Yin, W., Li, T., Mucha, P. J., Cohen, J. R., Zhu, H., Zhu, Z., & Lin, W. (2022). Altered neural flexibility in children with attention-deficit/hyperactivity disorder. Molecular Psychiatry, 27(11), 4673–4679.

# Early Detection of Mental Health Condition and Advising System Using Artificial Intelligence

**Minakshi Gogoi**

**Abstract** In India among the diverse community of population it is very hard to deal with their issues if any for their mental well being. Addressing their mental issues requires initiatives or promoting an advising system, where clinicians are needed to be educated on socio-cultural backgrounds and how they may impact mental health, illness presentation, and help-seeking behaviors. There can be a working area to deal self stigma, the negative perceptions that individuals with mental illnesses hold of their own conditions. Youth is in one group that faces the stigma associated with mental health due to their unconscious bias and stereotypes. These groups frequently encounter stigma in daily life, especially while seeking healthcare. Early identification and intervention can make a significant difference in managing mental health conditions, improving prognosis, and enhancing the quality of life for individuals at risk. Healthcare professionals can enhance patient outcomes, diagnose mental health disorders earlier, and tailor therapies by utilizing Artificial Intelligence. When these interventions are used in conjunction with conventional healthcare methods, mental health care can be given more quickly, efficiently, and individually.

**Keywords** Mental health · Healthcare · Psychiatric disorder · Artificial intelligence

## 1 Introduction

Due to the regional boundaries, distance to travel, and associated costs, not everyone in need of healthcare services can access every healthcare institution in the current healthcare system. This issue can be addressed to a large extent by the eHealthcare system, which offers everyone equitable access to all health-care facilities over the internet, regardless of geographic borders such as cities or countries.

M. Gogoi (✉)
Girijananda Chowdhury University, Assam, Guwahati, India
e-mail: minakshi_cse@gcuniversity.ac.in

 
T. P. Singh et al. (eds.), *Revolutionizing Healthcare: Impact of Artificial Intelligence on Diagnosis, Treatment, and Patient Care*, Studies in Computational Intelligence 1182,
https://doi.org/10.1007/978-3-031-80813-5_15

The North Eastern region of India are facing problems in the Health Sector due to several issues as mentioned in the report by Ministry of Activities in North East Region on "Activities in North East Region" ('Ministry of Activities in North East India' 2017). Issues mentioned are like providing access to sparsely populated, remote, far flung areas and improvement of Governance in the Health Sector and need for improved quality of health services rendered etc. It is the time to provide a behavioral health services across the people of North Eastern Region by providing tools for mental health screenings, initial intervention (Murthy 2017).

There is more to the idea of stigma around mental health than meets the eye. Youth is in one group that faces the stigma associated with mental health due to their unconscious bias and stereotypes. Self-stigma originates internally. Self-stigma sufferers may act victimized and push for others to treat them differently because they believe they are the victims of discrimination.

The significance of early detection of an individual at risk for the difficulties linked to emotional and behavioral disorders, or EBD. As per the Sustainable Development Goal (SDG)3 ('UN Sustainable Development Goal (SDG)3', 2015), which focuses on important targets that improve a nation's population's overall health in order to minimize needless suffering from preventable diseases and untimely death. Priority locations include those with the highest illness burden and underserved population segments.

There are several important factors to take into account while designing and implementing a comprehensive mental health screening program, such as the integrated support system and screening for mental health. First of all, screening is not a task that a single person can complete on their own. It ought to be included in Target Group's integrated services program. Now, it is the right moment for policymakers and healthcare professionals to collaborate and put an action plan into place to enhance care for the common peoples who suffer from mental illness. In addition, there is still substantial research to be done to validate the technical adequacy of the instruments and procedures employed in mental health screening procedures.

Keeping this as background, we present in this study a justification and explanation of screening and early identification for early intervention, an outline of screening instrumentation, procedural and ethical considerations for screening, and a comprehensive approach that integrates positive psychology assessments for mental health. Early identification and intervention can make a significant difference in managing mental health conditions, improving prognosis, and enhancing the quality of life for individuals at risk.

The work can be aligned with three folds objectives as follows keeping the aim of designing and developing an Android application for Mental Health Issues Screening Web enabled System.

1. Design of a user friendly android app for screening of mental health issues. The app will be able to
   (a) provide proper guidance to those who requires help.
   (b) create a platform for channelization of appropriate mental health intervention.

(c) help individuals who have already been diagnosed to quantify their mental health issues using appropriate questionnaire.
(d) provide scientific and necessary tips for positive mental help and stress management.

2. To design a web enabled Chatbot messaging system and
3. To create a database of health issues in aplenty normal population.

## 2 Early Intervention of Psychiatric Disorder and Strategies Using Artificial Intelligence

Managing mental health disorders, improve prognosis, and improve the quality of life for those who are at risk, early identification and intervention can be extremely beneficial. Psychiatric diseases should be identified and treated as soon as feasible by early intervention, preferably before to the full onset of symptoms or in the prodromal phase (early signs and symptoms).

In healthcare, particularly for mental health issues, early intervention techniques utilizing artificial intelligence (AI) concentrate on spotting any dangers or ailments before to their full expression, enabling prompt and efficient treatment. By utilizing behavioral analysis, prediction models, and data trends, AI-driven early intervention can greatly enhance results (Hasan and Alee 2018).

For early mental health intervention, the following are important AI-based strategies:

### *2.1 Predictive Analytics for Risk Detection*

AI is able to forecast the likelihood of acquiring a mental health disorder by analyzing vast volumes of patient data, including behavioral patterns, wearable device data, and medical records (Das et al. 2024). These forecasts enable medical professionals to take action prior to the emergence of a serious ailment. Both Machine Learning Algorithms and Deep Learning Models can be applied in this case.

- Machine Learning Algorithms: To estimate risk levels, classification models (such as random forests and support vector machines) can examine a variety of characteristics, including lifestyle information, past medical history, and demographics.
- Deep Learning Models: To identify early warning indicators, analyzing time-series data such as social media activity, electronic health records (EHRs), or repeated mental health assessments can be possible with various deep learning models such as Recurrent Neural Networks (RNNs), LSTMs, or GRUs.

In example, for depression risk analysis using AI, a model trained on patient PHQ (Colizzi et al. 2020) scores over time (e.g., PHQ-9, PHQ-12) can identify early indicators of depression based on minute variations in responses over several months Intervention strategies like Personalized Outreach can also be adopted. Individuals classified as high-risk may be notified in advance for follow-up appointments, lifestyle modifications, or counseling to prevent the illness from getting worse (Bhadra et al. 2024).

## 2.2 *Natural Language Processing for Early Mental Health Condition Detection*

In order to detect emotional discomfort or mental health problems in people, natural language processing (NLP) models can process and analyze text from a variety of sources, including Chatbots, social media, therapy session transcripts, and surveys (Bhadra and Kumar 2023).

Various techniques based on NLP for early detection of mental health conditions are:

- Sentiment analysis: evaluating mood or emotional states by analyzing the sentiment of spoken or written words.
- Topic Modeling: Recognizing recurrent themes in discussions (such as stress or hopelessness).
- Emotion Recognition: Using NLP and AI models, this technique identifies emotional indicators (such as sadness, anxiety, or rage) in voice or text.

As an intervention strategy, proactive support like Chatbot technology can provide early support, suggest mindfulness exercises, or escalate cases to human therapists if the AI detects concerning patterns.

## 2.3 *Wearable Technology for Continuous Monitoring:*

Wearable technology, such as fitness trackers or smart-watches, is used to continually gather physiological data, such as heart rate, sleep patterns, and activity levels. Artificial intelligence (AI) can then utilize this data to identify minute changes that may be associated to mental health (Bhadra and Kumar 2022).

Various wearable techniques for early detection of mental health conditions are:

- Time-Series Data Analysis: Artificial intelligence (AI) models, like long short-term memory (LSTMs), are able to analyse continuous data streams (heart rate variability, sleep quality, etc.) and identify pattern deviations.

- Anomaly detection is the process of identifying odd patterns in behaviour or physiological measurements that could point to the beginning of mental health conditions including stress, anxiety, or depression.

Artificial intelligence systems can recognize patterns that suggest mental health problems. For example a wearable technology can monitor anxiety and sleep, and it can also monitor sleep disturbances, which are closely associated with mental health problems including anxiety and depression.

### *2.4 AI-Powered Screening for Mental Health*

AI can improve on conventional mental health screening instruments (such as the PHQ9 and GAD7) by instantly evaluating responses and spotting trends or discrepancies that human clinicians might miss. AI also makes these screenings more automated and scalable. Techniques relevant to AI-Powered Screening for Mental Health are:

- Adaptive Questionnaires: AI-powered surveys that change their focus to pertinent risk factors and conditions in response to patient input.
- Cognitive Behavioral Markers: During evaluations, AI can detects minute variations in reaction times, answer consistency, and even voice intonation.

As a Mobile-based Screening, a smart phone based app can be designed to do AI-powered PHQ9 or GAD7 tests and offer real-time risk analysis, scoring, and feedback based on patient responses. AI models are fed data to determine whether or not early intervention is necessary.

Healthcare professionals can enhance patient outcomes, diagnose mental health disorders earlier, and tailor therapies by utilizing AI. When these interventions are used in conjunction with conventional healthcare methods, mental health care can be given more quickly, efficiently, and individually.

## 3 Design of an Android Application

The design and development of a user friendly android application can be used for screening of mental health issues which be used to provide proper guidance to those who requires help. An android application can be used to create a platform for channelization of appropriate mental health intervention. It will be useful for the individuals who have already been diagnosed to quantify their mental health issues using appropriate questionnaire and also to provide scientific and necessary tips for positive mental help and stress management. The methodology to design an Android application can be illustrated as mentioned in Fig. 1.

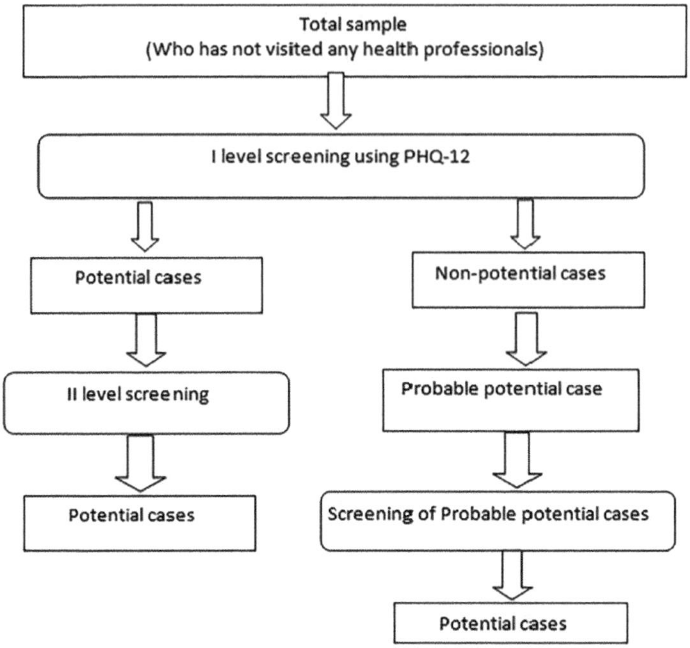


**Fig. 1** The methodology to design an Android application

# 4 Methodology

A simple model for the early detection of mental health conditions using Artificial intelligence (AI) can be designed with—

## *4.1 A Machine Learning Based Method Follows Various Steps as:*

1. *Dataset generation*: Needed a dataset of mental health questionnaires or symptoms; as an example, a dummy dataset is to create where the symptoms (anxiety, stress, depression, insomnia) are measured on a scale of 0 to 10, and the target (mental health condition) indicates whether the person has a mental health condition (0 or 1).
2. *Data Preprocessing*: here data need to preprocess followed by splitting the data into training and testing sets. Feature scaling or normalization can be applied to ensure all features contribute equally to the model and
3. *Feature Engineering*: using various algorithms (e.g., Logistic Regression, SVM etc.) to classify whether a person has a mental health condition or not. In this

case various Python libraries needed to install for this work e.g. are: pandas, scikit-learn, and numpy etc.
4. *Model Evaluation*: Model's performance is assessed using the classification report, confusion matrix, and accuracy after training during model evaluation.

In the next step, the dummy dataset is replaced with actual data (if available) and more advanced algorithm like Support Vector Machines (SVM), Random Forest, or Neural Networks for potentially better results.

## 4.2 Deep Learning Model with Recurrent Neural Networks (RNN):

Time-series or sequential data, such as the identification of mental health conditions from successive questionnaire answers or symptom data over time, are especially well suited for Recurrent Neural Networks (RNN). As an illustration of how to use an RNN model using Keras and Tensor Flow to identify mental health issues early, various steps to follow are:

1. Create a synthetic time-series dataset representing daily mental health checkups over 30 days for 100 individuals. The features include anxiety, stress, depression, and insomnia levels, while the target (mental health condition) represents whether a mental health condition is present (binary classification).
2. RNN Input Format: RNNs require input data in a specific 3D format [samples, time steps, features]. For example, for 100 people, the no. of samples can be n samples = 100, with time stamp of 30 as no. of days, i.e. n days = 30, and as no. of features considered as symptom categories with n features = 4.
3. RNN Model: A simple RNN layer with 64 units can be used, followed by a dropout layer to avoid over-fitting. For this case of binary classification (mental health condition), a Dense layer with a single neuron is considered as the output layer.
4. Training and Evaluation: As it's a binary classification problem, the model is trained for 10 epochs using the Adam optimizer and binary cross-entropy loss. Following the training, we assess the model using the test set and show the classification report, accuracy, and confusion matrix.

In the next step, for better prediction performance real time-series health data can be used. Hyper parameters like the number of RNN units, dropout rates, learning rate, etc also needed to train.

More sophisticated recurrent architectures, such as GRUs or LSTMs can be examined as they are frequently more effective at identifying long-term dependencies in time-series data.

### 4.3 Design of a Chatbot Messaging System

A Chatbot messaging system can be designed to create a database of health issues in aplenty normal population. The main objectives of the Chatbot will be

- It will asks questions and will response to the user.
- The Chatbot will calculate the users input and will give the result in normal, moderate and severe.
- If the user's condition is severe, the Chatbot will recommend available medical support and will help the user to get details of therapist available.
- The Chatbot will recommend related articles matching the user's input.

Dependencies or show stoppers of the method are: The various dependencies can be aligned are as mentioned in Fig. 2.

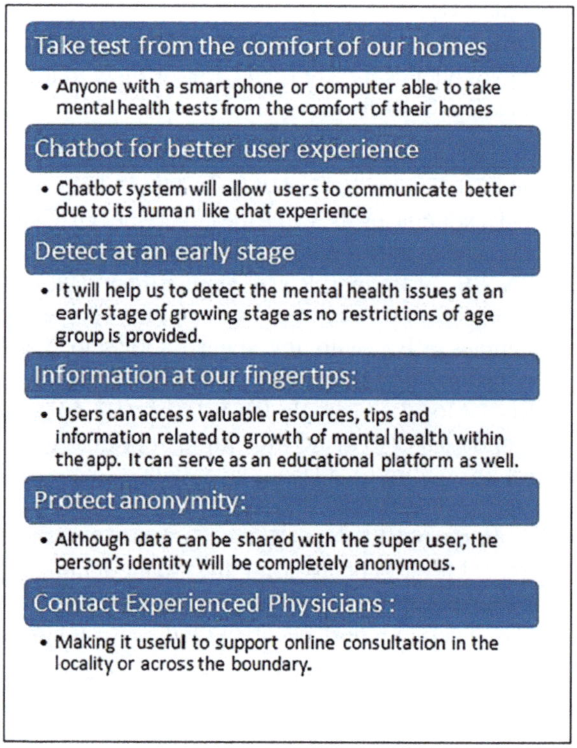


**Fig. 2** Dependencies of a Chatbot messaging system

## 5 AI Based PHQ-12 Screening for Early Detection of Mental Health Condition

In order to diagnose mental health disorders early with AI, the Patient-Health Questionnaire (PHQ-12), a modified version of the PHQ-9 can be intended to evaluate many facets of an individual's mental health (e.g., mood, sleep, anxiety etc.).

The following are required in order to use PHQ-12 scores in an early detection system utilizing a Recurrent Neural Network (RNN):

- Sequential information from several screening sessions (e.g., various patient visits throughout time).
- Design a classifier that uses this time-series data to determine if a mental health problem is present or not.

**The Steps to be followed are:**

I. ***Dataset generation***: Make use of a dataset in which every patient completed the PHQ-12 questionnaire several times during various time periods.
   - The shape of the input data(X_data) is [no. of patients(n_patients), no. of periods(n_periods), no. of features(n_features)]
   - In this case n_features = 12, representing the 12 PHQ questions.

II. ***Features***: The PHQ-12 scores are considered as the input features and every row represents a distinct patient visit.

III. ***Model***: A RNN-based architecture (Simple RNN, LSTM, or GRU) will be employed.

For example as a binary classification problem, the target is to indicate whether a mental health problem is present or not) and a RNN Layer with 64 units can be used, which processes sequential data (patient screening over time).

In the Dense layer, the sigmoid function as a activation function can be used for the binary classification problem.

IV. ***Training***: The sequential data is used to train the model. We employ binary cross-entropy loss and the Adam optimizer.

V. ***Model Evaluation***: Accuracy, confusion matrix, and a classification report are used to assess the model's performance on test data following the training phase.

In the next step, for better handling of longer-term dependencies in time-series data—

- the model can be switch to **LSTM/GRU by r**eplacing the Simple RNN with LSTM or GRU
- **PHQ-12 Data** (for the synthetic data) can be replaced with real patient PHQ-12 scores if the actual data are available.
- Hyper-parameter tuning: the batch size, learning rate, dropout rate, and number of units in the RNN layer can be adjusted.

## 6 Conclusion

Among the diverse community of population it is very hard to deal with their issues if any for their mental well being. Addressing their mental issues requires initiatives or promoting an advising system, where clinicians are needed to be educated on socio cultural backgrounds and how they may impact mental health, illness presentation, and help-seeking behaviors. A web enabled Chatbot messaging system can provide necessary help through a Clinical Mental Health professional who have adequate knowledge to deal with individuals stigmas surrounding mental health and thus reducing the stigmas surrounding mental health can be a vital step toward an individual seeking treatment. Early indicators of mental health disorders can be found by analyzing PHQ-12 responses over time using an LSTM or GRU network. Early interventions in clinical settings may benefit from these models' ability to capture the temporal patterns in a patient's responses.

## References

Bhadra, S.; Kumar, C.J.; Bhattacharyya, D.K. (2024). Multiview EEG signal analysis for diagnosis of schizophrenia: an optimized deep learning approach. *Multimedia Tools and Applications*. https://doi.org/10.1007/s11042-024-20205-y.

Bhadra, S., & Kumar, C. J. (2023). Enhancing the efficacy of depression detection system using optimal feature selection from EHR. Computer Methods in Biomechanics and Biomedical Engineering, 27(2), 222–236. https://doi.org/10.1080/10255842.2023.2181660

Bhadra, S.; Kumar, C.J. (2022). An insight into diagnosis of depression using machine learning techniques: a systematic review. *Current Medical Research and Opinion*, May, 38(5), 749–771. https://doi.org/10.1080/03007995.2022.2038487

Colizzi, M., Lasalvia, A. and Ruggeri, M. (2020) 'Prevention and early intervention in youth mental health: is it time for a multidisciplinary and trans-diagnostic model for care?', *International Journal of Mental Health Systems*, 14(1), p. 23. Available at: https://doi.org/10.1186/s13033-020-00356-9.

Das, P. R., Talukdar, R. R., & Kumar, C. J. (2024). Exploring the interplay of abstinence self-efficacy, locus of control, and perceived social support in substance use disorder recovery. *Current Medical Research and Opinion*, *40*(9), 1625–1635. https://doi.org/10.1080/03007995.2024.2390046

Hasan A., Alee N.T. (2018) *Mental health and psychological problems in North East India*. Hasan A., Alee N.T.

'Ministry of Activities in North East India'.

Murthy, Rs. (2017) 'National mental health survey of India 2015–2016', *Indian Journal of Psychiatry*, 59(1), p. 21. Available at: https://doi.org/10.4103/psychiatry.IndianJPsychiatry_102_17.

'UN Sustainable Development Goal (SDG)3' (no date). Available at: https://sdgs.un.org/goals/goal3.

# Virtual Health Assistants: AI in Patient Engagement

**Levi Adawari S. Damingo, Kingsley Theophilus Igulu, Nne Rena Saturday, Taibat Onome Asunogie, and Elliot Nkem Kizzy**

**Abstract** This chapter discusses automation in healthcare delivery services, including the use of artificial intelligence-driven healthcare solutions. Artificial intelligence and other underlying technological tools have evolved to provide more effective and efficient digital solutions in healthcare, such as virtual health assistants, which simulate human conversations and provide personalized responses based on user input. Similarly, we discussed AI-powered Virtual Health Assistants and how they are intended to assist healthcare professionals and patients in performing tasks such as accessing electronic health records, monitoring health conditions, delivering cognitive behaviour therapy for anxiety and depression, and so on. The fundamentals of virtual health assistants, their capabilities, the shift from virtual consultation to AI-enabled virtual health assistants, and the benefits and drawbacks of providing healthcare services were also examined.

**Keywords** AI · Virtual assistants · Healthcare · Intelligent systems

L. A. S. Damingo (✉)
Department of Computer Science, Ignatius Ajuru University of Education, Rumuolumeni, Port Harcourt, Rivers State, Nigeria
e-mail: levi.damingo@iaue.edu.ng

K. T. Igulu
Department of Computer Science, Kenule Beesor Saro-Wiwa Polytechnic, Bori, Rivers State, Nigeria
e-mail: igulu.kingsley@kenpoly.edu.ng

N. R. Saturday
Department of Computer Science, Rivers State University, Port Harcourt, Rivers State, Nigeria
e-mail: saturday.rena@ust.edu.ng

T. O. Asunogie
Department of Computer Science, Edo State University, Uzairue, Edo State, Nigeria
e-mail: asunogie.taibat@edouniversity.edu.ng

E. N. Kizzy
Department of Computer Science, Informatics and Cybersecurity, Federal University Otuoke, Otuoke, Bayelsa State, Nigeria
e-mail: elliotnk@fuotuoke.edu.ng



T. P. Singh et al. (eds.), *Revolutionizing Healthcare: Impact of Artificial Intelligence on Diagnosis, Treatment, and Patient Care*, Studies in Computational Intelligence 1182,
https://doi.org/10.1007/978-3-031-80813-5_16

# 1 Introduction

One of the key fundamentals of human existence is the state of health of an individual or group as the popular saying goes thus "health is wealth". This therefore suggests that the socioeconomic wellbeing of the society is directly proportional to the health and wellbeing of the individuals and groups that constitutes that society. The role and the importance of healthcare systems in the quality of life and social welfare in modern society, have been broadly well recognized as a major driver and indicator of economic growth.

The application of technology in enhancing the healthcare system in the twenty-first century has produced far reaching improvements in the area of diagnosis, treatment and patient-centered care across the sector. One prominent technological approach in enhancing the healthcare sector has been the introduction of artificial intelligence in healthcare. Artificial intelligence is rapidly evolving and becoming increasingly integrated into daily life. AI refers to the ability of a system to comprehend external data, learn from it, and adapt it to meet certain goals and activities geared towards informed decision making like humans (Kaplan & Haenlein 2019). Artificial intelligence (AI) has received a lot of support for its application in healthcare delivery, as systems are expected to enhance human capabilities in areas like patient care and system management, therapeutics, and diagnosis (Abbasgholizadeh et al., 2021). Through enhanced risk prediction, clinical decision support, faster and more accurate diagnosis, easier chart review and documentation, improved patient-physician relationships, and optimized operations and resource allocation, artificial intelligence (AI) systems have the potential to completely transform the landscape of the healthcare industry (Lin et al., 2019).

## *1.1 Concept of Virtual Assistants*

A virtual assistant is an independent contractor who provides administrative services to clients while working outside of their workplace. A virtual assistant usually works from home but can access important planning documents, such as shared calendars, remotely (Kenton 2024). Artificial intelligence-powered virtual assistants are ubiquitous in our daily lives. Most computer algorithms have humanized personal features, while being machine-generated (Yang et al., 2023). AI-powered virtual assistants are fast changing the way we engage with technology and complete activities. These intelligent systems use natural language processing (NLP) to interpret and respond to human requests, thereby automating operations and optimizing workflows (Uelsen 2021). Virtual assistants are typically characterized as digital systems that replicate human conversation and deliver personalized responses based on user input. They can be designed to engage in structured conversations or respond to user questions. Capabilities range from simple menu or multiple choice-based assistants to more advanced virtual assistants with natural language processing that recognize free speech or text.

Virtual assistants are currently frequently used in web-based banking and service contexts, decreasing dependency on people by responding to consumers' questions about products and services on-demand (Fitzpatrick et al., 2017).

The fundamental idea behind virtual assistants (VA) is that people want to communicate with Chatbots in the same natural language that they use to communicate with each other. This highlights the significance of human-like interactions and conversations, suggesting that people find it more enjoyable and better to engage with virtual assistants in a human-like manner. More specifically, a text input and text output mask is used to give virtual assistants a more human-like appearance by simulating real-world communication between the user and the assistant (Fei & Petrina 2013). Zumstein and Hundertmark (2018) identify five criteria that virtual assistants must possess. Virtual assistants should be viewed as team members, not just technology tools for human interaction. Virtual assistants should be built to provide human-like responses, such as not using bullet points. Virtual assistants adapt the personalities of the user, as individuals like to communicate with those that share their personality type. Expressing pleasant emotions, such as joy and happiness, improves the interaction between virtual assistants and users, increasing their credibility.

## *1.2 Overview of Virtual Health Assistance*

One key AI application that has revolutionized the healthcare sector is the use of virtual assistants (VA). Virtual assistants (VAs) are technology tools that facilitate natural language and other forms of human–machine interaction. It might be considered an intelligent tool that helps people with their everyday tasks. Voice assistants (VAs) are present in billions of devices worldwide and are predicted to number roughly 150 million by 2026 (Thormundsson, 2024). According to Bas (2024) virtual healthcare assistant is fundamentally a digital technology driven by artificial intelligence that aims to revolutionize the patient-care process. Take a look at an advanced software system that comprehends you're an individuals' health issues and provides prompt, tailored advice. The wonder of a virtual healthcare assistant is precisely that. This technological advancement is inexpensive to operate, scalable, and accessible. In order to be practical, it makes use of a number of techniques, including machine learning, speech recognition, neural networks, and natural language processing. These algorithms are taught using information gleaned from big data, patterns, and insights (Bhadra et al., 2024). It is now much more important in business, trade, production and manufacturing, healthcare systems, etc.

In the healthcare industry, these systems can recognize symptoms and recommend a suitable medical specialist. They can also generate medication dosage reminders for patients living in automated homes, send information to doctors, which always aids in managing patients' cases, and perform surgeries and other medical procedures. Additionally, it helps medical facilities by gathering information on insurance policies, medical histories of patients, demographics, finances, and procurement, as well as data mining and analysis of the gathered information. An increasing amount of research is being done to determine how the design features of virtual health assistants affect user experience. When considered collectively, there are findings indicate that user experience is impacted by the way virtual health assistants appearance and feel. This implies that they offer better user experiences as VHA demonstrates empathy, engage in nonverbal relational behaviours, and divulge personal information about themselves. Currently, there is a large body of evidence, and most research are small-scale and very diverse (Curtis et al., 2021). It is important to state that application of artificial intelligence driven virtual health assistants have improve healthcare delivery in recent time especially with remote care and virtual health assistants have come to state and will continue to evolve to provide move enhance patient-centered care.

Virtual Health Assistants (VHAs) bridge the communication gap between patients, healthcare institutions and professionals. It completely leveraged on artificial intelligence and natural language processing to provide customized and guided assistance to patients in the platform of interactive digital interfaces (Kolluri 2024). Unlike traditional healthcare facilities, the capacity of VHA to offer patients ongoing assistance and education is one of its main advantages. It can communicate with patients using Chatbots, mobile apps, etc. The availability of this technology gives people a sense of empowerment over their health, which in turn encourages positive outcomes and adherence to medical recommendations.

It is important also to note that virtual health assistants (VHA) can help alleviate healthcare disparities, particularly in poor or remote locations as it deploys digital platforms to provide health information and services to persons who cannot access traditional healthcare due to geographical, financial, or logistical barriers (Dille & Dille 2022). Ensuring equal access to healthcare aligns with global efforts to reduce inequities in patient care in line with the sustainable development goals (SDG) 3. VHAs assist both patients and healthcare workers by automating routine chores, streamlining administrative processes, and promoting improved clinical decision-making. It can create and generate appointment calendars, classify patient queries, and collect data to help healthcare providers prepare treatment plan for their patients (Thayananthan 2023). Virtual health assistants supports healthcare providers with decision-making and prioritizes improving care quality and patient-provider relationships.

The outbreak of COVID-19 pandemic led to the increase use of this technology, launch of more new products; with the mix of artificial intelligence in developed countries. However, there are some factors that influences the rise of these technological products which include: rise in consumer preference on technological products, population rise in geriatrics, rise in the number of users of digital devices, etc. Globally major organizations that operate in the VA healthcare industry are Amazon, CSS Corporation, Microsoft Corporation, ADA Digital Health, Verint Systems, Orbita, Babylon Holdings, MedRespond LLC, Kognito Solutions, Nuance Communications, Sensely, Intelligent Digital Avatars, Egain Corporation, Virgin Plus, Healthtrap, etc.

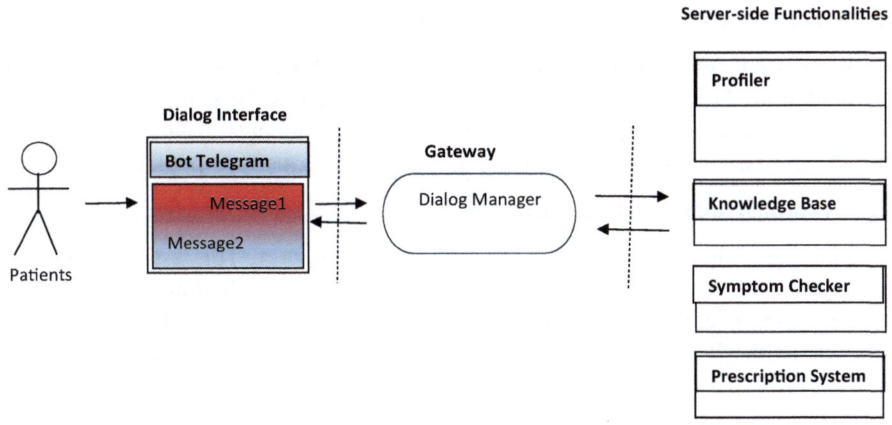


**Fig. 1** Basic design of a virtual health assistant

### *1.3 Fundamental Design of Virtual Health Assistants*

Figure 1 shows the basic design of a regular virtual health assistant. It is made up of the Dialog interface, Gateway, Server-side functionalities which comprises of the knowledge base module, system checker, prescription system and profiler. This can serve as a springboard for other designs of a virtual health assistant. From the basic design architecture the patient and user can access the VHA remotely to get medical assistance from the dialog interface module by sending message to the system stating their present health condition with respect to feelings and symptoms and the message will go through the gateway to the VHA server side where the message will be analyzed based on the information available in the knowledge base (Sharma, 2024). The diagnosis from the VHA is matches with the symptoms from the message sent with dataset in the knowledge base and further make prescription for the patient via the prescription system embedded in the VHA. This function make it easy for patients that cannot readily and easily access healthcare facility to get medical attention and care on the go.

## 2 Categories of Virtual Health Assistant

There is no clear categorization of virtual health assistants, but it will be based on the distinct services they provide and by implication. According to Asad (2023), a virtual assistant can improve patient care and results by bringing pertinent information to the notice of healthcare personnel and allowing them to do their duties more efficiently. The following are some of the areas in which virtual assistant is being implemented in the healthcare sector.

(i) **Virtual Assistant for Clinical Support**

These assistants play a key role in the delivery of healthcare; as they help to reduce the burden on the medical doctors and specialists, thereby focusing on core patient care. Their functions comprise of scheduling of appointments, navigating complex health records, exchange of critical patient's health information, medical billing, etc.

(ii) **Administrative Virtual Assistant**

This can be regarded as a higher level of clinical support. This assistant is more intelligent and drives on in-depth professional knowledge. It transforms medical offices into efficient models with its unique versatility in the medical field and other relevant fields. It joggles between e-learning, legal assistance in order to give a wholesome experience to patients and satisfying health outcomes.

(iii) **Care Coordinators for Patients**

These category of virtual health assistants plays a major in the treatment process of patients. Its usefulness is applied in ambulatory care unit, larger healthcare institutions and hospitals. They bridge the communication difference between the patient and healthcare professionals. Amongst there several duties are planning unique healthcare needs for patients in order to follow the care prescription. They ensure patients religiously follow through with complex healthcare journeys in order to achieve an unhindered wellness. Care coordinators are professionally proficient with prerequisite knowledge in nursing or social works and this makes them very important in the tele-health services.

## 3 Advantages of AI-Powered Virtual Health Assistants (VHA)

AI-powered virtual health assistants have transformed the traditional healthcare landscape, providing significant benefits for promoting healthy living habits through ongoing, personalized support. Chatbots, which act as virtual health assistants (VHAs), can engage users in real-time chats, provide specialized information, and encourage them. VHAs are capable of adapting to an individual's health objectives and settings. They can also offer quick support for someone attempting to enhance their health routine (Maher et al., 2024). The importance of the application of artificial intelligence based virtual health assistants in the sector cannot be overemphasized as it has gained prominence in the various ways with respect to timely and ready intervention strategies with high prescription accuracy. The deployment of bot and other tools provide numerous benefits to the user such as:

(i) **Reduced wait times**: Virtual assistants in healthcare provide immediate replies to patients, reducing long hospital waits.
(ii) **Reduced care costs**: Virtual assistants dramatically cut company overheads associated with desk and call centre staffing.

(iii) **Timely medical advice**: In an emergency, virtual health assistants can help by providing prompt information that could save a person's life.
(iv) **Round the Clock support**: On a tight schedule, it is practically difficult for doctors to see every patient; nevertheless, medical virtual assistants are available around-the-clock.
(v) **Privacy and confidentiality**: AI-powered medical chat is intended to protect the patients' privacy and ensure confidentiality which is advantageous in delicate situations.
(vi) **Timely monitoring**: A virtual healthcare assistant can offer treatment recommendations, follow-up reminders, and health monitoring services. This service could be specifically designed using IoT based smart watches and device to send reminder to the user.
(vii) **Real-time interaction**: Virtual health assistants deliver rapid responses in real time and adjust the response based on the input, similar to engagement with a doctor.
(viii) **Scalability**: Virtual health assistants may handle a large number of patients concurrently and in real time without compromising quality or increasing costs.
(ix) **Patient satisfaction**: Virtual health assistant provides safe, timely, and intelligent solutions, improving patient experience and satisfaction.

## 4 Challenges of AI-Powered Virtual Health Assistants

Some of the perceived challenges and limitations of the application of virtual health assistant are;

(i) **Ethical Consideration**

Ethics refers to the ideas and values that govern human behaviour, and it involves the study of what is good and wrong, as well as how people should behave in respect to others and their surroundings. In the context of scientific research and development, ethical considerations are crucial to ensuring that technology is designed and developed responsibly and ethically. The Ethics by Design approach incorporates ethical concepts throughout the design and development process, allowing ethical issues to be addressed as early as feasible while minimizing negative impacts and hazards to users and society.

(ii) **Data Security and Privacy Concern**

Privacy concerns are raised by the use of AI in any industry, and the healthcare sector is no exception with regard to virtual health assistance. These assistants rely significantly on human input to improve their performance, raising concerns about data privacy and security. There is concern that personal and sensitive information exchanged with virtual assistants may be exposed to unauthorized access or exploitation, resulting in potential privacy violations and identity theft (Martín et al.,

2022). The implementation of Artificial Intelligence for Virtual Health Assistance will require healthcare systems to establish suitable protocols for managing and preserving confidential patient information.

(iii) **Constant Maintenance Cost**

AI virtual health assistants, in contrast to the notion of a "one-time setup," need regular routine maintenance and updates to function at their best. There may be ongoing costs and resource requirements associated with this. Inadequate execution may result in malfunctions that impair the overall performance of the system.

(iv) **Limited Availability of Quality Clinical Data**

Access to high-quality data is required for developing an AI-powered virtual healthcare assistant. It can be limited in specific areas, causing the assistant to make less accurate forecasts and recommendations. This disadvantage can provide a considerable impediment to its efficacy.

(v) **Error Prone**

Though AI powered virtual health assistants are adept at carrying out duties precisely, mistakes can nevertheless happen. They could break down due to a small glitch in the system or an unanticipated circumstance. These interruptions have the potential to cause user discontent leading to wrong diagnosis and prescriptions as well as an accumulation of incomplete tasks.

## 5 Conclusion

Health and wellbeing is a critical component of individual and organizational growth and development. A healthy nation is said to be a wealthy nation therefor the need for continuous improvement in the health and wellbeing of the people. Technological innovations in healthcare delivery experienced a transformation with the integration of artificial intelligence in the healthcare sector and this has also enhanced the wellbeing of people across the globe. The healthcare sector globally is undergoing a radical transformation due to the implementation and deployment of AI-powered virtual healthcare assistants. They have the potential to lower operating costs for healthcare systems and increase patient access to healthcare services. This is probably going to introduce new capabilities that will keep coming as technology advances. From there, AI-enabled virtual healthcare assistants could contribute to the advancement of revolutionary personalized healthcare. These evolution in the sector has improved healthcare delivery and patient interaction through round-the-clock assistance, customized attention, ongoing observation, and enhanced correspondence, these improve patient results and optimize healthcare procedures.

## References


Abbasgholizadeh, R.S., Légaré, F., Sharma, G., Archambault, P., Zomahoun, H.T.V., Chandavong, S., Rheault, N.T, .... Légaré, J. (2021). Application of Artificial Intelligence in Community-Based Primary Health Care: Systematic Scoping Review and Critical Appraisal. Journal of Medical Internet Research, 23(9), 234–247.

Asad, S. (2023). 23 Types of Virtual Assistants (Virtual Assistants & Their Tasks). Retrieved from ASL: https://www.aslpreservationsolutions.com/types-of-virtual-assistants (12/09/2024).

Bas, A. (2024). A Blueprint to Craft AI-powered Virtual Healthcare Assistant. Retrieved from Uptech: https://www.uptech.team/blog/how-to-build-ai-powered-virtual-healthcare-assistant (27/08/24).

Bhadra,S and Kumar,C.J. and Bhattacharyya,D.K. (2024), Multiview EEG Signal Analysis for Diagnosis of Schizophrenia: An Optimized Deep Learning Approach., Multimedia Tools and Applications, Springer.

Curtis, R.G., Bartel, B., Ferguson, T., Blake, H.T., Northcott, C., Virgara, R & Maher, C.A. (2021). Improving User Experience of Virtual Health Assistants: Scoping Review. Journal of Medical Internet Research, 23(12).

Dille, F & Dille, Z. (2022). Virtual Assistants and Storytelling. In R. Berger, Storytelling. for New Technologies and Platforms (pp. 11–22). CRC Press.

Fei, Y & Petrina, S. (2013). Using Learning Analytics to Understand the Design of an Intelligent Language Tutor-Chatbot Lucy. International Journal of Advanced Computer Science and Applications, 4(11), 124–131.

Fitzpatrick K.K., Darcy, A & Vierhile, M. (2017). Delivering cognitive behavior therapy to young adults with symptoms of depression and anxiety using a fully automated conversational agent (Woebot): A Randomized Controlled trial. Journal of Medical Internet Research (JMIR) Mental Health, 4(2), 1–11.

Kaplan, A & Haenlein, M. (2019). A brief history of artificial intelligence: On the past, present, and future of artificial intelligence. California Management Review, 61(4), 5–14.

Kenton, W. (2024). What Is a Virtual Assistant, and What Does One Do? Retrieved from Investopedia: https://www.investopedia.com/terms/v/virtual-assistant.asp (12/09/2024).

Kolluri, V. (2024). A Cutting-Edge Research on Virtual Health Assistants: AI-Based Technology Platform for Engaging and Supporting Patients. International Journal of Creative Research Thoughts (IJCRT), 12(2), 48–52.

Kumar,C.J. and Das,P.R.(2021). The diagnosis of ASD using multiple machine learning techniques, International Journal of Developmental Disabilities, Jun 10;68(6):973–983. https://doi.org/10.1080/20473869.2021.1933730.

Lin, S.Y., Mahoney, M.R & Sinsky, C.A. (2019). Ten ways artificial intelligence will transform primary care. Journal of general internal medicine, 34(8), 1626–1630.

Maher, C., Singh, B., Wylde, A & Chastin, S. (2024). Virtual health assistants: a grand challenge in health communications and behavior change. Frontiers in Digital Health, 6(1), 1–4.

Martín, A.P., Mateo, C.G., Fernández, L.D & Pérez, MCL. (2022). Ethical Challenges in the Development of Virtual Assistants Powered by Large Language Models. Preprint, 1–13.

Sharma, V. (2024). Comprehensive Guide on How to Create a Virtual Health Assistant. Retrieved from Idea Usher: https://ideausher.com/blog/how-to-create-a-virtual-health-assistant/ (26/08/2024).

Thayananthan, V. (2023). AI-Based Digital Health Communication for Securing Assistive Systems. IGI Global.

Thormundsson, B. (2024). Virtual Assistant Technology - statistics & facts. Retrieved from Statista: https://www.statista.com/topics/5572/virtual-assistants/#topicOverview (27/07/2024).

Uelsen, A. (2021). Usefulness vs. Uncanniness: Exploring the perceived usefulness of virtual assistants in the customer journey. University of Twente. University of Twente. Retrieved from http://essay.utwente.nl/86548/1/vanUelsen_BA_BMS.pdf (12/09/2024).

Yang, N., Palma, M & Drichoutis, A.C. (2023). Humanization of Virtual Assistants and Delegation Choices. Munich Personal RePEc Archive(119275), pp. 1–24.

Zumstein, D & Hundertmark, S. (2018). Chatbots: an interactive technology for personalized communication and transaction. ADIS International Journal, 15(1), 96–109.

# Revolutionize Electronic Health Records with AI in Hospitals for Patient Benefit

**Manmohan Mishra, Bireshwar Dass Mazumdar, Prashant K. Gupta, and Abhishek Tripathi**

**Abstract** This study investigates the transformative potential of artificial intelligence (AI) in revolutionizing electronic health records (EHRs) and compares the adoption rates and impacts in both public and private hospital settings. With the rapid advancement of AI technologies, there exists a significant opportunity to enhance the efficiency, accuracy, and accessibility of EHR systems, thereby improving patient care and outcomes. Through a comparative analysis of public and private hospitals, this research assesses the extent to which AI-driven EHR solutions have been implemented and integrated into healthcare practices. By examining factors such as infrastructure, investment, and regulatory frameworks, this study seeks to identify the drivers and barriers influencing AI adoption in each setting. Furthermore, this research evaluates the tangible benefits that patients derive from AI-enabled EHR systems. These benefits may include streamlined appointment scheduling, personalized treatment recommendations, predictive analytics for disease prevention, and enhanced communication between healthcare providers and patients. By leveraging AI algorithms, EHRs can efficiently process vast amounts of patient data, leading to more informed decision-making and proactive healthcare management. The findings of this study aim to provide valuable insights for policymakers, hospital administrators, and healthcare professionals seeking to optimize EHR systems through

M. Mishra (✉)
Department of Master of Computer Application, United Institute of Management, Prayagraj, Uttar Pradesh 211010, India
e-mail: mishramanmohan@gmail.com

B. D. Mazumdar · P. K. Gupta
School of Computer Science Engineering and Technology, Bennett University, Greater Noida, India
e-mail: bireshwar.mazumdar@bennett.edu.in

P. K. Gupta
e-mail: guptaprashant1986@gmail.com

A. Tripathi
Department of Accounting and Information Systems, School of Business, The College of New Jersey, Ewing Township, NJ, USA
e-mail: tripatha@tcnj.edu

 
T. P. Singh et al. (eds.), *Revolutionizing Healthcare: Impact of Artificial Intelligence on Diagnosis, Treatment, and Patient Care*, Studies in Computational Intelligence 1182,
https://doi.org/10.1007/978-3-031-80813-5_17

AI integration. By understanding the comparative advantages and challenges faced by both public and private hospitals, stakeholders can develop strategies to maximize the potential of AI in revolutionizing healthcare delivery and improving patient outcomes. Ultimately, this research contributes to the ongoing discourse on leveraging technology to address the evolving needs of modern healthcare systems.

**Keywords** Artificial intelligence · Expletory research · e-health · e-governance · Competitive advantages · Adoption · Acceptance

## 1 Introduction

This article provides a comprehensive overview of the objective, importance, methodology, hypothesis, scope, limitations, and literature review related to the study. The initial review of the literature highlights a clear gap in research on e-Health. To effectively implement e-Health in India, practitioners and researchers need a better understanding of why people resist using e-Health systems. Incorporating Artificial Intelligence (AI) into these studies could offer deeper insights and innovative solutions to overcome this resistance (Bhadra and Kumar 2023).

## 2 Research Questions

The concentration of this research is based on the adaptation of e-Health in India, with a focus on incorporating Artificial Intelligence (AI) to enhance outcomes. An AI-driven study is carried out to explore the possibility of increasing the level of acceptance and adoption of e-Health in different public hospitals in India. The following AI-informed Research Questions have been formulated after conducting an AI-based literature study of e-Health across various states of India. Some AI-related literatures were also found in this context from Asian and European perspectives.

1. What is the role of AI in enhancing ICT in the health sector of India?
2. What is the status of health sectors in terms of efficiency and effectiveness after the adoption of AI-integrated ICT?
3. What is the role of AI in e-governance within the health sector of India?
4. How can AI increase the acceptance of e-Health among different users across the health sector in India?
5. Identifying the different AI-driven factors responsible for the adaptation of e-Health systems in India

## 3 Research Objectives

To answer the above research questions, the following AI-enhanced objectives are set for this research:

1. To study the current usages of AI-integrated ICT in delivering Health Services to the citizens.
2. To study the potential applications of AI within ICT in e-Health sectors.
3. To identify the AI-driven factors affecting users' adoption of e-Health Services.
4. To identify the Competitive Advantages acquired by the health sector after the use of AI-enhanced ICT.
5. To develop and validate an AI-based model empirically for e-Health Services adoption in India.

## 4 Issues

Through an initial examination of the literature and engaging with users, certain challenges pertaining to the acceptance and execution of information and communication technology (ICT) in healthcare services have been discovered. These factors significantly influence the adoption and acceptability of e-Health. Integrating Artificial Intelligence (AI) can successfully tackle these problems (Das et al. 2024).

(i) Technological Infrastructure: AI-enhanced technology refers to modern methods of performing tasks, often at lower costs and with increased efficiency, thereby setting a competitive platform for e-Health. If updated and advanced AI technology is available for e-Health, the adoption of ICT will be smoother (Paul et al. 2006; Alexander 2011; Al Ghamdi and Moussa 2011; Reddy et al. 2018; Garg et al. 2023).

(ii) Research and Development Cell: AI-driven research and development involve efforts to find innovations in e-Health. It measures expenditures in research and development areas. e-Health initiatives investing in AI research and development require high ICT adoption (Czerwinski et al. 2009; Balamuralikrishna and Dugger 2009; Buchanan et al. 2011; Ekeland et al. 2012; Amann et al. 2020).

(iii) Government Support: AI can enhance government support in terms of infrastructure, which is essential for the smooth adoption of ICT in e-Health. Since e-Health is relatively new at every level, AI-supported government assistance in infrastructure, tax rebates, and adoption of new systems can be highly beneficial (Beedasy et al. 2008; Narayana 2009; Uryan et al. 2010; Obry et al. 2011; Arnold 2021).

(iv) Citizen Awareness: AI can improve the dissemination of basic information to citizens provided by the government regarding policies and programs in the field of e-Health. Government announcements of supportive policies and programs through AI-powered e-governance portals can enhance the smooth adoption of ICT in e-Health (Subrahmanyam 2008; Narayana et al. 2008; Czerwinski et al. 2009; Nores and José 2011; Anttila 2012; Vaart et al. 2012; Pande et al. 2012; Amann et al. 2020).

(v) Literacy: AI applications in enhancing education for healthcare providers are essential. Given that India's literacy rate is lower compared to developed countries and it is a multilingual nation, AI can provide user-friendly and cost-effective tools, applications, and content in various Indian languages. AI technologies such as optical character recognition, text-to-speech, speech recognition, cross-lingual information retrieval, and machine translation can address linguistic data resource issues, content creation, and language processing (Lituchy and Rail 2000; Czerwinski et al. 2009; Raj 2011; Anttila 2012; Vaart et al. 2012; Davenport and Kalakota 2019).

(vi) Socio-Cultural: AI can help address socio-cultural factors that pose challenges to the adaptation of e-Health in India. These factors are critical components of health policy analysis and e-Health adaptation. There is a persistent dichotomy between respectable economic growth rates and marginal improvements in social indicators, which AI can help to bridge (Duggal 2007; Valeri et al. 2010; Uryan et al. 2010; Burcu et al. 2011; Pandel et al. 2011; McLaughlin et al. 2012; Sharma et al. 2024).

## 5 Research Scope

The research scope is a statement outlining the assumed boundary of the research in terms of its coverage and focus. This study is concentrated on Delhi and the National Capital Region (NCR), specifically focusing on CGHS Empanelment government hospitals, which have proven to be fertile ground for the implementation of emerging technologies, including Artificial Intelligence (AI). Government healthcare centers and their AI-enhanced information accessibility facilities for citizens regarding different schemes and policies have been taken into consideration (Bhadra et al. 2024). The data will be collected from Delhi and NCR-based CGHS Empanelment government hospitals and will be used to create a model for the successful implementation of e-Health. This will be achieved through an in-depth analysis of questionnaires, with AI playing a crucial role as the prime source of data collection and analysis.

## 6 The Research Model and Hypotheses

The study is designed with the objective to identify AI-driven factors that determine the implementation of e-Health adoption or rejection by citizens. An AI-enhanced research model is designed according to Fig. 1, considering the importance of these variables. This AI-based research model is grounded in a comprehensive literature study.

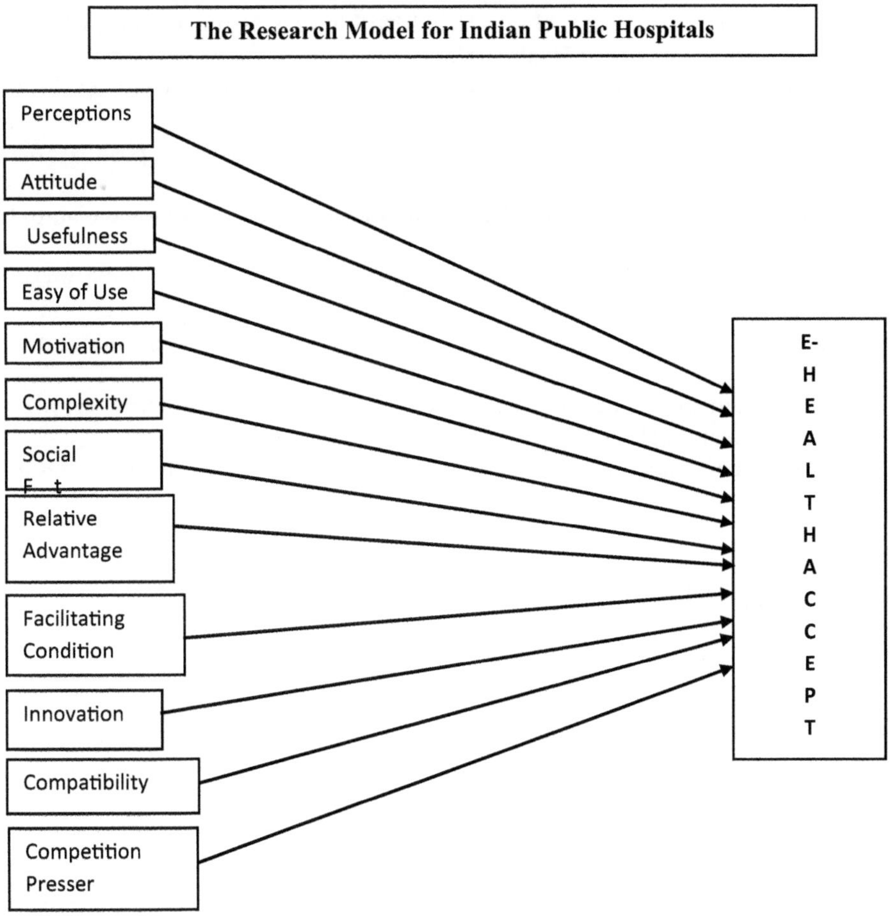


**Fig. 1** AI-enhanced research model for the adaptation of e-Health in Indian government hospitals

## 7 Research Hypothesis

Research hypothesis development is based on the existing AI-enhanced research model. These AI-driven research hypotheses are:

1. Perceptions towards using AI in e-Health will have a significant positive effect on the use of e-Health. [H1]
2. Attitude toward using AI in e-Health will have a significant positive effect on the intention to use e-Health. [H2]
3. Perceived usefulness of AI will have a significant positive effect on the intention to use e-Health. [H3]
4. Perceived ease of use of AI will have a significant positive effect on the attitude toward using e-Health. [H4]

5. Motivation toward using AI in e-Health will have a significant positive effect on the intention to use e-Health. [H5]
6. Complexity of AI-related risk will have a significant negative effect on the attitude towards using e-Health. [H6]
7. Social factors toward using AI in e-Health will have a significant positive effect on the intention to use e-Health. [H7]
8. Relative advantage of AI will have a significant positive effect on the attitude towards using e-Health. [H8]
9. AI-facilitation conditions will have a significant positive effect on the attitude towards using e-Health. [H9]
10. AI-driven innovation will have a significant positive effect on the attitude towards using e-Health. [H10]
11. Compatibility of AI will have a significant positive effect on the attitude toward using e-Health. [H11]
12. AI-induced competitive pressure will have a significant positive effect on the attitude towards using e-Health. [H12]

## 8 Research Methodology

The research study is based on questionnaire survey method, data collected from secondary sources. The secondary data is collected from published as well as unpublished sources of the organizationals internal records, manuals, files and brochures. Data published by various books, journals, organizationals manuals on the subject along with the number of newspaper clippings and journals.

The successful investigation of the research objectives is dependent upon analysis of a large number of responses, and consequently survey research is the most suitable form of data collection. This section describes how a research methodology is set for the proposed research.

Understanding the totality of the issues of the research may require the ingenious application of several tools that are both qualitative and quantitative.

### *8.1 Questionnaire Survey*

The AI-enhanced questionnaire-based approach is a well-established technique in obtaining data for social sciences research. Numerous IS research projects aimed at gathering data from user groups have successfully utilized this AI-driven method. The objectives of AI-supported questionnaire design are:

- Obtaining quantitative input through AI for testing the hypotheses of the study.
- Saving time in collecting voluminous data with AI-assisted data management and analysis.

## *8.2 Questionnaire Design*

A precise, structured multiple-choice questionnaire will be designed keeping in mind the need for eliciting the requisite information. The questionnaire format proposed to be worked out in targeted variables of the Research Model. Generalized questions will be avoided to protect the information from biases of the respondents. Questionnaire was based on English language for respondent convince & Descriptions of Questionnaire is given below.

(a) Questionnaires are being designed for two types of respondents.
   - Patients for public hospitals and government hospitals
(b) Questionnaires are designed with the objective to collect data which is necessary to test the hypotheses formulated for the study.
(c) Questions related to the test of hypotheses were in the statement format, which seeks opinion in terms of level of acceptance or rejection of the respondent.
(d) Basically two types of questions are being used i.e. dichotomous, multiple choice along with seven point likert scale.
(e) The likert scale has been divided in seven stages with different meanings of each. The 1st stage indicates 'strongly disagree', 2nd indicates 'disagree', 3rd indicates 'partially disagree', 4th 'neutral', 5th indicates 'partially agree', 6th indicates 'agree', and 7th indicates 'strongly agree'.
(f) This justifies that the answers of the categories is mutually exclusive so that respondents have to select only one choice against a question. In the beginning part of the questionnaire based on the personal information such as name (optional), age, gender, education and geographical area.

## *8.3 Questionnaire Validation*

- **The questionnaire** validation is based on pretesting of data. The validations of questionnaire are divided into four parts viz. face validity, criterion validity, and content validity and construct validity.
- **Face Validity verifies** the degree of fitness between the researcher's perception and concept of variables, which are being operationalized through the instrument. In this process, twenty experts selected as judge. Among them ten judges are general Citizen, five experts are from hospitals and five experts are from the academic field with great research experience. Their opinion/suggestions are being incorporated in the questionnaire (Hair et al. 2009)
- **Criterion Validity** refers to the degree to which the measurement with the questionnaire is meaningfully related to the objective of the research. This validation completed with the involvement of the experts who gave significant input during the face validity. The suggestions for improvement are incorporated in the questionnaire (Hair et al. 2009)

- **Content Validity** consists essentially in judgment. For this questionnaires along with the definition of the variables used were given to the experts (Hair et al. 2009)
- **The Construct Validity** aims to validate the theory behind the test. Construct validity has been established on statistical basis (Hair et al. 2009).

### *8.4 Sample Design*

- Sampling Technique—Convenience Technique.
- Sample Size: 312 Samples.
- Sample Place: Delhi and National Capital Region (NCR) based CGHS Empanelment.
- Multi Specialist Hospitals (Private and Government hospitals both).
- Nature of Elements: Patients from above mentioned hospitals.
- Sampling duration: from January 2024 to June 2024.

### *8.5 Hospital's Selections*

- **Government Hospitals**: Delhi and National Capital Region (NCR) CGHS Empanelment Multi Specialist Hospitals

### *8.6 Questionnaire Validation*

The testing of the Questionnaire will done in a group of respondents or expert of the health services with a pointed question as to whether the questionnaire used is relevant and covers all the research questions under enquiry. This can be explained by the flow chart as given Fig. 2.

## 9 Research Tools and Technique

Mathematical tools and technique are used to study the response SPSS software is used to analyze the response. To validate the questionnaires Chron back alfa and Confirmatory factors analysis will be used. Regression Analysis, Structural Equations Modeling and ANOVA/t-test have to be done to test the hypothesis.

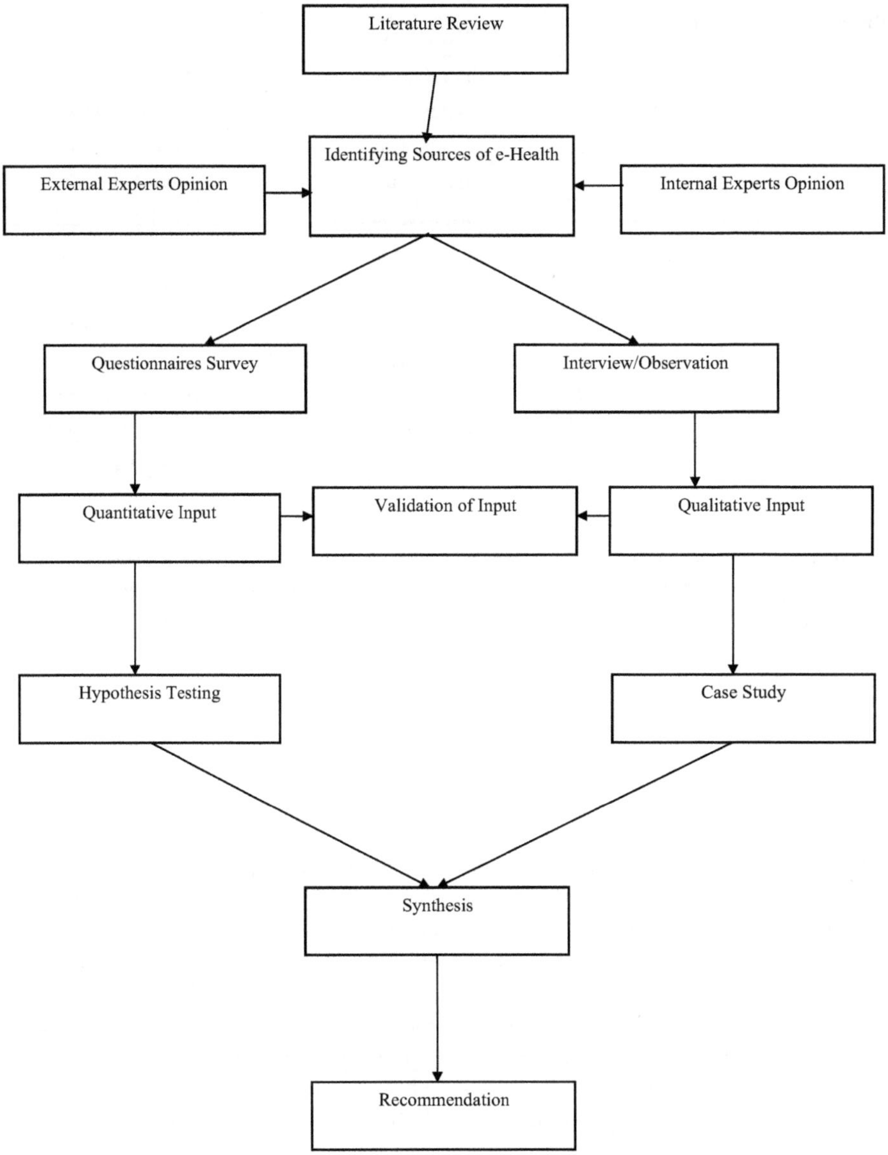


**Fig. 2** Research model flow chart

## *9.1 Questionnaire Testing*

- The questionnaire undergoes pre-testing by administering a survey to a confined group of respondents. The respondents for the pre-testing phase were chosen from the same demographic that would later be surveyed.

**Table 1** Questionnaire features

| Sl. No. | Variable name | No. of questions |
|---|---|---|
| 1 | Perception | 5 (pre1–pre5) |
| 2 | Attitude | 5 (att6–att10) |
| 3 | Usefulness | 8 (use11–use18) |
| 4 | Easy of use | 6 (easy19–easy24) |
| 5 | Motivation | 5 (moti25–moti29) |
| 6 | Complexity | 7 (com30–com36) |
| 7 | Social factors | 5 (sf37–sf41) |
| 8 | Relative advantage | 5 (ra42–ra46) |
| 9 | Facilitation condition | 4 (fc47–fc50) |
| 10 | Innovation | 6 (inno51–inno56) |
| 11 | Compatibility | 5 (comp57–comp61) |
| 12 | Competition pressure | 5 (cp62–cp66) |

- During this phase, a total of 312 participants were provided with the questionnaire to complete. Subsequently, each responder was questioned using a questionnaire in order to identify any deficiencies in the questionnaire. The questionnaire was revised according to the recommendations in order to enhance the respondents' comprehension.

## 9.2 *Questionnaire Format*

The questionnaire aims to gather personal information and opinions on the uses of AI in e-health within public and private hospitals on the basis of questionnaire features (see in Table 1). It includes multiple-choice questions to collect demographic data such as age, gender, and occupation. Additionally, it employs ranking methods to evaluate respondents' views on various AI applications, such as diagnostics, patient monitoring, administrative tasks, and personalized treatment. By comparing preferences and perceived benefits between public and private hospital settings, the survey seeks to identify key areas where AI can enhance healthcare services and pinpoint potential differences in AI adoption and effectiveness between the two sectors.

## 9.3 *Survey Data*

**Collection Flow chart**

Flow chart is the pictorial presentation of study, according to Fig. 3 fallowing flow chart represents the pilot survey data collection methodology during the survey.

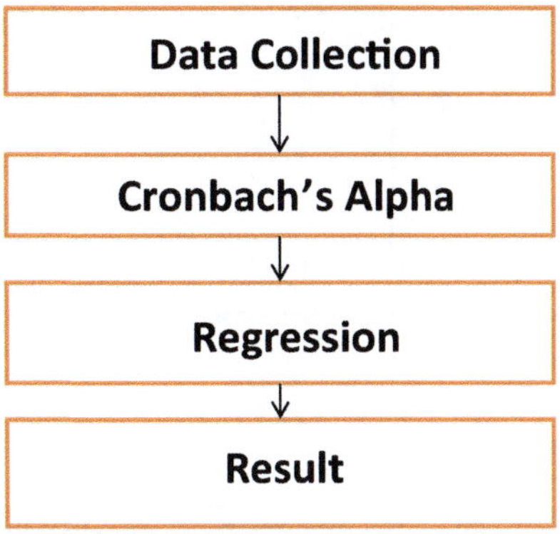


**Fig. 3** Flow chart is the pictorial presentation of study

**Gender of Respondent**

**Description: Gender profile**: According to Fig. 4, it was found in the analysis that out of total 54 respondents 80% i.e. 43 are males and 20% i.e. 11 were females, so it can be understood that maximum are male who prefer to use e-Health services.

**Residents of Respondent**

**Description: Gender profile**: According to Fig. 5, it was found in the analysis that out of total 54 respondents 93% i.e. 50 are Urban and 07% i.e. 04 were rural, so it can be understood that maximum respondent are belong to urban area & only few respondent are belong to rural area who aware with the e-Health services.

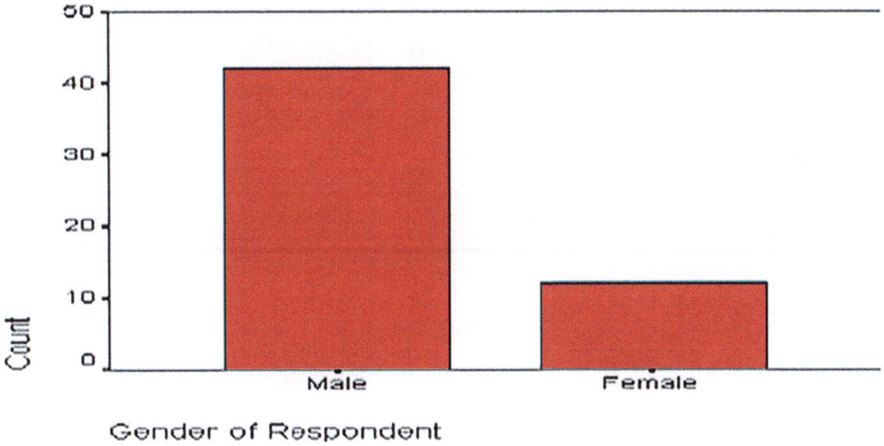


**Fig. 4** Gender of respondent

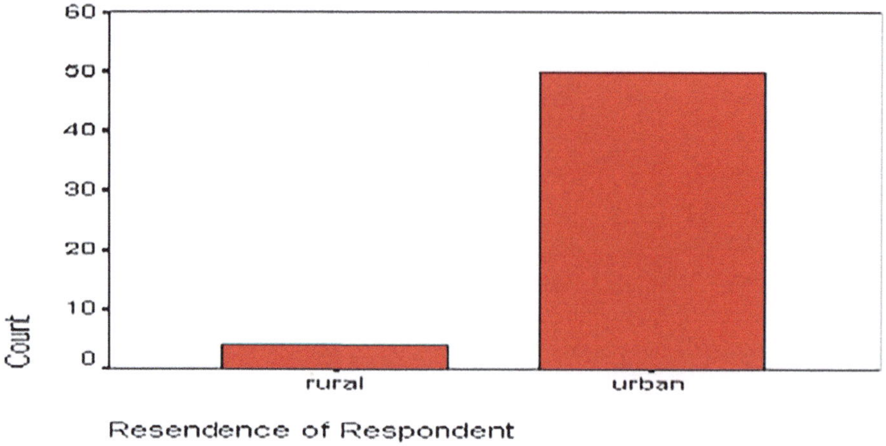


**Fig. 5** Residents of respondent

## Age of respondent

**Description: Age profile**: According to Fig. 6, it was found in the analysis that the respondents are of different age groups 02% (upto 20), 44% (21–30), 47% (31–40), 7% (41–50). So, it can be understood that majority of the respondents are middle age group between (31–40) years, the detail is presented in graph.

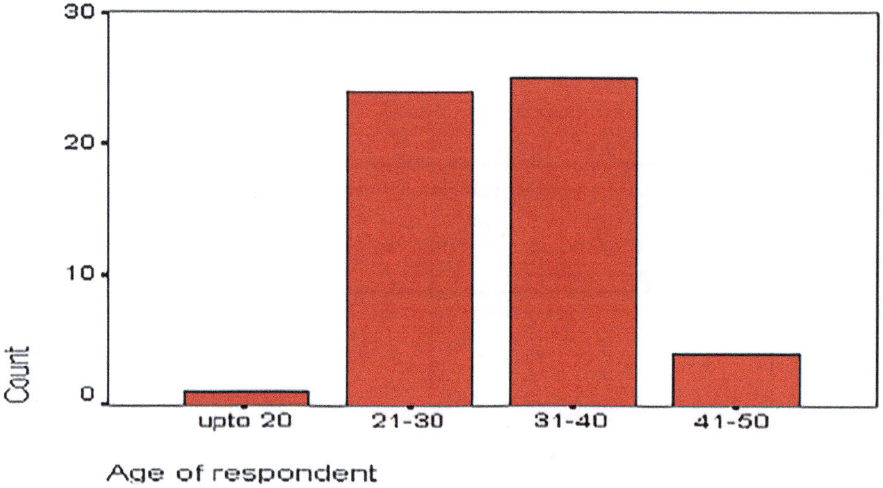


**Fig. 6** Age of respondent

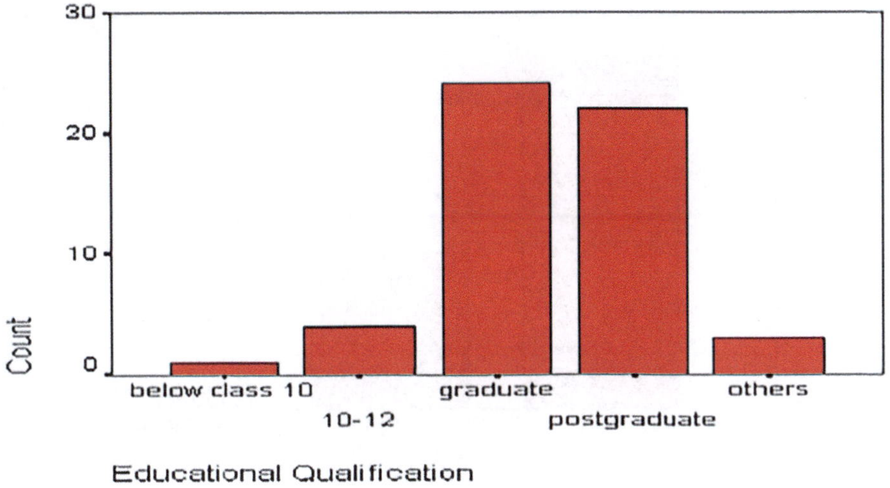


**Fig. 7** Educational qualification

**Educational Qualification**

**Description: Education profile**: According to Fig. 7, as shown in the analysis, the education of the respondents below class 10th is 2% i.e. 1, 10–12 is 7% i.e. 04, graduate is 47% i.e. 25, post-graduation is 39% i.e. 21, and others is 5% i.e. 3, so it can be interpret that majority of the people were graduate and above, the detail is presented in table.

**Uses of Internet**

**Description: Gender profile**: It was found in the analysis that out of total 54 respondents 68% i.e. 37 are daily users, 30% are weekly users i.e. 16 and 2% not at all i.e. 01.so it can be interpret that majority of the people are using daily Internet and weekly are second number (Fig. 8).

## 10 Reliability Analysis Table (Chron Back Alpha Value)

The Cronbach's Alpha values for the majority of variables are more than 0.7, suggesting that the survey questions associated with each variable have strong internal consistency and reliability. Several variables, including "Attitude," "Usefulness," and "Innovation," have Alpha values that are notably high (over 0.9), indicating a high level of dependability.

The dependent variable, "Acceptance of e-Health," has a high Alpha value of 0.9801, suggesting a very reliable measurement of this variable (from in Fig. 9).

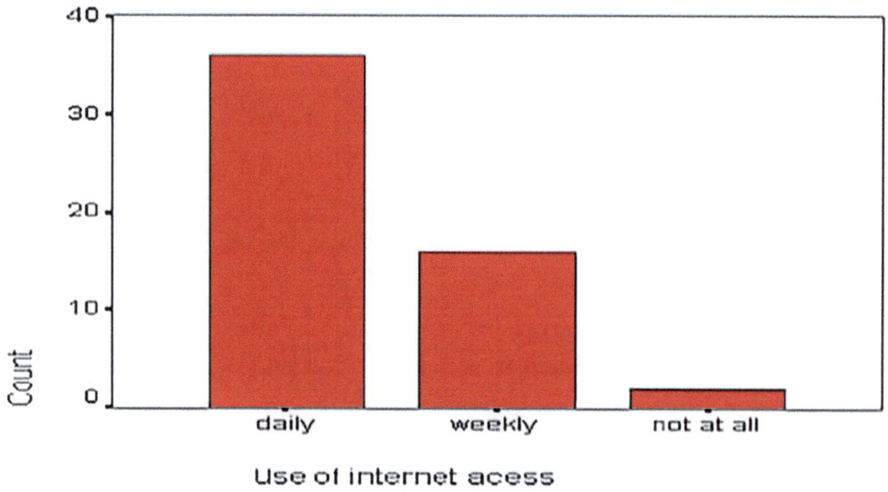


**Fig. 8** Use of Internet

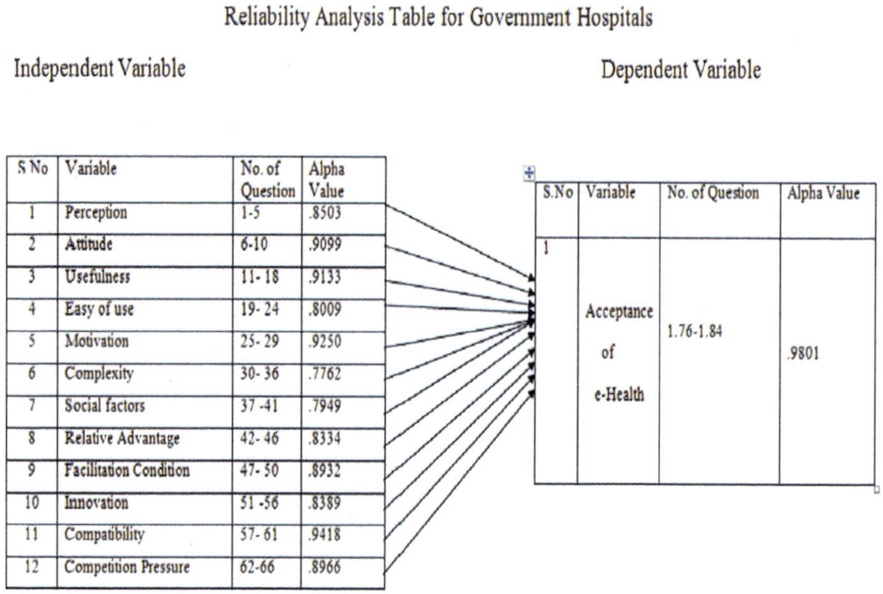


| S No | Variable | No. of Question | Alpha Value |
|---|---|---|---|
| 1 | Perception | 1-5 | .8503 |
| 2 | Attitude | 6-10 | .9099 |
| 3 | Usefulness | 11- 18 | .9133 |
| 4 | Easy of use | 19- 24 | .8009 |
| 5 | Motivation | 25- 29 | .9250 |
| 6 | Complexity | 30- 36 | .7762 |
| 7 | Social factors | 37 -41 | .7949 |
| 8 | Relative Advantage | 42- 46 | .8334 |
| 9 | Facilitation Condition | 47- 50 | .8932 |
| 10 | Innovation | 51 -56 | .8389 |
| 11 | Compatibility | 57- 61 | .9418 |
| 12 | Competition Pressure | 62-66 | .8966 |

| S.No | Variable | No. of Question | Alpha Value |
|---|---|---|---|
| 1 | Acceptance of e-Health | 1.76-1.84 | .9801 |

**Fig. 9** Reliability Analysis for Government Hospitals

**Table 2** Model summary of regression analysis

| Model | R | R Square | Adjusted R Square | Std. error of the estimate |
|---|---|---|---|---|
| 1 | 0.538(a) | 0.289 | 0.076 | 1.977 |

**Table 3** Model summary of ANOVA

| Model | | Sum of squares | Df | Mean square | F | Sig. |
|---|---|---|---|---|---|---|
| 1 | Regression | 63.608 | 12 | 5.301 | 1.357 | 0.227(a) |
| Residual | | 156.281 | 40 | 3.907 | | |
| Total | | 219.889 | 52 | | | |

### *10.1 Regression Analysis*

**Summary of Regression Analysis for Government Hospitals**

**Model Summary**

The regression analysis conducted on government hospitals indicates a moderate connection between the predictors and the dependent variable, as shown by a R value of 0.538. The R Square score of 0.289 indicates that 28.9% of the variability in the dependent variable can be accounted for by the model. Nevertheless, the Adjusted R Square value of 0.076 indicates a substantial loss in the explanatory power of the model when accounting for the number of factors. The standard error of the estimate is 1.977, which is the average deviation of the observed data from the regression line (see in Table 2).

**ANOVA**

The ANOVA Table 3 displays the analysis of variance for the regression model, specifically for the dependent variable UIGH. The regression sum of squares is 63.608, and the residual sum of squares is 156.281, resulting in a total sum of squares of 219.889. These calculations were performed using 52 degrees of freedom. The regression has a mean square of 5.301, whereas the residual has a mean square of 3.907. The F-statistic is 1.357 with a p-value of 0.227, indicating that there is no statistically significant difference.

## 11 Finding

The AI-enhanced notional model of public hospitals was verified using a regression approach. By using AI methods, this study helped with the prediction of changes in the dependent variable by examining its association with all independent variables. The regression analysis was performed with a significance level of 95%. The AI analysis used covariance between the dependent variable and all independent factors to predict the variability in the dependent variable. The value of the coefficient of determination (R Square) suggests that the AI-based research model accounts for just 28.9% of the variability in the adoption of e-Health among Indian hospital patients. Regression coefficients, often known as beta values, show the significance of the relationship between the dependent variable and independent factors. The corrected R square value was determined to be 0.076, while the standard error of estimate, which is the square root of the residual mean square in ANOVA, is 1.977. The standard error of estimate quantifies the extent of the residual error.

The analysis found support for ten (10) out of twelve (12) hypotheses related to the use of e-Health in private hospitals. The hypotheses of 'Usefulness' and 'Motivation' were not shown to be valid predictors of e-Health adoption in Indian institutions. The research, which used AI technology, identified 'Perception,' 'Attitude,' 'Ease of Use,' 'Complexity,' 'Relative Advantage,' 'Facilitation Condition,' 'Innovation,' 'Social Factors,' 'Compatibility,' and 'Competition Pressure' as factors that might predict the adoption of e-Health.

## 12 Limitation

(1) Less Number of Respondents.
(2) Time constraints.
(3) Lack of Technological Awareness.
(4) Less or limited Literacy in computer.
(5) Perceptions on e-Health acceptance varies from Respondent to Respondent.

## 13 Conclusion

To sum up, the incorporation of artificial intelligence (AI) into electronic health records (EHRs) has great potential to improve hospital healthcare delivery. The research emphasizes how AI has the ability to revolutionize healthcare by enhancing the effectiveness, precision, and usability of EHR systems. It is clear from examining public and private hospital settings that several elements, including infrastructure, legal frameworks, and technological investment, have an impact on the adoption of AI. The study emphasizes how crucial it is to remove obstacles to AI integration in order to fully reap its advantages, which include improved communication, predictive analytics, and tailored treatment suggestions. Leveraging AI in EHRs offers a crucial chance to enhance patient outcomes and advance healthcare delivery as healthcare systems continue to change. In order to enhance EHR systems via AI and eventually contribute to the continuous transformation of healthcare, legislators, administrators, and healthcare professionals may benefit greatly from the insights this study offers.

## References

Alexander J. A. M. D. (2011). Internet skill-related problems in accessing online health information. International Journal of Medical Informatics, 81(1), 61-72.
Al Ghamdi, K. M., & Moussa, N. A. (2011). Internet use by the public to search for health-related information. Journal of Health Communication, Elsevier, 11, 219–236.
Amann, J., Blasimme, A., Vayena, E., Frey, D., & Madai, V. I. (2020). Explainability for artificial intelligence in healthcare: A multidisciplinary perspective. BMC Medical Informatics and Decision Making, 20(1). https://doi.org/10.1186/s12911-020-01332-6.
Alsheibani, S., Cheung, Y., & Messom, C. (2018). Artificial intelligence adoption: AI readiness at firm-level. In M. Tanabu & D. Senoo (Eds.), Proceedings of PACIS2018: Pacific Asia Conference in Information Systems (PACIS) [37] Association for Information Systems. https://aisel.aisnet.org/pacis2018/37/.
Anttila, M. (2012). Use of web-based patient education sessions on psychiatric wards. International Journal of Medical Informatics, 81(6), 424-433.
Arnold, M. H. (2021). Teasing out artificial intelligence in medicine: An ethical critique of artificial intelligence and machine learning in medicine. Journal of Bioethical Inquiry, 18(1), 121-139. https://doi.org/10.1007/s11673-020-10080-1.
Balamuralikrishna, R. & Dugger, J. C. (2009). SWOT analysis: a management tool for initiating new programs in vocational schools. Journal of Vocational and Technical Education, 12(1).
Beedasy, J., Ramloll, R., Piland, N., Hudnall Stamm, B., Cunningham, M., & Ikeyo, B. (2008). A blended approach to emergency preparedness training. Telemedicine and e-Health, 14(1), 117.
Bhaskaranarayana, A., Satyamurthy, S. L., & Remilla, Murthy, L. N. (2009). Indian space research organization and telemedicine in India. Telemedicine and e-Health, 15(6), 586–591.
Bhadra, S. & Kumar, C. J. (2023). Enhancing the efficacy of depression detection system using optimal feature selection from EHR. Computer Methods in Biomechanics and Biomedical Engineering, 27(2):222–236. https://doi.org/10.1080/10255842.2023.2181660.
Bhadra, S.; Kumar, C.J.; Bhattacharyya, D.K. (2024). Multiview EEG signal analysis for diagnosis of schizophrenia: an optimized deep learning approach. Multimedia Tools and Applications.
Burcu, C., Arnrich, B., Roberto, L. M., & Gerhard, T. (2011). Monitoring of mental workload levels during an everyday life office-work scenario. Springer-Verlag London Limited.
Czerwinski, M., Cutrell, E., & Horvitz, E. (2009). Cyberchondria: Studies of the escalation of medical concerns in web search. Microsoft Research Redmond WA, HI Publication, 75–182.
DACAR Platform for eHealth Services Cloud, L. Fan, W., Buchanan, C., & Thümmler, C. (2011). IEEE, computer society of India, pp. 219–226. https://doi.org/10.1109/CLOUD.2011.31.
Davenport, T., & Kalakota, R. (2019). The potential for artificial intelligence in healthcare. Future Healthcare Journal, 6(2), 94–98. https://doi.org/10.7861/futurehosp.6-2-94.
Das, P. R., Talukdar, R. R., & Kumar, C. J. (2024). Exploring the interplay of abstinence self-efficacy, locus of control, and perceived social support in substance use disorder recovery. Current Medical Research and Opinion, 40(9), 1625–1635. https://doi.org/10.1080/03007995.2024.2390046.
Ekeland, A. G., Bowes, A., & Flottorp, S. (2012). Methodologies for assessing telemedicine: A systematic review of reviews. Journal of Medical Informatics, Elsevier, 81(1), 1–11.
Garg, R., Aggarwal, K., & Arora, A. (2023). Applications of augmented reality in medical training. In Mathematical Modeling, Computational Intelligence Techniques and Renewable Energy: Proceedings of the Third International Conference, MMCITRE 2022 (pp. 215–228). Singapore: Springer Nature Singapore.
Hair, J. F., Blake, W. C., Babin, B. J., Anderson, R. E., & Tatham, R. L. (2009). Multivariate Data Analysis, Eighth Edition, Pearson Education Inc.
Jens H. Weber-Jahnke & Christina Obry. (2011). Protecting privacy during peer-to-peer exchange of medical documents. Published online: 12 April 2011, Springer Science+Business Media, LLC.

Lituchy, T. R., & Rail, A. (2000). Bed and breakfasts, small inns, and the Internet: The impact of technology on the globalization of small businesses. Journal of International Marketing, 8(2), 86–97.

McLaughlin, M., Nam, Y., Gould, J., Pade, C., Kathleen, A., Meeske, K., Ruccione, S., & Janet F. (2012). A video-sharing social networking intervention for young adult cancer survivors. Journal of Computers in Human Behavior, 28(2), 631–641.

Nores, M. L., & José J. P. A. (2011). Property-based collaborative filtering for health-aware recommender systems. Computers & Education, 39(8), 7451–7457.

Pande, A., Ramanujam, R., & Ramaswamy, S. (2012). Distributed processing and Internet technology to solve challenges of primary healthcare in India. Springer Berlin Heidelberg publication, 7154, 188–199.

Pradeep, Paul, G., Raman, R., Padmajakumari, R. K., Deshmukh, H., & Sharma, Tarun. (2006). Patient satisfaction levels during teleophthalmology consultation in rural South India. Telemedicine and e-Health, 12(5), 571–578.

Reddy, S., Fox, J., & Purohit, M. P. (2018). Artificial intelligence-enabled healthcare delivery. Journal of the Royal Society of Medicine, 112(1), 22-28. https://doi.org/10.1177/0141076818815510.

Rosalie van der Vaart, C. H. C. Drossaert, E. Taal, & M. A. F. J. van de Laar. (2012). Giving rheumatology patients online home access to their electronic medical record (EMR): advantages, drawbacks and preconditions according to care providers. Springerlink.com.

Salem, R. M., Pandey, B., Richard, E., Maple, M. F., Garcia, E. P., Brophy, V. H., Brophy, N. J., Daniel, C., O'Connor, T., & Bhatnagar, V. (2010). The VA hypertension primary care longitudinal cohort: electronic medical records in the post-genomic era. Journal of Health Informatics, 16, 274–286.

Sharma, M., Kumar, C. J., Singh, T. P., Talukdar, J., Sharma, R. K., & Ganguly, A. (2024). Enhancing disease region segmentation in rice leaves using modified deep learning architectures. Archives of Phytopathology and Plant Protection. 18(1):20220689. https://doi.org/10.1515/biol-2022-0689.

Valeri, L., Giesen, D., Jansen, P., & Klokgiieters, K. (2010). Business models for eHealth Study Report. ICT for Health Unit DG Information Society and Media European Commission, 1–42.

# Ethical Considerations in AI-Enabled Healthcare

**Vinaytosh Mishra, Yotam Lurie, and Shlomo Mark**

**Abstract** Integrating Artificial Intelligence (AI) in healthcare has revolutionized patient care and operational workflows, yet it introduces significant ethical considerations. This chapter explores the impact of AI on key ethical principles—beneficence, nonmaleficence, autonomy, and justice. AI systems, while enhancing patient outcomes and operational efficiency, must be designed to promote patient welfare, minimize harm, and ensure equitable access to care. However, algorithmic bias presents a critical challenge, often resulting from biased data and flawed algorithms, leading to disparities in diagnosis and treatment, particularly for marginalized communities. The importance of diverse training data is emphasized to mitigate bias and foster fair and inclusive AI systems. This chapter underscores the necessity of ethical frameworks in AI development and implementation, addressing privacy, security, and transparency issues to build trust and ensure responsible AI use in healthcare.

**Keywords** Ethical considerations · Explainability and trust · Algorithmic bias · Healthcare AI ethics

V. Mishra (✉)
Thumbay Institute for AI in Healthcare, Gulf Medical University, Ajman, UAE
e-mail: dr.vinaytosh@gmu.ac.ae

Y. Lurie
Ben-Gurion University of the Negev, Beersheba, Israel
e-mail: yotam@bgu.ac.il

S. Mark
Shamoon College of Engineering, Ashdod, Israel
e-mail: marks@sce.ac.il

 
T. P. Singh et al. (eds.), *Revolutionizing Healthcare: Impact of Artificial Intelligence on Diagnosis, Treatment, and Patient Care*, Studies in Computational Intelligence 1182,
https://doi.org/10.1007/978-3-031-80813-5_18

# 1 Introduction

AI has significantly revolutionized the healthcare sector. The practice of applying AI to healthcare is currently manifested by manifold application implementation in various domains (Chen and Decary 2020). Currently, the use of AI can be seen in the diagnosis and prediction of diseases, personalization of treatments, analysis of medical images, drug discovery, decision support systems, monitoring of the state of patients, robotics, personalized medicine, clinical trials, and organizational workflow management (Agafonov 2024; Poalelungi 2023). Hence, integrating AI technologies into healthcare is the key to enhancing patient care and improving operational workflows; this changes healthcare practices. AI has emerged as the transforming force that opens opportunities for enhanced patient outcomes, process simplification, and decreased costs across all healthcare settings. AI-driven technologies such as Generative AI, Natural Language Processing, Artificial Neural Networks, Machine Learning, Deep Learning, physical robots, and rule-based expert systems are redefining a new era for healthcare delivery. The AI healthcare market is growing exponentially and is set to grow to $6.6 billion by 2021 (Collier 2020). Even with all these advancements, some people say there will be a possible displacement of human medical professionals by AI technologies. Research has held that AI is the hope for better healthcare efficiency, patient outcomes, and transformative practice in care. The ethical issues in using AI for healthcare are increasingly gaining prominence, whereby a balance is sought between technological development and ethical, equitable, and human-centred healthcare practices. Key challenges to AI implementation in health include trust issues, actual clinical utilization, and its perception by healthcare professionals and patients.

## *1.1 Applications of Artificial Intelligence in Healthcare*

Artificial Intelligence is increasingly integrated into health care, including diagnosis, treatment planning, patient monitoring, and health administration. AI can help doctors make accurate diagnostic results, predict the course of diseases, create personalized patient treatment paths, and improve medical imaging for better analysis (Secinaro et al. 2021). It is important for disease diagnosis, treatment personalization, medical imaging, drug discovery, and clinical data management (Parija and Padmavathi 2022). AI can also substantially improve the care and safety of patients, public health, research, and healthcare professionals' daily workflow and labour in a clinical setting. AI applications have been seen in drug discovery, diagnostics, and administrative workflows to improve health operations and resource allocation (Parija and Padmavathi 2022). AI technologies also optimize hospital administrative processes, insurance collection, clinical reporting, and medical billing, including medical records management, to achieve improved healthcare operations and resource use effectiveness (Waheed 2024). Clinical decisions among nurses and associated documentation processes also benefit from AI, including those tasks that are remote from direct patient interaction (Seibert et al. 2021; O'Connor et al. 2022). In addition, AI tools in behavioural health care may improve services using process efficiency, administrative automation, and objective data collection on patient progress (Bhadra et al. 2024).

Artificial Intelligence in healthcare has transformed the industry, improving diagnostic accuracy and personalization of treatments, patient monitoring, and administrative efficiency. Moreover, AI technologies have modernized existing healthcare practices towards better patient care, optimally increasing workflow efficiencies and enabling innovation in health delivery. In this article, healthcare and medicine have been used interchangeably, but both concepts have distinct differences. Healthcare broadly encompasses the overall system and services provided to maintain or improve health, including preventive care, treatment, and managing chronic conditions. On the other hand, medicine typically refers to diagnosing, treating, and preventing disease and is often focused on the clinical aspects of care. AI adoption in healthcare is not only limited to medicine but has touched various aspects of service delivery, bringing efficiency to healthcare systems.

## *1.2 Benefits of AI in Healthcare Services*

Integrating artificial Intelligence (AI) into health can actualize efficiency and promptness in various systems. According to Mishra and Islam (2021), AI technologies will further enhance healthcare by streamlining workflows, enabling personalized approaches to treatment, increasing the degree of diagnosis, and improving overall patient outcomes. In these ways, human intervention should be reduced, productivity should be increased through automation, and real-time data analytics and insights should be provided in large measure to improve efficiency in healthcare with the technologies of AI (Brey and Dainow 2023). The application of AI in healthcare can bring about a healthcare delivery paradigm in which expenses decrease, efficiency rises, and patient outcomes improve. AI techniques generally enhance work efficiency, automate processes, reduce administrative burdens, and improve the quality of the services delivered within healthcare settings (Navath 2023). According to Lee and Yoon (2021), AI in healthcare can further predict, diagnose, and treat diseases, enhancing efficiency and reducing overcrowding in healthcare resource utilization. The amalgamation of advanced machine learning algorithms with big data analytics and modern healthcare technologies promises an increased potential for improved patient outcomes, more efficient healthcare delivery, and superior resource allocation in the future (Wu et al. 2023). AI, in return, with the speedy developments under it and related technologies, will become a source of new value to patients and improvement in operational processes in healthcare. AI, through the automation of digital health systems, significantly impacts diagnoses, treatment, and management of illnesses and supports claims and payments. Artificial Intelligence in healthcare is pitched to improve the quality of healthcare services and address the dimensions of healthcare

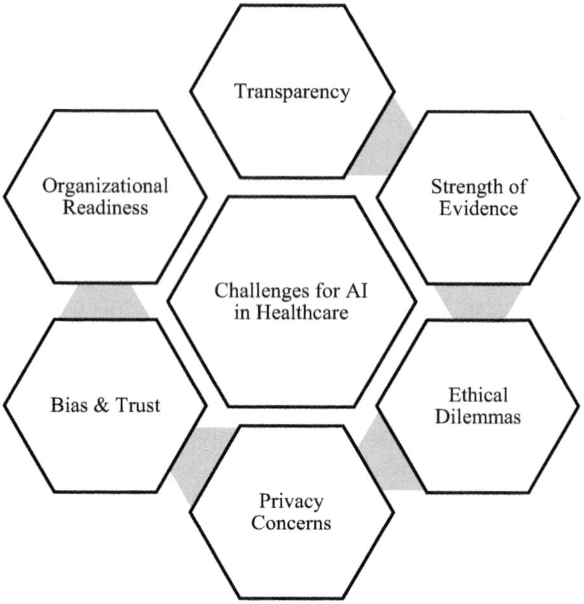


**Fig. 1** Challenges for AI integration in healthcare. *Source* Author's compilation

quality defined globally. By integrating AI techniques, health institutions improve electronic care service with a better management strategy, benefiting improved patient outcomes. Integration of AI in healthcare introduces several complex aspects: transparency, strength of evidence, ethical dilemmas, privacy concerns, bias, trust, and organizational readiness, Fig. 1. Such challenges require thorough comprehension so that their implementation is feasible and associated issues can be competently addressed.

## 2 Ethical Framework for Artificial Intelligence

In the governance of AI in healthcare, an ethical framework is of paramount importance. Ethics forms the cornerstone of the proper development and implementation of AI technology in healthcare setups. It provides guidance and principles in handling complexities brought about by AI technology, including concerns about transparency, fairness, accountability, and privacy (Hindocha and Badea 2021; Akinrinola 2024; Ijiga 2024). It is a compass that directs the decision-making process so that the application of AI abides by the set ethical standards and values. Ethical frameworks must address moral dilemmas and ensure transparency and fairness in developing, using, or deploying AI technologies (Dhawan 2024; Elendu 2023). These frameworks help a healthcare organization address associated risks, build trust, and alleviate bias in the AI system under concrete and ethical guidelines.

The ethical considerations related to AI applications in healthcare have attracted considerable debate. International organizations such as the World Health Organization (WHO), the Organization for Economic Cooperation and Development (OECD), and the European Union (EU) have come up with guidelines related to the ethical use of AI in healthcare (Van Noordt and Misuraca 2022). These entities have worked on regulations underlining ethical values, including fairness, accountability, transparency, and privacy. The adherence to these values and principles across many of these regulations highlights the importance of ethical considerations in using AI in healthcare.

The central features of healthcare practice are that the following ethical principles are most recognized: respect for autonomy, beneficence, nonmaleficence, and justice. Put the ethical theories and principles at the core and understand the values needed for complete ethical consideration (Manderius et al. 2023). Ethical principles in applying AI for mortality prediction in the emergency department are five, including autonomy, beneficence, nonmaleficence, justice, explicability, and professional governance. The ethical theories these principles relate to cover virtue, deontology, and consequentialism. Healthcare ethics include four bioethical principles: respect for autonomy, nonmaleficence, beneficence, and justice. Beauchamp and Childress summarized these principles decades ago, forming the foundation of healthcare ethics (Newnham and Kirkham 2019). In this regard, to structure the design, development, implementation, and assessment of drones used in public healthcare within an ethical framework around the four bioethics principles—beneficence, nonmaleficence, autonomy, and justice—and a fifth principle coming from artificial intelligence ethics, explicability (Cawthorne and Wynsberghe 2020), an analytical overview shall be provided. The description of these five ethical principles is depicted in Table 1.

## 3 Impact of AI Technologies on Ethical Principles

The effect of AI on beneficence and nonmaleficence in healthcare is considered in a discourse setting for ethical values governing the development and use of AI technologies to serve the patient's best interest while minimizing harm. Balasubramanian (2023) thus focuses on the point that AI systems should work to augment the well-being of patients by reducing physical and psychological harm. The ethical considerations of AI-augmented decision-making in end-of-life care underline that AI systems must contribute positively to the benefits of patients while avoiding harm. According to Wang et al. (2021), autonomy has a great positive impact on the intention to use by healthcare practitioners; therefore, it plays a determining role, acting like a signal of a different kind, leading to the decision to be involved in workplace AI Therefore, concerning AI-enabled health, autonomy could be characterized as the ability of individuals, including patients and health professionals, to make decisions with proper consent and under their control concerning their health.

**Table 1** Ethical principles for AI in healthcare

| Ethical principle | Description | Key considerations |
|---|---|---|
| Beneficence | The principle of doing good and acting in ways that promote the welfare of others | – Actively contributing to the well-being of patients and communities<br>– Providing benefits while balancing against risks<br>– Ensuring beneficial interventions |
| Nonmaleficence | The obligation to avoid causing harm | – Avoiding unnecessary harm<br>– Preventing harm where possible<br>– Weighing potential harms against benefits |
| Autonomy | The right of patients to make informed decisions about their health and lives | – Respecting patient decisions and choices<br>– Providing all necessary information for informed consent<br>– Supporting patient independence and self-governance |
| Justice | Fair and equitable treatment, ensuring that benefits and burdens are distributed fairly | – Ensuring fair access to healthcare resources<br>– Avoiding discrimination<br>– Promoting social justice and addressing inequalities |
| Explicability | The principle of ensuring that decisions and processes are transparent, understandable, and justifiable | – Providing clear explanations of AI and healthcare decisions<br>– Ensuring transparency in data usage and decision-making processes<br>– Making AI systems understandable to users and stakeholders |

In the context of providing healthcare justice and equitability, Gurevich et al. (2022) state that AI can advance health equity through access to care, reduced disparities, and increased fairness in the delivery of healthcare services. AI technologies may be especially important in reducing healthcare disparities and ensuring all citizens have equal access to excellent care. Thus, AI technologies could impact beneficence and nonmaleficence by the increased welfare of the patients and minimization of injury; their impact on autonomy is seen in higher individual power exercised over making healthcare decisions, and they could play a role in justice and fairness in eliminating disparities in healthcare. The impact of AI technologies on ethical principles is depicted in Table 2.

**Table 2** Impact of AI technologies on ethical principles

| Ethical principle | Impact of AI technologies | Source |
|---|---|---|
| Beneficence | AI systems should augment the well-being of patients by reducing physical and psychological harms, ensuring positive contributions to patient benefits | Balasubramanian (2023) |
| Nonmaleficence | AI technologies must minimize patient harm, focusing on reducing potential physical and psychological harm | Balasubramanian (2023) |
| Autonomy | AI enables healthcare practitioners and patients to make informed decisions with proper consent and under their control, enhancing individual power in healthcare decisions | Wang et al. (2021) |
| Justice and Equity | Artificial Intelligence can contribute to equity in healthcare by improving accessibility, reducing disparities, and ensuring fairness in healthcare | Gurevich et al. (2022) |

*Source* Author's compilation

## 4 Privacy, Security and Algorithmic Bias

Patient consent and their right to protected health information (PHI) are essential concepts in medical ethics. Frameworks such as the General Data Protection Regulation (GDPR) from the European Union and the Health Insurance Portability and Accountability Act (HIPAA) from the US have made many provisions to ensure that. The ethical considerations of data privacy and security extend beyond simple adherence to these regulations. It includes a commitment to protecting patient data from breaches and illegal access, crucial for preserving trust in healthcare systems (Ahammed 2024). With increasing digitalization, it is imperative to implement stringent protocols to safeguard PHI. HIPAA Journal summarises data privacy and security breaches under various categories in Fig. 2 (Joe 2024; Mishra et al. 2024).

The ethical considerations in using AI in healthcare are driven by ensuring patient's rights, promoting their well-being, and safeguarding their privacy and security. Healthcare organizations must abide by the ethics and regulations related to protected health information. AI technologies should ensure that activities are rooted in beneficence, which is about doing good (Petersson et al. 2022; Formosa et al. 2021).In healthcare, Artificial Intelligence can be used to ensure that it is more personalized and better performing for the interests and needs of patients regarding set ethical standards. Nonmaleficence principles focus on avoiding harm, which includes mitigating any risks linked with using technologies, such as data breaches and unauthorized access to protected health information (Jobin et al. 2019). Data security, privacy, and moral artificial intelligence development are important safeguards for nonmaleficence or causing no harm towards patients or the healthcare system. As far as autonomy is concerned, the system will respect the rights of a person to make decisions about their health in an AI-enabled healthcare system (Bruxvoort and Keulen 2021). This entails effective communication with patients, encompassing

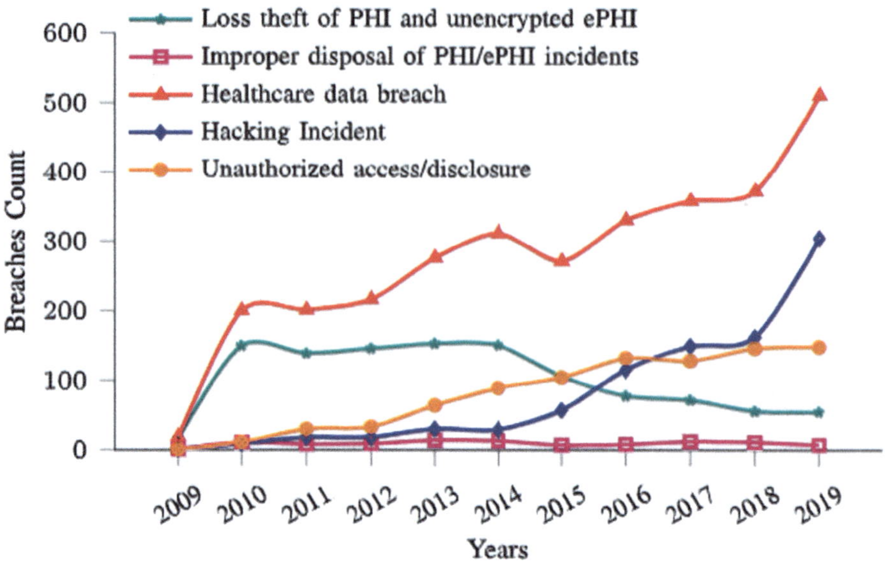


**Fig. 2** Reported data privacy issues in the USA in last decade. *Source* Mishra et al. (2024)

obtaining informed consent and enhancing individuals' ability to engage in health-related decision-making processes. The notion of justice, encompassing fairness and equity, will be essential for ensuring that AI technologies are deployed to facilitate equal access to quality healthcare services for all societal members (Wang et al. 2021).

Algorithmic bias in healthcare denotes the systematic and unjust discrimination that may occur in AI systems due to biased data or defective algorithms, leading to erroneous or inequitable results. The underrepresentation of specific minority groups in medical databases may result in erroneous diagnoses or treatment recommendations for those populations (Thompson et al. 2021; Zhang and Zhang 2023; Seyyed-Kalantari et al. 2021). Furthermore, algorithmic bias can profoundly impact healthcare outcomes among various communities, leading to discrepancies in diagnostic and treatment recommendations. Algorithmic bias can exacerbate health disparities and result in inequitable healthcare outcomes for vulnerable populations (Cerrato et al. 2022). Ensuring diversity in training data is essential for reducing algorithmic bias and promoting equitable and inclusive AI systems. Incorporating diverse and representative datasets enables healthcare professionals to mitigate bias and improve the accuracy and efficacy of AI algorithms (Mittermaier et al. 2023; Nazer 2023). Studies suggest that deep learning algorithms for CT lung nodule identification exhibited differences in performance affected by patient demographics, implying that biases in training data can directly influence clinical outcomes (Sourlos et al. 2022). An illustrative scenario involves an algorithm that predicts patient care requirements based on healthcare expenses rather than health status. This results in the erroneous

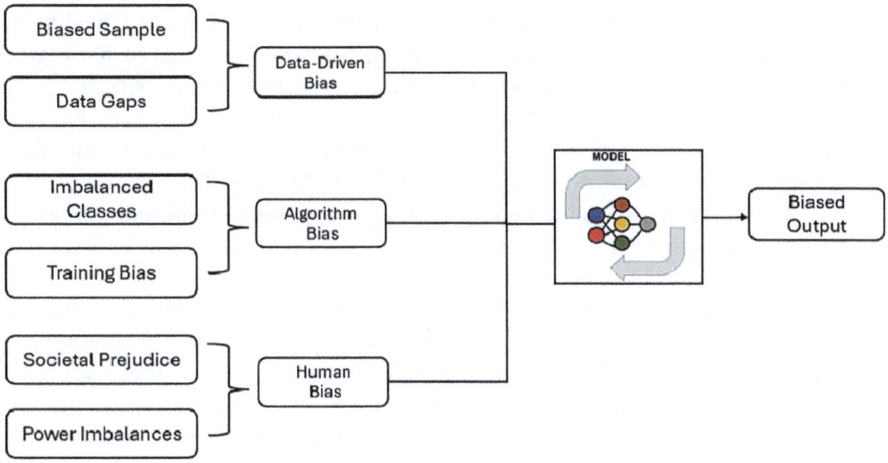


**Fig. 3** Types of bias in AI-enabled healthcare. *Source* Author's compilation

conclusion that Black patients exhibit superior health due to reduced expenditure on their care (Zhang and Zhang 2023).

For example, the use of AI in medical image analysis is affected by continuous racial inequities as it frequently mirrors previous prejudices in healthcare access and treatment (Aquino et al. 2023). The world has witnessed increased adoption of digital technologies in healthcare in post-pandemic settings. For example, healthcare systems adopted Generative AI rapidly after the launch of ChatGPT in December 2022. These models are trained on humongous data from certain geography or sections of society subject to representation bias. These issues necessitate careful management to ensure fair results and maintain confidence in AI systems. The bias related to the use of artificial Intelligence can fall under three categories, namely (1) data-driven bias, (2) algorithmic bias, and (3) human bias, as shown in Fig. 3. Addressing these ethical dilemmas is crucial for the responsible deployment of AI technology in healthcare settings (Tilala 2024; Anyanwu 2024).

## 5 Conclusion

Artificial Intelligence in healthcare has great potential in providing superlative care, improving operational efficiencies, and making healthcare affordable. Although AI has great potential in healthcare, its implementation involves substantial ethical dilemmas that must be addressed to ensure responsible and fair implementation of AI in healthcare. This book chapter analyses fundamental ethical principles such as beneficence, nonmaleficence, autonomy, and justice in the context of the implementation of AI in healthcare. The article highlights the need for a strong ethical framework to guide the implementation of AI in healthcare.

This article has three implications for the practice. First and foremost, healthcare practitioners and AI developers must work together to guarantee that technological products are developed and put into practice in a way that prioritizes patients' well-being, reduces potential harm, and upholds patient autonomy. Secondly, diversity in the training data is required so that every population is well-represented and algorithmic bias can be minimized. Finally, to build confidence in these AI-augmented systems, it is necessary to ensure that technology is explainable and transparent.

Similarly, this chapter provides four implications for future research directions. There is a need to develop a comprehensive ethical framework for implementing AI in healthcare which can address challenges in AI-enabled healthcare. Secondly, it is imperative to engage in interdisciplinary research to comprehensively comprehend and alleviate the effects of AI on different populations, with a specific focus on underprivileged communities, intending to foster health equity. Third, future studies can help develop stringent regulations that impose ethical norms in developing and implementing Artificial Intelligence (AI) for further use in healthcare systems. Finally, future studies can examine the role of education in implementing ethical AI in healthcare. Providing ongoing education and training to healthcare workers regarding AI technologies and their ethical implications is crucial to safely incorporating AI into healthcare procedures. Thus, this article concludes that although AI has transformative potential, its ethical implementation necessitates thoughtful deliberation and proactive actions to guarantee that it benefits all patients, thereby fostering a fairer and more efficient healthcare system.

## References


Agafonov, O. (2024). Editorial: Trustworthy AI for healthcare. Frontiers in Digital Health, 6. https://doi.org/10.3389/fdgth.2024.1427233.

Ahammed, M. (2024). Privacy-preserving data sharing in healthcare: advances in secure multiparty computation. Journal of Medical and Health Studies, 5(2), 37–47. https://doi.org/10.32996/jmhs.2024.5.2.4.

Anyanwu, E. (2024). Artificial Intelligence in healthcare: a review of ethical dilemmas and practical applications. International Medical Science Research Journal, 4(2), 126–140. https://doi.org/10.51594/imsrj.v4i2.755.

Aquino, Y., Carter, S., Houssami, N., Braunack-Mayer, A., Win, K., Degeling, C., … & Rogers, W. (2023). Practical, epistemic and normative implications of algorithmic bias in healthcare artificial Intelligence: a qualitative study of multidisciplinary expert perspectives. Journal of Medical Ethics, jme-2022–108850. https://doi.org/10.1136/jme-2022-108850.

Akinrinola, O. (2024). Navigating and reviewing ethical dilemmas in AI development: strategies for transparency, fairness, and accountability. GSC Advanced Research and Reviews, 18(3), 050–058. https://doi.org/10.30574/gscarr.2024.18.3.0088.

Balasubramanian, E. (2023). Ethical considerations in ai-assisted decision-making for end-of-life care in healthcare. Power System Technology, 47(4), 167–182. https://doi.org/10.52783/pst.168.

Bhadra, S., Kumar, C.J. & Bhattacharyya, D.K. (2024). Multiview EEG signal analysis for diagnosis of schizophrenia: an optimized deep learning approach. Multimed Tools Appl. https://doi.org/10.1007/s11042-024-20205-y.

Brey, P., & Dainow, B. (2023). Ethics by design for artificial Intelligence. AI and Ethics, 1–13.

Cawthorne, D., & Robbins-van Wynsberghe, A. (2020). An ethical framework for the design, development, implementation, and assessment of drones used in public healthcare. Science and Engineering Ethics, 26(5), 2867–2891.

Cerrato, P., Halamka, J., & Pencina, M. (2022). A proposal for developing a platform that evaluates algorithmic equity and accuracy. BMJ Health & Care Informatics, 29(1), e100423. https://doi.org/10.1136/bmjhci-2021-100423.

Chen, M., & Decary, M. (2020, January). Artificial Intelligence in Healthcare: An essential guide for health leaders. In Healthcare Management Forum (Vol. 33, No. 1, pp. 10–18). Sage CA: Los Angeles, CA: SAGE Publications.

Collier, M. (2020). AI: An engine for growth. Accenture. https://www.accenture.com/au-en/insights/health/artificial-intelligence-healthcare.

Dhawan, S. (2024). Ethical implications of AI in healthcare. International Journal of Scientific Research in Engineering and Management, 08(03), 1–13. https://doi.org/10.55041/ijsrem29006.

Elendu, C. (2023). Ethical implications of AI and robotics in healthcare: a review. Medicine, 102(50), e36671. https://doi.org/10.1097/md.0000000000036671.

Formosa, P., Wilson, M., & Richards, D. (2021). A Principlist Framework for Cybersecurity Ethics. Computers & Security, 109, 102382.

Gurevich, E., Hassan, B., & Morr, C. (2022). Equity within AI systems: what can health leaders expect? Healthcare Management Forum, 36(2), 119–124. https://doi.org/10.1177/08404704221125368.

Hindocha, S. and Badea, C. (2021). Moral exemplars for the virtuous machine: the clinician's role in ethical artificial Intelligence for healthcare. AI and Ethics, 2(1), 167–175. https://doi.org/10.1007/s43681-021-00089-6.

Ijiga, A. (2024). Ethical considerations in implementing generative AI for healthcare supply chain optimization: a cross-country analysis across India, the United Kingdom, and the United States of America. International Journal of Biological and Pharmaceutical Sciences Archive, 7(1), 048–063. https://doi.org/10.53771/ijbpsa.2024.7.1.0015.

Jobin, A., Ienca, M., & Vayena, E. (2019). The global landscape of AI ethics guidelines. Nature Machine Intelligence, 1(9), 389–399.

Joe Robinson. US Healthcare Data Breach Statistics. Accessed on August 26, 2024. Available from: https://www.privacyaffairs.com/healthcare-data-breach-statistics/.

Lee, D., & Yoon, S. N. (2021). Application of artificial intelligence-based technologies in the healthcare industry: Opportunities and challenges. International journal of environmental research and public health, 18(1), 271.

Manderius, C., Clintståhl, K., Sjöström, K., & Örmon, K. (2023). The psychiatric mental health nurse's ethical considerations regarding the use of coercive measures–a qualitative interview study. BMC nursing, 22(1), 23.

Mishra, V., & Islam, S. M. S. (2021). Framework for the Adoption of Healthcare 4.0–An ISM Approach. In Disruptive Technologies for Society 5.0 (pp. 235–247). CRC Press.

Mishra, V., Gupta, K., Saxena, D., & Singh, A. K. (2024). A Global Medical Data Security and Privacy-Preserving Standards Identification Framework for Electronic Healthcare Consumers. IEEE Transactions on Consumer Electronics, 70(1): 4379–4387.

Mittermaier, M., Raza, M., & Kvedar, J. (2023). Bias in AI-based models for medical applications: challenges and mitigation strategies. NPJ Digital Medicine, 6(1). https://doi.org/10.1038/s41746-023-00858-z.

Navath, S. (2023). Transforming Healthcare: the Impact and Future of Artificial Intelligence in Healthcare. JAIM, 1(1), 16–21. https://doi.org/10.55124/jaim.v1i1.234.

Nazer, L. (2023). Bias in artificial intelligence algorithms and recommendations for mitigation. Plos Digital Health, 2(6), e0000278. https://doi.org/10.1371/journal.pdig.0000278.

Newnham, E., & Kirkham, M. (2019). Beyond autonomy: Care ethics for midwifery and the humanization of birth. Nursing Ethics, 26(7–8), 2147–2157.

O'Connor, S., Yan, Y., Thilo, F., Felzmann, H., Dowding, D., & Lee, J. (2022). Artificial Intelligence in nursing and midwifery: a systematic review. Journal of Clinical Nursing, 32(13-14), 2951–2968. https://doi.org/10.1111/jocn.16478.
Parija, S. and Padmavathi, S. (2022). Artificial Intelligence in health care. Annals of SBV, 10(2), 23–23. https://doi.org/10.5005/jp-journals-10085-9112.
Petersson, L., Larsson, I., Nygren, J., Nilsen, P., Neher, M., Reed, J., ... & Svedberg, P. (2022). Challenges to implementing artificial Intelligence in healthcare: a qualitative interview study with healthcare leaders in Sweden. BMC Health Services Research, 22(1). https://doi.org/10.1186/s12913-022-08215-8.
Poalelungi, D. (2023). Advancing patient care: how artificial Intelligence is transforming healthcare. Journal of Personalized Medicine, 13(8), 1214. https://doi.org/10.3390/jpm13081214.
Secinaro, S., Calandra, D., Secinaro, A., Muthurangu, V., & Biancone, P. (2021). The Role of Artificial Intelligence in Healthcare: a Structured Literature Review. BMC Medical Informatics and Decision Making, 21(1). https://doi.org/10.1186/s12911-021-01488-9.
Seibert, K., Domhoff, D., Bruch, D., Schulte-Althoff, M., Fürstenau, D., Bießmann, F., ... & Wolf-Ostermann, K. (2021). Application scenarios for artificial Intelligence in nursing care: rapid review. Journal of Medical Internet Research, 23(11), e26522. https://doi.org/10.2196/26522.
Seyyed-Kalantari, L., Zhang, H., McDermott, M., Chen, I., & Ghassemi, M. (2021). Underdiagnosis bias of artificial intelligence algorithms applied to chest radiographs in under-served patient populations. Nature Medicine, 27(12), 2176–2182. https://doi.org/10.1038/s41591-021-01595-0.
Sourlos, N., Wang, J., Nagaraj, Y., Ooijen, P., & Vliegenthart, R. (2022). Possible bias in supervised deep learning algorithms for CT lung nodule detection and classification. Cancers, 14(16), 3867. https://doi.org/10.3390/cancers14163867.
Tilala, M. (2024). Ethical Considerations in the use of Artificial Intelligence and Machine Learning in health care: a comprehensive review. Cureus. https://doi.org/10.7759/cureus.62443.
Thompson, H., Sharma, B., Bhalla, S., Boley, R., McCluskey, C., Dligach, D., ... & Afshar, M. (2021). Bias and Fairness Assessment of a Natural Language Processing Opioid Misuse Classifier: Detection and Mitigation of Electronic Health Record Data Disadvantages Across Racial Subgroups. Journal of the American Medical Informatics Association, 28(11), 2393–2403. https://doi.org/10.1093/jamia/ocab148.
van Bruxvoort, X., & van Keulen, M. (2021). Framework for assessing ethical aspects of algorithms and their encompassing socio-technical system. Applied Sciences, 11(23), 11187.
Van Noordt, C., & Misuraca, G. (2022). Artificial Intelligence for the public sector: results of landscaping the use of AI in government across the European Union. Government Information Quarterly, 39(3), 101714.
Waheed, M. (2024). Perceptions of family physicians about applying AI in primary health care: a case study from a premier health care organization. Jmir Ai, 3, e40781. https://doi.org/10.2196/40781.
Wang, W., Chen, L., Xiong, M., & Wang, Y. (2021). Accelerating AI adoption with responsible AI signals and employee engagement mechanisms in health care. Information Systems Frontiers, 25(6), 2239–2256. https://doi.org/10.1007/s10796-021-10154-4.
Wu, H., Lu, X., & Wang, H. (2023). The application of artificial Intelligence in health care resource allocation before and during the COVID-19 pandemic: scoping review. Jmir Ai, 2(1), e38397.
Zhang, J. and Zhang, Z. (2023). Ethics and governance of trustworthy medical artificial Intelligence. BMC Medical Informatics and Decision Making, 23(1). https://doi.org/10.1186/s12911-023-02103-9.

# Overcoming Challenges in the Integration of AI in Healthcare

**Ubrurhe Ogheneochuko, Okpu Okpomo Eterigho, and Eluemunor Kizito**

**Abstract** AI continues to make inroads into the healthcare sector, as public and private-sector actors grapple with challenges that are inhibiting wide preadoption of this disruptive technology, which stands poised to level up patient outcomes, boost care quality and shrink health service delivery costs. It has a huge potential to change the dynamics of how patients get treated, diagnosed and hospital operation in healthcare sector by making them faster accurate and cost-effective. By understanding the barriers and possibilities for AI in healthcare, regulators can support patient health also promoting truly revolutionary changes within an industry. One noticeable problem with using AI is that it requires a high-quality and widely varied dataset to train from. For patient data to be able to move safely and ethically, it will need robust frameworks for managing the different lines of accountability (data governance), agreements about how data should or can be shared between institutions (data-sharing agreements) and sets of standards that describe what is meant by a "record" when medical records are transferred as one unit across any arbitrary pair of systems in semantic interoperability. The other major issue is the public's distrust of AI algorithms. Accountability, interpretability and bias in the AI algorithms driving decision-making have also been raised by patients and healthcare professionals. In this chapter, we investigate different downsides in integrating AI into the medical domain generalizing their possible usage and issues while covering fundamental success factors for deploying it. It includes successful integration case studies, and a forward-looking conversation on the future of AI along with its challenges. An outcome may be fullest realization of AI technologies will also address disparities related to AI in healthcare, preparing for a future where all people have

U. Ogheneochuko (✉) · O. O. Eterigho
Delta State University of Science and Technology, Ozoro, Nigeria
e-mail: ubrurheo@dsust.edu.ng

O. O. Eterigho
e-mail: okpuoe@dsust.edu.ng

E. Kizito
Department of Information Systems and Technology, Delta State University of Science and Technology, Ozoro, Nigeria
e-mail: anaziake@dsust.edu.ng



T. P. Singh et al. (eds.), *Revolutionizing Healthcare: Impact of Artificial Intelligence on Diagnosis, Treatment, and Patient Care*, Studies in Computational Intelligence 1182,
https://doi.org/10.1007/978-3-031-80813-5_19

access to equitable and patient-centered care involving these tools. The chapter concludes with recommendations for ongoing research in AI integration, highlighting the significance of responsible advancement and the use of AI for the benefit of the society.

**Keywords** Artificial intelligent · Healthcare · Machine language · Patient

## 1 Introduction

In 1956, Artificial Intelligence (AI) was first initiated, where algorithm was developed by examining existing data and self-learning (Lushan et al. 2021) and (Hamet and Tremblay 2017). Subsequent years of improvement, AI has progressively make change in health services across industrialized markets, and has the prospect to drive game-changing developments for underserved societies in global health. Through assisting health givers to better serve patients in remote areas to supporting governments avert deadly disease epidemics before they arise (USAID 2019). There is increasing acceptance of the huge prospect of AI tools to break core trade-offs in health access such as quality and cost. However, AI usually denotes to a computerized system such as hardware or software, equipped with the capability to execute tasks or reasoning processes that we regularly linked with the intelligence level of humans (Union 2018). AI is not viewed as a single form of technology but comprises of diverse sorts within numerous application sections which includes inpatient and outpatient care, pharmed and MedTech, and care services (Lena et al. 2019; USAID 2019). Cases of AI been applied to supplement patient care comprises of systems that provide treatment based on prescription, codes for insurance billing and application that reads vital signs to monitor patients in emergence unit (Micheal et al. 2019). These AI systems include machine-learning applications example pattern recognition, natural language processing, and expert systems (WHO 2024).

The prospect of AI is firmly connected to the accessibility of relevant data which is fairly expensive to gather, for instance in the case of longitudinal studies and clinical trials and, there is abundant of data in the field of healthcare hence, it tends to be closely guarded once it is collected (Dolores 2017). Sadly, the obtainability and, the quality of the data remain a key challenge. Besides, the gathering and allocation of this data makes it clumsy compared to other forms of data, this is as a result of privacy issues. Similarly, patients and healthcare givers may be unwilling to trust AI-driven decision-making because of worries about interpretability, and bias in the algorithms. Further, the lack of data governance frameworks, ethical flow of data and environmental data in current systems leaves a main set of variables out of information flow for individual well-being (Das et al. 2024).

Artificial Intelligent can provide a sole opportunity to enhance and increase the capabilities of strained healthcare providers. Thus, the introduction of AI for medical care, as in several other fields of the society, could have a major negative impact on the medical care personnel. It can lead to reduction in the number of healthcare personnel or diminish the skills of medical care providers, and compel them to reorient to acclimatise to the utilisation of AI. Years of working experience in healthcare sector are based on association between healthcare professional and patient, and certain care have to be considered when presenting AI technologies so that they do not interrupt such associations. Patient rights which was included in the Universal Declaration of Human Right (UDHR) for example dignity, confidentiality and informed consent, may radically be reviewed or compromised as natural language processing technologies takes hold and evolve. The efficiency of natural language processing relies among other parameters such as the type and volume of information, and the circumstances under which such information were collected.

The integration of Artificial Intelligence (AI) into healthcare is transforming the industry by enhancing patient care, diagnosis, treatment, and operational efficiency. However, several challenges impede the widespread adoption of AI technologies. This chapter aims to analyze these challenges, explore potential solutions, and present successful case studies to provide a comprehensive understanding of AI integration in healthcare.

## *1.1 Artificial Intelligent*

Artificial intelligent sometimes called machine language was founded on the knowledge that computers could be utilised to simulate human intelligence. Initial AI technologies were usually instructions based systems whereby a system is programmed, centred on expert knowledge, and based on algorithms which are decoded into machine code that conveys commands for swift analysis and conversion of data into outputs. It is an aspect of computing with the support of digital technology that highlights the development of robot that function and respond like human beings (Reagan 2018). The term is regularly used to the project of developing systems through data been collected in large amount, and the system train itself by utilising the data to improve accuracy of its forecasts. Increased accessibility of big data sets and processing power has allowed current machine knowledge growth for example natural language processing, and computer vision.

Researchers use several methods to train machine learning algorithms, including:

Machine learning also known as natural language processing has been used to train this learning algorithms, which includes supervised learning, unsupervised learning and reinforced studying, deep learning (WHO 2024; Mobeen et al. 2021). In supervised learning, in which the trainer knows the correct dataset used to train the model, and the two common types are classification and regression. Unsupervised learning does not involve known data, which we wish to examine and find pattern within, and the significant examples includes dimension reduction and clustering. Reinforcement learning, in which an agent such as a robot or controller seeks to study the ideal actions to take based on the state of the world, and hence the drawbacks of past actions (Aaron and David 2023). Example of enforcement learning is similar to offering and withholding treats when training a dog a fresh manoeuvre. Deep Learning is current the principal model for machine learning and its extensive applications (Sharma et al. 2024a, b). Deep Learning is a group of machine learning built on the use of multi-layer robotics as a techniques for solving a spectrum of supervised and unsupervised or semi-supervised machine learning problems. Deep learning usually needs huge quantities of data to be fed into the robot.

However, Mobeen et al. (2021), further summarize deep learning into Artificial Network (ANN), and ANN further divided into Convolutional Neutral Network (CNN), Recurrent Neutral Network (RNN) and Multilayer Perception (MP) as presented in Fig. 1. The ANN is a method stimulated by the human brain which resolves the problems that would have been very difficult by statistical standard. CNN are developed in several array and is frequently used in signal processing among others (Nina and Brian 2020). For complex signaling procedures in which the input data have a massive number of input attributes and enormously abstract linkages, the use of CNN could avoid the stress of attribute chosen by directly introducing the input data into the system (Yankang et al. 2018). The RNN, represent another form of structural design in deep learning with the key goal of handling sequential data. However, in RNN each hidden layer with targeted cycles could be unfolded and processed as a conventional neural networks allocating equal weight in every equivalent layer. The MLP came into existence due to the failure against the heterogeneity of the objects and the connections of classical network representation used for century, and this gave birth to a richer structure capable of hosting objects and connections of diverse scales known as multilayer network (Kivelä et al. 2014).

Numerous machine learning methods are data thirsty and rely on huge quantities of precise data, to generate desirable results. Huge data also known as big data are intricate data that are swiftly gathered in such extraordinary amounts such as terabytes, and petabytes or even zettabytes of storage space may be required as well as unconventional approaches for their management. The distinctive attributes of big data are defined by the following: volume, velocity, veracity and variety (Hao 2017; WHO 2024).

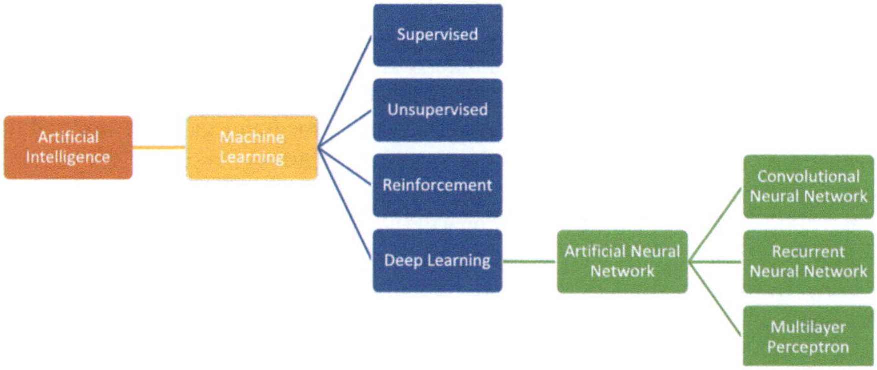


**Fig. 1** Artificial intelligent (Mobeen et al. 2021)

Integrating Artificial Intelligent in health services marks a major landmark, indicating the beginning of a new era in accurate medicine. This metamorphic change holds enormous potential for transfiguring medical care, providing improvements that were once considered innovative. Through exploiting natural language processing capabilities (Steven and Victor 2024), medical care systems stand on the edge of a blueprint swing distinguished by improved diagnostic precision, prevention, personalised treatment approach, treatment of diseases and increase competency in medical care administration (Sharma et al. 2023).

## 2 Potential Applications of AI in Healthcare

Section 2 deliberates on AI technologies designed and applied in medical field, however cases of such automations are evolving in global healthcare. Digital medical automations are already deployed in regions, including for data gathering, dissemination of health details via cell phones and increased use of electronic health records on open software platforms and cloud based computing (Brian et al. 2018). The following are potential applications of AI in Healthcare.

### *2.1 Diagnostics and Imaging*

AI systems are really good at spotting many diseases and conditions by looking at medical images like X-rays, MRIs, and CT. When it comes to healthcare, artificial intelligence (AI) has come a long way and we have seen significant improvements specifically in detecting complex diseases like cancer, neurological disorders or heart conditions. Sometimes, AI even outperforms human radiologists. Today, this technology is used in a variety of medical applications including disease tracking, patient triaging and drug discovery. For example, early detection of disease through AI powered picture analysis. Examples of this include AI systems that can analyze retinal images and pull out early signs of diabetic retinopathy, one the most common causes for blindness. Still, it has been more than 30 years since Andreas et al. (2020) published their seminal study showing that medical imaging informatics played a critical role in the implementation of clinical research projects which were instrumental to convert discoveries into viable healthcare solutions. Recently, Haridimos et al. (2023) mentioned that a hybrid method is best for managing data efficiently in AI medical imaging, by combining both federated and centralized systems. Medical image processing covers important tasks like improvement, segmentation, classification, localization, and detection to make the data easier to understand (Sistaninejhad et al. 2023). Utilizing automated image analysis, healthcare workers can breathe a little easier, because they get more time to focus on patient care instead of being bogged down with pictures.

## 2.2 Predictive Analytics

AI-driven predictive models are super helpful, because they can predict the condition of the patients and if diseases will get worse by using data from wearables, electronic health records (EHRs), and more. Predictive algorithms dive into EHRs to find out likely issues with chronic diseases like diabetes or heart problems. Using this Predictive algorithms doctors can step in earlier when needed. To do this, they use all kinds of methods like neural networks, deep learning, support vector machines, and natural language processing for both organized info and unstructured data (Jimma 2023). The insights gained can shape personalized care plans and keep people healthier. Predictive analytics also helps spot patients who may be at risk of getting certain diseases so that early action can be taken.

## 2.3 Operational Efficiency

Scheduling and managing resources in healthcare processes, can be made much better with AI. Imagine AI-powered scheduling systems, they can generate awesome appointment timetables that can help decrease the wait time, which will lead to happier patients. They investigate who needs to see a doctor, when staff are available, and even check for any limits on resources. AI is truly helpful in handling supplies, automating paperwork, and making clinical workflows smoother. All these things can improvement how well an operation runs and even save some money and time.

## 2.4 Patient Monitoring

Did you know that wearable gadgets using AI can keep track of patients' vital signs and activity levels? These nifty remote monitoring tools can watch over health parameters all the time. When something unusual happens, like a sudden change in heart rate or breathing. AI algorithms send real-time alerts to healthcare providers. According to Chen and Wu (2023), there are contactless ways to monitor patients continuously using radio frequency (RF) technology. Dubey and Tiwari (2023) note that AI makes remote patient monitoring easier in the healthcare field. It helps cut down on hospital stays and saves money as well.

## 2.5 Personalized Medicine

AI has a prodigious ability to create personalized treatment plans. By examining information such as a patient's genes, medical history, and even their lifestyle choices, it crafts recommendations just for them. AI-powered systems can suggest the best treatments, dosages of medications, and lifestyle changes by learning about each person's particular traits and risk factors. This improves treatment outcomes and lowers the possibility of negative reactions. AI-powered personalized medicine has the potential to provide more focused and effective treatment, which will eventually improve patient outcomes and save expenses.

### *2.6 Telemedicine*

Telemedicine also known as telehealth or e-health is a remote provision of healthcare services over the internet infrastructures, and this consist of a larger change from hospital to home based care, with use of AI technologies to expedite the change. They comprises of remote monitoring systems and virtual assistance to support patient care. Example of this was observed before the COVID-19 epidemic, with over 50 health care devices in the USA were using e-medicine services (Judd et al. 2020). Also, search engines such as Google and HotBot rely on algorithms for medical data as well as by the increase in quantity of text or speech application for medicare care, the performance of which has enhanced with enhancement in machine language processing.

In addition, machine language application can play extra function in the supervision of patient well-being outside traditional clinical situations such as hospitals and doctors offices, and this depend on sensors to offer patients with precise interventions according to present and preceding data collected (Sharma et al. 2024a, b).

## 3 Challenges in Integrating AI into Healthcare

We identified numerous challenges faced when integrating AI applications for healthcare and this was drawn on data from experts, stakeholders, and the scientific literature in this section. These challenges affect technology designers and health care practitioners, and may slow the acceptance of such tools. Some of the challenges in integrating AI into Healthcare systems are;

### *3.1 Data Requirements and Governance*

To train AI models effectively, we need a lot of high-quality, diverse data. There are some challenges to this.

#### 3.1.1 Availability of Data

Collecting and arranging extensive and diverse datasets in the healthcare field can be a tough task. It can be particularly challenging when dealing with rare illnesses or smaller groups of patients due to the difficulty in accessing data that comes in various formats. Furthermore, when it comes to AI in healthcare, there are strict regulations in place to protect patient privacy and security. However, these rules can sometimes limit the data available for AI advancements. Factors like human errors, data variations, regulatory challenges, and tech constraints can make it tough to successfully integrate AI into healthcare (Zemplényi et al. 2023).

#### 3.1.2 Information Management

It's really vital to a have strong systems in place for sharing data safely and legally. This means creating rules around consent, who owns the data, and how to use patient information ethically. To address the ethical and legal challenges related to collecting, storing, and using patient data in AI, we need clear governance structures. Good data governance helps build trust and reduces risks tied to using AI in healthcare.

#### 3.1.3 Interoperability

When AI is brought into healthcare systems, it brings along challenges like bias, transparency, safety, accountability, fairness, and autonomy (Pasricha 2022). In the healthcare sector, various Electronic Health Record (EHR) systems and databases are in use, but they don't always sync well. Connecting these systems can be tricky as there is a need for common standards and methods to share data. Problems with interoperability can complicate the process of gathering and utilizing information from different origins.

### *3.2 Algorithmic Bias and Fairness*

AI models can sometimes show biases found in their training data, leading to unfair or biased results. These biases might come from not having enough representation of certain patients or past inequalities in healthcare access or even just society's biases that slip into the data itself. Tackling these algorithmic biases is super important to ensure fairness in AI-based healthcare and avoid worsening health gaps while promoting equal access to care.

#### 3.2.1 Interpretability and Explainability

AI models are sometimes referred to as "black boxes" since it might be challenging to understand the reasoning behind their choices. Decisions made in the healthcare industry are vital and have a lasting effect, therefore this lack of openness is concerning. AI systems must be clear and simple to comprehend in order to foster trust and adhere to medical norms. In order to successfully incorporate AI into healthcare, concentrate on creating models that provide concise justifications.

#### 3.2.2 Regulatory Approval and Liability

AI-powered medical software needs to go through hard approval processes, which can be tricky and take time. Developers and healthcare workers often feel confused as regulatory bodies are still thinking out rules for evaluating safety and ethical issues tied to AI in health settings. Worries about liability could hold back the use of AI in making medical decisions, especially if things don't go well. By addressing these points effectively, while being aware of our responsibilities, we can better support a fairer healthcare system. The implementation of AI-driven medical decision-making in clinical settings may potentially be hampered by liability concerns, particularly in cases of unfavorable outcomes.

### 3.3 *Ethical and Legal Considerations*

Data regulation and ethical use of data raises significant concerns by differing and uncertain regulatory across regions and countries. Since AI applications require access to huge amounts of health data, stakeholders fear that this would increase the danger for anticompetitive attitude if a single player is involved. Besides, AI companies freely provide their product directly to users including health care practitioner and patients, and the users regularly assumes the product for the company monetary gain, and it raises concern around ethical and regulatory policies. The following are many difficulties in navigating the ethical and legal landscape of AI in healthcare.

#### 3.3.1 Confidentiality and Assent

Since AI systems frequently rely on enormous volumes of sensitive personal health data, protecting patient privacy is a critical problem when using AI in healthcare. Informed consent from patients and the establishment of strong data governance procedures are necessary to guarantee the moral and lawful use of personal data. This entails creating precise policies for gathering, storing, and exchanging data as well as systems that let patients limit how their information is used and withdraw their consent at any time. Following data protection laws, like the General Data Protection Regulations (GDPR) and Health Insurance Portability and Accountability Act (HIPAA), wonderful important for creating and using AI-powered medical software responsibly.

#### 3.3.2 Equity and Access

It's important that dissimilar groups, whether having varying socioeconomic backgrounds, demographics, or locations, get to enjoy the aids of AI healthcare technology similarly. There is need to tackle the digital divide so that underprivileged and marginalized folks can access healthcare services driven by AI. Addressing hurdles like computer skills, language issues, and cultural differences is key. Some helpful approaches might include outreach programs, offering financial help, or teaming up with public organizations.

#### 3.3.3 Liability

To know the legal responsibility when AI causes harm or makes errors is important. The different players, which are AI developers, healthcare workers, and institutions, should know what they are responsible for. Clear legal frameworks will help define accountability for errors or harmful results produced by AI systems. It's all about setting standards for testing, validation, and monitoring these systems to make sure they are harmless and dependable.

## 4 Essential Elements of a Successful Integration

These are some essential factors for an effortless integration of AI in healthcare.

### *4.1 Cooperation and Involvement of Stakeholders*

A study illustrates that trust and good communication are vital for effective AI integration in healthcare using the Non-adoption, Abandonment, Scale-up, Spread, Sustainability (NASSS) approach (Alami et al. 2023). It's critical to include stakeholders at every step of generating and rolling out AI solutions to guarantee they consider everyone's needs and concerns. Healthcare workers really should be part of the planning, design and testing process so that everything aligns with clinical workflows and patient care priorities. Discussing with patients and advocacy groups is key too. Policy-makers need to team up with healthcare workers and make rules that control ethical AI use in medicine. Sharing knowledge and working together across different parties can help shape trust and support successful AI solutions.

### *4.2 Continuous Education and Training*

There is need to provide strong education and training programs for everyone in healthcare, from admin staff to doctors for smooth AI integration. Linking stakeholders boosts practice and training while tackling issues like bias (Nyberg and Morris 2023). Everyone employed in healthcare must recognize what AI can do and its boundaries. Training of workers in healthcare should cover basics like how AI works, categories of algorithms used, pros and cons of technologies and best practices for clinical decisions with AI visions. Giving staff chances to continue learning confirms they stay updated on new advances and remain skilled at using these knowledge.

### *4.3 Regulation and Ethical Issues*

Producing strong ethical standards and rules is important in applying AI considerately and reliably in healthcare. When setting these ethical standards and rules, we should deliberate on things like data privacy and transparency. It is significant for policymakers to work together with healthcare providers, AI developers and other stakeholders to certify fairness in applying AI in medicine. Providers should follow data privacy laws like GDPR and HIPAA closely while ensuring there are strong governance practices in place to protect patient information.

## 5 Case Studies of Successful AI Integration

Here are some astonishing cases of effective AI use.

### *5.1 Diagnostic Imaging*

In a hospital, an AI system figured out how to spot breast cancer using deep learning models and loads of mammogram images (Kumar et al. 2022). Lin et al. (2022) sought to make breast cancer diagnosis better and cut down on mistaken calls by tinkering with how the models were set up. This system could check mammograms in a second and catch tricky signs that human radiologists might overlook. Thanks to this workflow enhancement, radiologists could emphasis on tougher cases while the

AI managed routine screenings successfully. Because the malignancies were discovered at earlier, more curable stages, this method resulted in speedier diagnosis and treatment as well as better patient outcomes. The utilization of diagnostic imaging has experienced a notable surge in fee-for-service healthcare regimes (Kumar et al. 2022). Members of integrated health care systems' imaging patterns are not well understood (Smith-Bindman et al. 2012). Estimating trends in imaging use and related radiation exposure among people who are a part of integrated health care systems is the aim.

## 5.2 *Predictive Analytics in Chronic Disease Management*

The medical professional identified patients with chronic illnesses like diabetes or heart disease who were most likely to experience complications by using AI-powered predictive analytics. Medical team can use predictive analytics driven by AI to identify chronic illness patients likely facing complications. They looked at a lots of patient data, like medical history and then created personalized risk profiles with their system's help. In Singhania and Reddy's study (2024), big data analytics provided correct insights allowing defensive measures against chronic illnesses through targeted programs like regular check-ups or tailored treatment plans. By catching high-risk patients early on, this positive strategy improved chronic disease management while reducing hospital visits.

## 5.3 *Operational Efficiency in Hospitals*

Diving into historical patient flow data, staffing information, AI systems employed cutting-edge algorithms to streamline various hospital tasks, from scheduling patients to allocating resources. The AI systems discovered trends which will help to sharpen schedules. For illustration, forecasting patient arrival times helped improve staffing levels while lowering wait times. Mr and Ab (2021) emphasized on how combining science with tech creates improved care by minimizing human error while refining health outcomes.

## 5.4 *AI in Drug Discovery*

In drug discovery land, a pharmaceutical company analyzed tons of data with machine learning algorithms searching for new treatments. The speed at which the system processed this information was much faster than humans could manage alone. By automating initial steps of finding medication candidates, they sped up drug development processes significantly. In addition, it suggested safer options based on expected interactions.

## 6 Future Prospects and Challenges

The following are opportunities and challenges we might face as we push forward with integrating AIs into healthcare.

### 6.1 Technological Advancements in AI

As time goes on we'll see some big changes, which will improve its applications in healthcare:

1. More powerful machine learning algorithms: Progresses here will make dependable tools explicit for healthcare needs.
2. Amazing processing speed: Better hardware will let us crunch larger datasets faster and helping us make decisions quickly.
3. Multimodal approaches: Merging various data sources will offer a stronger picture of health trends.
4. Explainable AIs: New strategies will make it clearer how decisions are made by these systems, which will be able to build trust with patients.
5. Continuous learning: Systems evolving from knowledges form continuous usage, will become gradually effective over time.

### 6.2 Addressing Ongoing Challenges

Some of the few big hurdles need sorting out:

1. Data security is crucial; we must find a way to protect sensitive information carefully
2. Algorithm biases could widen gaps and even can cause troubles.
3. Integrating within clinical settings must happen seamlessly, which can be achieved using teamwork.
4. Regulatory clarity is important, just as we want security while tech keeps being developing.
5. Ethical discussions are important, they need to focus on right levels of human monitoring and transparency.

### 6.3 Worldwide Application

Bringing this tech globally introduces potential but also challenges:

1. Global health structures vary widely; each area needs tailored approaches.
2. Different regulations mean smart navigation skills are required.

3. Resource accessibility is so important and helping low-resource communities access tech.
4. Local talent matters, which will help in building capacities among local health workers.
5. Collaborative research amongst different countries drives good solutions forward.

## 7 Conclusion

It is imperative to solve the issues of data requirements, public trust, and ethical considerations in order to fully achieve the promise of AI in the healthcare industry. To reveal the potential benefits from AIs within health care we must tackle relevant challenges about ethics sustainability trustworthiness. These will shape our next steps towards success together. Achieving better collaboration with stakeholders means making things more engrossed on patients while being accessible equitably too. The final goal here would be continual innovation ensuring social benefits for All.

It is critical to recognize that numerous of these items mentioned require fundamental changes in policymaker oversight of medical-care and care delivery system. Though, approaching innovative applications of AI in medicine with a crucial but with open mind would enable users to be the forerunner in ushering in the next generation of medical care advancement.

## 8 Recommendations

The following are some recommendations for you to consider:

### *8.1 Build Strong Data Governance Foundations*

It's vital to establish rules for sharing data and guaranteeing it's used in an accountable and ethical way. This contains creating rules for who can access the data, who owns it, and following regulations. Security actions like encryption, restricted access, and keeping a record of who accesses the data are used to keep data safe and prevent unauthorized use. Develop standardized data formats and interoperability rules to facilitate smooth data exchange between AI systems and existing healthcare IT systems. Authorize patients to control the usage of their health data by engaging them in data governance decisions. Collaborate with regulatory bodies to ensure that data governance frameworks align with recent privacy regulations such as GDPR and HIPAA.

## 8.2 *Establish Public Confidence*

Transparency and accountability matter for patient trust of health care providers, AI Systems That means making it easier to comprehend which AI algorithms are guiding decisions, and creating a framework for who is responsible if things go wrong. To develop a framework to outline obligations of health workers, artificial intelligence coders and contributors that ensures the predictable use of AI in healthcare. Provide explicit mechanisms for AI errors or failures, accountability processes and corrective actions. Open communication and effective education are key to dispelling misperceptions and fears attached with AI in healthcare. For AI solutions in healthcare to be a success, they must address patient needs and align with the values of their communities, which is why involving patients advocacy groups alongside community representatives during design and implementation are core pillars for deployment on these platforms.

## 8.3 *Promote Continuous Education*

Ongoing training will also be necessary to ensure that healthcare workers can use AI effectively. This involves knowing AI well, what it can do and cannot, as also being able to 'interpret' the data produced by this machine (Sharma et al. 2023). We must include AI as a part of the curriculum for medical and nursing courses in order to enable future healthcare professionals with skills required to use these instruments more efficiently. Organize workshops, webinars and on-site training for your medical staff to ensure that it is updated with the latest of AI-driven healthcare solution which can ultimately help in providing a boost to their career. Facilitate mutual learning in this domain by encouraging healthcare professionals, data scientists and AI experts to learn from each other on the challenges and opportunities of implementing solutions. Offer user friendly and intuitive AI tools/interfaces. This will help you learn faster and seamlessly apply them in your regular clinical practice.

## 8.4 *Foster Collaboration*

Encouraging teamwork is crucial. Bringing together various parties like patients, lawmakers, healthcare workers, and tech experts is key to solving issues and developing AI solutions that meet the healthcare industry's needs. By involving all these stakeholders, you can ensure that AI solutions are effective and beneficial for everyone involved. Ensuring that AI-powered solutions in healthcare meet the specific needs and processes of the industry involves fostering communication and sharing information between AI developers and healthcare providers. To promote innovation in AI healthcare, it is essential to back collaborative research and development initiatives that capitalize on the combined expertise of academic institutions, healthcare entities, and tech companies. Establish innovation hubs and collaborations between public and private sectors to accelerate the transition of AI technology from research to practical use in healthcare. Engage community groups and patient advocates in creating and deploying AI healthcare solutions to gather their feedback and perspectives.

### *8.5 Promote the Development of Ethical AI*

These efforts are important considering with great power comes the need to ensure AI is used appropriately, and that it benefits society as a whole. This involves things like addressing algorithmic fairness, privacy regarding data set use and some sort of ethical consideration should be used when considering the results that come out as lays. We need strong oversight and standards to promote fair practice throughout every stage of the AI lifecycle from data collection, through model development and deployment. Ethics should be at the heart of designing and reviewing AI-assisted health tools. This requires that we examine whether there is bias in models, the after-effects of such tools and equitable access to them. Collaborate with the regulatory bodies to develop standards and guidelines, embedding deliberate check and balance for AI-Based services. That way you can still find new ground in an ever-changing field.

## References

Aaron H. and David J. F. (2023). Introduction to Machine Learning. Available: https://www.cs.toronto.edu/~fleet/courses/C11/Handouts/Introduction.pdf.

Alami, H., Lehoux, P., Papoutsi, C., Shaw, S. E., Fleet, R., & Fortin, J. P. (2023). Understanding the integration of artificial intelligence in health systems through the NASSS framework: A qualitative study in a leading Canadian academic centre.

Andreas, S., Panayides., Amir, A., Amini., Nenad, Filipovic., Ashish, Sharma., Sotirios, A., Tsaftaris., Alistair, A., Young., David, J., Foran., Nhan, Do., Spyretta, Golemati., Tahsin, Kurc., Kun, Huang., Konstantina, S., Nikita., Ben, P., Veasey., Michalis, Zervakis., Joel, H., Saltz., Constantinos, S., Pattichis. (2020). AI in Medical Imaging Informatics: Current Challenges and Future Directions, *IEEE Journal of Biomedical and Health Informatics*, https://doi.org/10.1109/JBHI.2020.2991043.

Brian W., Aline C., Stefan G., and Nina R. S. (2018). Artificial intelligence (AI) and global health: how can AI contribute to health in resource-poor settings? BMJ global health. Available: https://gh.bmj.com/content/bmjgh/3/4/e000798.full.pdf.

Chen, Y., & Wu, M. (2023). Artificial Intelligence-enabled contactless sensing for medical diagnosis. *Medical Review*, *3*(3), 195–197.

Das, P. R., Talukdar, R. R., and Kumar, C. J. (2024). Exploring the interplay of abstinence self-efficacy, locus of control, and perceived social support in substance use disorder recovery. Current Medical Research and Opinion,_ _Taylor and Francis, 2024.

Dolores D. (2017). Artificial Intelligence for Health and Health Care. Available: https://www.healthit.gov/sites/default/files/jsr-17-task-02_aiforhealthandhealthcare12122017.pdf.

Dubey, A., & Tiwari, A. (2023). Artificial intelligence and remote patient monitoring in US healthcare market: a literature review. *Journal of Market Access & Health Policy*, *11*(1), 2205618.

Hamet P and Tremblay J. (2017). Artificial intelligence in medicine. Metabolism, 69s: S36–S40.

Hao K. (2017).What is machine learning? Machine-learning algorithms find and apply patterns in data. And they pretty much run the world. MIT Technology. (https://www.technologyreview.com/2018/11/17/103781/what-is-machine-learning-we-drew-you-another-flowchart/.

Haridimos, K., Varvara, K., Stelios, S., Kostas, M., Manolis, T., Ana, J., Eduardo, C., Ignacio, B, J., Damian, S., Sergio, L., Magdalena, K., Gianna, T., Nikolaos, S., Paschalis, B. V., Drossou., Antonios, L., Konstantinos, V., Pedro, M., Luis, M, L., Cerda, A., Karine, S., Esther, C., Lauren, A. F., Jordi, R., Alexander, H. A., Gutierrez., Martijn, P. A., Starmans., Fred, W., Prior., Karim, L. (2023). Data infrastructures for AI in medical imaging: a report on the experiences of five EU projects, *European Radiology Experimental*, https://doi.org/10.1186/s41747-023-00336-x.

Jimma, B. L. (2023). Artificial intelligence in healthcare: A bibliometric analysis. *Telematics and Informatics Reports*, *9*, 100041.

Judd E. H., M.D., and Brendan G. Carr, M.D. (2020). Virtually perfect? Telemedicine for COVID-19. New England Journal of Medicine. Vol. 382, No. 18, pg 1679–1681. https://doi.org/10.1056/NEJMp2003539.

Kivelä M, Arenas A, Barthelemy M, Gleeson JP, Moreno Y, Porter MA. (2014). Multilayer networks. Journal of Complex Networks, Vol. 2, Issue 3, pg 203–271. Available: https://doi.org/10.1093/comnet/cnu016.

Kumar C.J, Das P.R., Hazarika A (2022). Autism spectrum disorder diagnosis and machine learning: a review. International Journal of Medical Engineering and Informatics, Inderscience, vol. 14(6).

Lena Petersson, Ingrid Larsson, Jens M. Nygren, Per Nilsen, Margit Neher, Julie E. Reed, Daniel Tyskbo and Petra Svedberg (2019). Challenges to implementing artificial intelligence in healthcare: a qualitative interview study with healthcare leaders in Sweden. BMC Health Services Research. Available at: https://doi.org/10.1186/s12913-022-08215-8.

Lin, R. H., Kujabi, B. K., Chuang, C. L., Lin, C. S., & Chiu, C. J. (2022). Application of deep learning to construct breast cancer diagnosis model. *Applied Sciences*, *12*(4), 1957.

Lushun Jiang, Zhe Wu, Xiaolan Xu, Yaqiong Zhan, Xuehang Jin, Li Wang and Yunqing Qiu (2021). Opportunities and challenges of artificial intelligence in the medical field: current application, emerging problems, and problem-solving strategies. Journal of International Medical Research. 49(3) 1–11, https://doi.org/10.1177/03000605211000157.

Micheal M., Sonoo T. I., Mahnoor A., and Danial W. (2019). Artificial Intelligence in Health Care: The Hope, the Hype, the Promise, the Peril. NAM Special Publication. Washington, DC: National Academy of Medicine.

Mobeen N., Muhammad M. A. and Elad Y. (2021). A systematic review of human-computer interaction and explainable artificial intelligent in healthcare with artificial intelligence techniques. In *IEEE Access*, vol. 9, pp. 153316–153348, https://doi.org/10.1109/ACCESS.2021.3127881.

Mr, N. S. Z., & Ab, N. (2021). The Application of Artificial Intelligence (AI) In Highly Intelligent Hospitals. *International Journal of Public Health & Clinical Sciences (IJPHCS)*, *8*(6).

Nina S. and Brian W. (2020). Artificial intelligence and the future of global health. National Library of medicine, pg 1579–1586, https://doi.org/10.1016/S0140-6736(20)30226-9.

Nyberg, C. C., & Morris, E. (2023). "Revolutionizing clinical education: opportunities and challenges of AI integration". *European Journal of Physiotherapy*, *25*(3), 127–128.

Pasricha, S. (2022). AI ethics in smart healthcare. *IEEE Consumer Electronics Magazine*, *12*(4), 12–20.

Reagan N. R (2018). Artificial Intelligence: Its Importance, Challenges and Applications in Nigeria Available at https://directresearchpublisher.org/drjeit/files/2018/12/Robinson.pdf.

Sharma M, Kumar CJ, Talukdar J, Singh TP, Dhiman G, Sharma A., Identification of rice leaf diseases and deficiency disorders using a novel DeepBatch technique, Open Life Sci, De Gruyter, 2023.

Sharma M., Kumar C. J and Bhattacharyya D. K. (2024a). Machine/deep learning techniques for disease and nutrient deficiency disorder diagnosis in rice crops: A systematic review, Biosystems Engineering, Elsevier.

Sharma M., Kumar C. J., Singh, T. P., Talukdar, J., Sharma, R. K. and Ganguly, A. (2024b). Enhancing disease region segmentation in rice leaves using modified deep learning architectures, Archives of Phytopathology and Plant Protection, Taylor and Francis.

Singhania, K., & Reddy, A. (2024). Improving preventative care and health outcomes for patients with chronic diseases using big data-driven insights and predictive modeling. *International Journal of Applied Health Care Analytics*, *9*(2), 1–14.

Sistaninejhad, B., Rasi, H., & Nayeri, P. (2023). A review paper about deep learning for medical image analysis. *Computational and Mathematical Methods in Medicine*, *2023*(1), 7091301.

Smith-Bindman, R., Miglioretti, D. L., Johnson, E., Lee, C., Feigelson, H. S., Flynn, M., ... & Williams, A. E. (2012). Use of diagnostic imaging studies and associated radiation exposure for patients enrolled in large integrated health care systems, 1996–2010. *Jama*, *307*(22), 2400–2409.

Steven M. W. and Victor P. (2024) Balancing Privacy and Progress: A Review of Privacy Challenges, Systemic Oversight, and Patient Perceptions in AI-Driven Healthcare. Vol. 14, no 2, pg 675 https://doi.org/10.3390/app14020675, Switzerland.

Union E. (2018). A definition of AI: Main capabilities and scientific disciplines. https://ec.europa.eu/futurium/en/system/files/ged/ai_hleg_definition_of_ai_18_december_1.pdf.

US Agency for International Development (USAID), (2019). Artificial Intelligence in Global Health: Defining a Collective Path Forward. Available at: https://www.usaid.gov/sites/default/files/2022-05/AI-in-Global-Health_webFinal_508.pdf.

WHO (2024). Ethics and governance of artificial intelligence for health, pg 1–69, **ISBN:** 978–92–4–008475–9.

Yankang J., Yuemin B., Ziheng H., Lirong W. and Xiang-Qun S. X. (2018) Deep Learning for Drug Design: an Artificial Intelligence Paradigm for Drug Discovery in the Big Data Era. The AAPS Journal. Pg 1–10, https://doi.org/10.1208/s12248-018-0210-0.

Zemplényi, A., Tachkov, K., Balkanyi, L., Németh, B., Petykó, Z. I., Petrova, G., ... & Kaló, Z. (2023). Recommendations to overcome barriers to the use of artificial intelligence-driven evidence in health technology assessment. *Frontiers in Public Health*, *11*, 1088121.

# AI's Potential: Embracing AI in the Future of Healthcare

**Vijay Prakash, Kirtan Dua, Carl James Debono, and Lalit Garg**

**Abstract** Artificial intelligence (AI) has not just revolutionised but transformed medical operations, opening up new frontiers in treatment. This transformative power of AI has empowered healthcare professionals to overcome major obstacles in cost, accessibility, illness management, and therapy optimisation. Advanced AI technologies, such as Deep Learning (DL) and Machine Learning (ML), have significantly impacted telemedicine, diagnosis, new drug development, and patient monitoring. Great strides have been achieved in improving intervention and disease detection with the help of AI-based algorithms utilised in clinical decision support systems and predictive models for disease identification. The landscape of drug trials has transformed with AI's rapid advancement in researching therapeutic effectiveness, adverse reactions and the development of potential molecular structures. This chapter delves into the role of intelligence in healthcare, exploring its capabilities, opportunities and applications within the realm. Modern AI algorithms like the da Vinci surgical system have made minimally invasive surgeries possible, reducing hospital stays and recuperation periods and improving surgical precision and results. AI is being rapidly used to streamline supply chains and improve hospital patient flow, reducing wait times and enhancing patient satisfaction.

**Keywords** Patient monitoring · Telemedicine · Diagnostics · AI in healthcare · Surgical precision

V. Prakash (✉) · C. J. Debono · L. Garg
Faculty of ICT, University of Malta, Msida, Malta
e-mail: vijay.prakash@um.edu.mt

C. J. Debono
e-mail: carl.debono@um.edu.mt

L. Garg
e-mail: lalit.garg@um.edu.mt

K. Dua
School of Computer Science, University of Petroleum and Energy Studies, Dehradun, India



T. P. Singh et al. (eds.), *Revolutionizing Healthcare: Impact of Artificial Intelligence on Diagnosis, Treatment, and Patient Care*, Studies in Computational Intelligence 1182,
https://doi.org/10.1007/978-3-031-80813-5_20

# 1 Introduction

According to AI's vision, the healthcare landscape of tomorrow goes beyond disease treatment; it focuses on foreseeing and averting ailments before they manifest. This proactive stance, advocated by AI, was succinctly captured by Topol (2019). In a village where medical services are scarce, an elderly gentleman grapples with overseeing his persistent ailment. A lot of such cases exist and they all have a common element: the requirement for economical and easily accessible healthcare solutions. With its potential to revolutionise healthcare, AI offers hope and optimism for the future.

AI can help overcome these obstacles. AI's role in overcoming significant healthcare challenges, such as cost, disease management, and accessibility, is evident in its accurate analysis of genetic, environmental, and lifestyle characteristics to tailor treatment plans for each patient, with a strong focus on prediction and prevention. Figure 1 captures the constructive view of AI in the medical sector. During the COVID-19 pandemic, AI-assisted telemedicine established a standard for patient follow-up and distant healthcare delivery. Furthermore, wearable technology and AI-generated apps have made checking and monitoring vital indicators possible during everyday activities. In addition to being incredibly revolutionary, AI's impact on healthcare is constrained by legal and ethical issues (Tripathi et al. 2021). AI-enabled data security and algorithm openness ought to strictly conform to moral standards. Strong governance structures must be established before using AI to improve medical education, manage chatbots, diagnose disease or provide virtual reality training for mental health interventions. Practical expertise is crucial because AI has a minimal role in medical decision-making. Finding the right harmony between the technology's capabilities and ethical concerns is crucial to guaranteeing impartial outcomes in healthcare settings and increasing understanding of AI's possibilities in healthcare.

The study aims to show how AI is revolutionising medical procedures like diagnostics, telemedicine, and patient monitoring, demonstrating AI's huge healthcare transformational potential. AI optimises medicine development, surgery, patient flow, and resource management, improving therapy and operational efficiency. This improves patient outcomes and system efficiency. This chapter also focuses on the importance of governance, transparency, and data security for ethical and responsible AI-driven medical technology implementation.

The chapter is divided into five sections. Section 1 summarises the current role of AI in healthcare, focusing on accessibility, cost, and treatment optimisation. It explores how AI might improve hospital efficiency, patient flow, and surgery preoperative planning. Section 2 focuses on different applications of AI in the healthcare sector, including advanced robotic tools used in surgery and monitoring of patients. Section 3 focuses on AI-driven diagnostics and imaging, administering clinical decision support systems, personalised Medicine and Precision Treatment, Drug Discovery and Development, etc. Section 4 discusses the future of AI in healthcare, focusing on medicine discovery, illness treatment, and remote patient monitoring. The section emphasises the importance of using AI to improve healthcare services

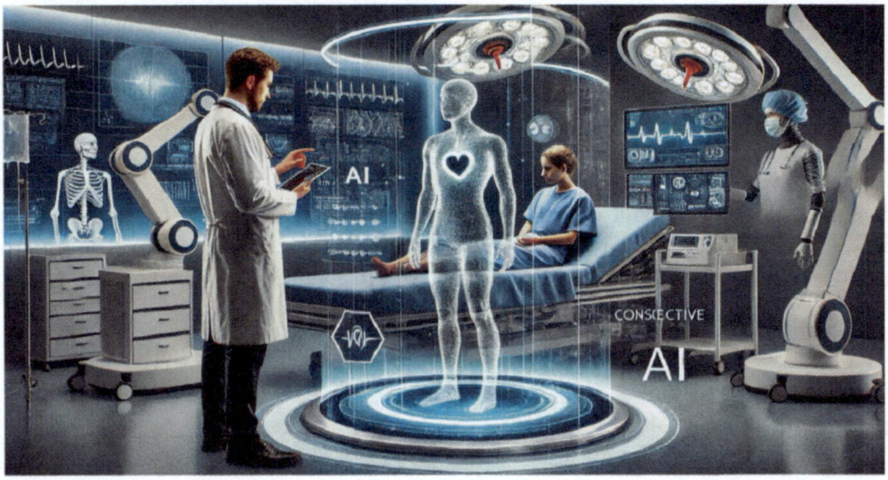


**Fig. 1** Artificial intelligence constructive view in the medical sector (generated using AI)

while addressing data protection and ethics. Section 5 concludes that AI has the potential to transform healthcare by providing sophisticated diagnostics, tailored therapies, and effective care delivery. Still, it must overcome ethical concerns, data security, and integration into current systems to realise its benefits fully. Responsible AI integration in healthcare requires strict control and continual research.

## 2 Application and Potential of Artificial Intelligence

The use of AI has the potential to enhance abilities and tackle intricate medical challenges with remarkable precision and efficiency. This has sparked a transformation in the healthcare industry. Through data analysis to generate 3D models of anatomical structures, AI algorithms are revolutionising preoperative planning in surgery, aiding surgeons in meticulously organising procedures and reducing risks (Khan et al. 2023). Advanced robotic tools with AI improve precision and eliminate operational errors. A successful AI medical application is the da Vinci surgical system. This AI-guided platform has performed minimally invasive procedures worldwide. Therefore, patients have shorter hospital stays and faster recovery times. AI is reshaping not just environments but all facets of healthcare operations. ML algorithms are optimising supply chains to ensure the delivery of prescription medications and other medical essentials by analysing data to forecast demand patterns, pinpoint potential bottlenecks, and enhance inventory management. These advancements lead to cost savings and increased productivity. Additionally, AI is being utilised to assess hospital patient flow, identify inefficiencies, and provide insights for process enhancement, ultimately reducing waiting time and enhancing satisfaction.

The integration of intelligence in healthcare shows promise for the future. Cutting-edge AI concepts such as "Brain Intelligence" have the potential to inspire approaches to healthcare methods. This may involve using AI algorithms to analyse patient data, estimate medication dosage efficacy, or propose new pharmacological research goals to create individualised treatment programs. Thanks to technological advances, AI can improve precision, proactivity, and personalisation to match each patient's demands. While challenges such as protecting data privacy and addressing issues persist, ongoing research and development offer hope that AI can effectively complement expertise to enhance health outcomes for all (Dileep and Gianchandani Gyani 2022).

## *2.1 Potential of AI*

AI has made its way into many sectors, reshaping industries like telecommunications, healthcare, education, and employment. It is no longer a concept found in science fiction. In addition to streamlining operations, AI enhances what humans can do, handling tasks that used to need input and changing how things are done in areas such as manufacturing, logistics, and finance (Dileep and Gianchandani Gyani 2022). AI can uses data and advanced analysis to revolutionise the healthcare industry entirely. Remarkable progress in identifying, diagnosing and treating diseases across fields has driven the development of machine-learning algorithms for detecting complex patterns within intricate data sets (Chandrashekar et al. 2020). Moreover, AI is hastening the quest for drugs by speeding up their development through analysing molecular structures and forecasting their therapeutic capabilities. While AI holds potential for the healthcare field, it is crucial to recognise the societal impacts of this technology. Safeguarding information from misuse or unauthorised access necessitates prioritising data security and privacy. Additionally, with AI algorithms playing a role in decision-making, it is essential to address potential biases that could lead to unfair outcomes for specific groups. Robust regulatory frameworks are vital to ensure that AI in healthcare is developed and utilised to align with values and ethical norms (Raddo et al. 2021).

## *2.2 Healthcare Sector*

The healthcare industry faces a difficult moment as it strives to offer fair care in a rapidly evolving world. With challenges like the shortage of professionals, changing patient demands and rising healthcare expenses, there is an urgent call for creative approaches. AI stands out as a player in promoting sustainability in healthcare by optimising resource utilisation, improving results, and addressing longstanding disparities in access to top-notch medical services. Globalisation has revealed the glaring gap between wealthy and developing countries regarding healthcare access and quality,

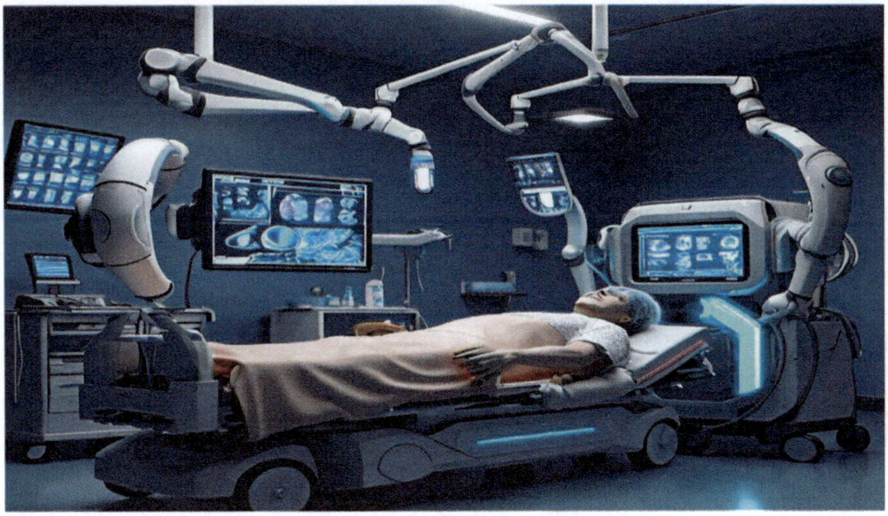

**Fig. 2** AI-driven diagnostic on medical image (generated using AI)

even as it promotes innovation by sharing information and resources (Li et al. 2020). AI can close this gap by making specialised knowledge and diagnostics more accessible, especially in underprivileged areas. For example, AI-driven diagnostic tools can accurately analyse medical images, as shown in Fig. 2, even in resource-constrained situations. At the same time, AI-powered telemedicine platforms may link patients in remote places with specialists.

In the healthcare industry, constant innovation is crucial to improve processes, find new disease cures, and support healthcare workers. The demand for continuous adaptation and development is driven by changing patient expectations, workforce demographics, and increasing cost pressures (Olive-Gadea et al. 2020). In this constant progression, AI is positioning itself at the epicentre of modern systems as it improves clinical decision-making, streamlines the administration of patient care, and advances diagnostics (Lin et al. 2022). Additionally, new business models centred on efficiency, value-based care delivery, and better outcomes are emerging due to AI. To fully utilise AI in healthcare, several obstacles must be overcome. Implementation challenges and reluctance may arise from the need to make substantial process changes and staff training to integrate AI into the current workflows. In the ensuing subsections, we shall discuss eHealth and mHealth in more detail.

Information and communication technology had a significant role in the late 1990s eHealth revolution, transforming patient empowerment and healthcare delivery (Iqbal et al. 2022). As a result of this change, telehealth platforms and Electronic Health Records (EHRs) were integrated, making eHealth a vital component of contemporary healthcare. eHealth aims to streamline healthcare procedures while promoting patient-centred care and collaboration. AI significantly improves eHealth by evaluating EHR data to create personalised medicine strategies and customising treatment plans for each patient. AI also transforms diagnosis in eHealth systems by helping radiologists spot anomalies in medical imagery. Additionally, AI-driven chatbots improve patient communication and expedite administrative tasks, while AI-driven predictive analytics pinpoint high-risk patients for targeted preventive care.

Through the analysis of data from wearables and sensors, real-time insights, and the facilitation of timely treatments, AI integration is also revolutionising mobile health (mHealth). AI powers virtual health aides like Woebot for mental health assistance and other solutions that improve glycaemic control in diabetes treatment and allow early identification of abnormal heart rhythms (Nguyen et al. 2018). Furthermore, early detection of worsening medical problems and rapid management are made possible by AI-powered remote patient monitoring, ultimately improving patient outcomes (Mostafa et al. 2022). AI has enormous potential to change healthcare as it develops, opening the door to more proactive, individualised, and efficient healthcare solutions in the future.

## *2.3 Artificial Intelligence and the Healthcare Sector*

The healthcare industry is quickly changing due to AI, which presents numerous opportunities to improve patient care, expedite procedures, and cut costs (Comito et al. 2022). AI applications are becoming increasingly prevalent across the healthcare ecosystem, driven by the exponential growth of healthcare data and improvements in analytical methodologies (Brinker et al. 2019). AI algorithms are showing great promise in the early diagnosis of chronic conditions, including diabetes, Alzheimer's disease, and many types of cancer. For example, Google Health's AI-powered technology has outperformed human radiologists when identifying breast cancer from mammograms (Santosh and Gaur 2021). Early identification made possible by AI-powered technologies can result in prompt therapies and greatly enhance patient outcomes. AI is improving patient support, communication, and diagnostics, employing virtual assistants and intelligent robots, as depicted in Fig. 3. These AI-powered solutions can respond to questions from patients, provide emotional support, and deliver personalised health information. As a result, patient engagement and treatment plan adherence are increased. For instance, Woebot, an AI-powered chatbot, has demonstrated encouraging results in lowering users' feelings of anxiety and depression.

AI is also improving workflows, automating tedious jobs, and strengthening cybersecurity safeguards, all changing healthcare operations (Tran et al. 2021). According to an Accenture analysis, by 2026, AI may reduce healthcare costs in the United States by $150 billion annually through automation and optimisation. AI is used in various healthcare applications, including remote patient monitoring, clinical trials, medication development, and even nanotechnology-based treatments (Hameed and Krishnan 2022). AI has even outperformed humans in several fields, like medical image processing and diagnostic tasks, highlighting its enormous potential to transform healthcare delivery (Akkus et al. 2018). However, more than merely creating

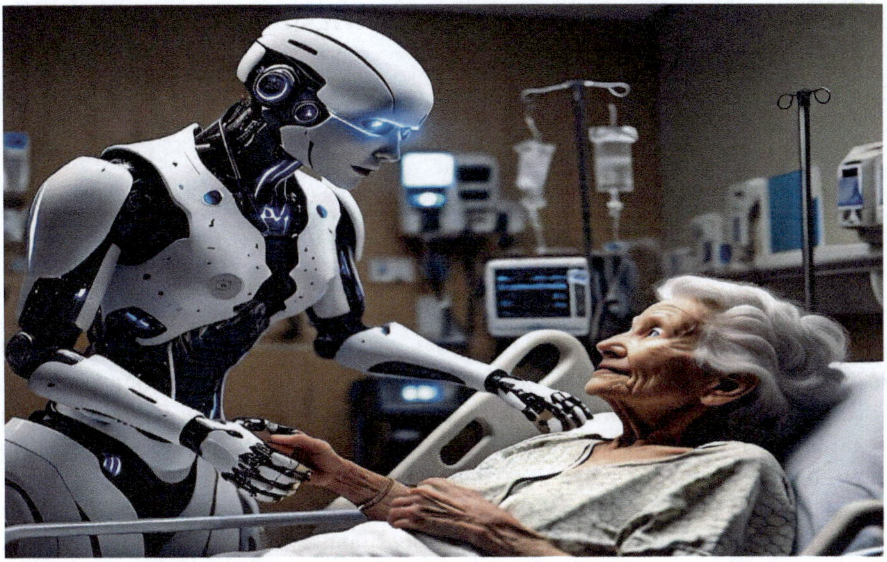

**Fig. 3** Patient support through AI-driven robots (generated using AI)

AI algorithms is required to realise AI's full potential in the healthcare industry. For successful implementation, it is necessary to prioritise design thinking, process improvement strategies, and implementation science (Hawkins et al. 2016). To achieve this goal, it is essential to comprehend the requirements of patients and healthcare providers to create AI solutions that seamlessly integrate into processes and carefully assess the impact of these systems on outcomes and patient contentment.

AI technology holds the promise of transforming the healthcare sector. Some challenges must be addressed. Technical constraints like the need for high-quality data and computer capacity to train AI models present significant obstacles. The ethics of prejudice, data privacy, and AI worsening pre-existing health inequities must be considered. AI in healthcare needs accountability and transparency frameworks. The potential impact of AI in addressing healthcare workforce shortages through the automation of diagnostic decision-making and administrative tasks is noteworthy (Andreeva et al. 2022). To ensure that AI complements and then replaces skills, deliberate planning and strategic implementation are crucial in integrating AI into healthcare systems. The future of intelligence in healthcare is promising but ethical and legal concerns and the requirement for continuous research and advancement remain essential to guarantee AI's suitable and just utilisation for the good of all.

# 3 Review of the Applications of AI in Healthcare

This section discusses some of the exciting uses of AI techniques in different health sector applications.

## 3.1 *Artificial Intelligence-Driven Diagnostics and Imaging*

AI can rapidly diagnoses diseases using ML and DL methods. As a result, issues are reduced, and illnesses can be diagnosed earlier (Fabrizio et al. 2021). Clinical decision support systems (CDSS) aid clinicians in diagnosing and treating disorders. For example, manual or automatically generated rules are applied by diagnostic CDSs to aid in diagnosis. Experiments carried out in electronic medical records (EMRs) show how successful this strategy is (Chang et al. 2020), as depicted in Fig. 4. Recent advances in DL algorithms can generate contrast images from non-contrast data. This reconstruction can be done using a two-dimensional cycle Generative Adversarial Network (cycleGAN). The disadvantages of employing contrast can be addressed by refining and enhancing this approach (Battineni et al. 2020).

Timely outbreak detection and management are essential. Digital Q-learning Auxiliary Diagnostics (DQAD) from AI identified epidemics faster than previous approaches (Zhang et al. 2021). COVNet, a neural network based on chest CT imaging, can distinguish COVID-19 from community-acquired pneumonia (Currie et al. 2019). A powerful shock advisory algorithm (SAA) is needed to diagnose and operate an automated external defibrillator. The new SAA used DL and ML

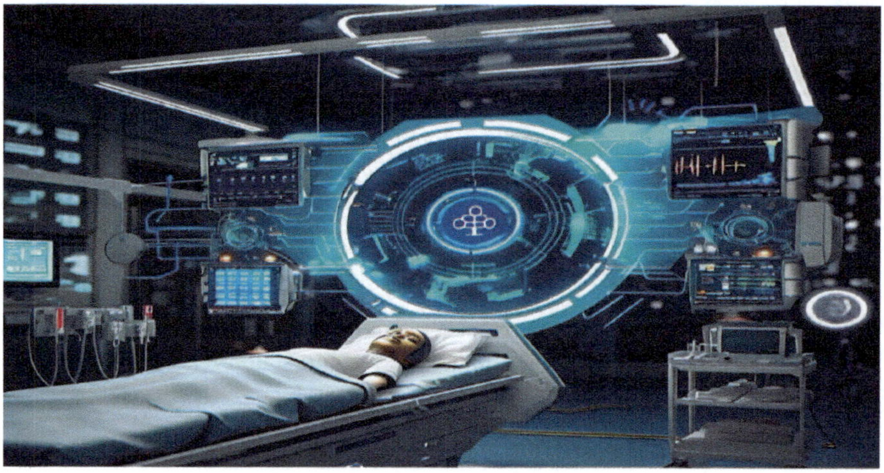

**Fig. 4** AI-driven EMR and ECG diagnostic (generated using AI)

algorithms and had better sensitivities and specificities compared to other solutions. Researchers developed AI algorithms with enhanced specificity, sensitivity, and operator repeatability for precisely detecting tumour volumes using the peaks in Magnetic Resonance Imaging (MRI) histograms. By integrating radionics and clinical markers, machine-learning models have effectively identified medulloblastoma molecular subgroups and distinguished between chondrosarcomas and skull base chondromas (Nomura et al. 2021).

It is imperative to diagnose strokes early because delaying endovascular thrombectomy (EVT) reduces their effectiveness. Using non-contrast computed tomography (NCCT), Methinks LVO (large vascular obstruction) is an ML-based technique that locates the LVO. As a result, patients can be sent to EVT facilities more quickly, which enhances the results (Kratz et al. 2005). Reconstructing a volumetric brain from CT head films with fully Convolutional Neural Networks (fCNN) is fast and efficient (Chen et al. 2021). AI-powered smartphone apps can track health data and diagnose illnesses (Gedefaw et al. 2023). AI classifies heart rhythms in the Kardia Mobile Cardiac Monitor (KMCM) into three categories: regular, perhaps Afib, and unclassified. Comparing KMCM to the 12-lead ECG revealed similar sensitivities and specificities. The sensitivities and specificities increase when a physician interprets the KMCM (Bokhari et al. 2022).

The use of AI in the diagnosis of some diseases, such as pancreatic cancer, which is frequently found in its advanced stages, has the potential to have a profound effect on the lives of several individuals. Multiple problems arise throughout the diagnosis procedure, including non-specific symptoms, delayed presentation of symptoms, and the lack of molecular markers. Scientists are presently investigating the application of AI methods, such as Support Vector Machine (SVM), Artificial Neural Network (ANN), and K-Nearest Neighbour (KNN), to detect slightly aberrant imaging signals that can assist in the early diagnosis of pancreatic cancer (Shiwlani et al. 2024). After the low-dose CT scan screening, the AI-powered ML system was used to assess the likelihood of acquiring lung cancer. In terms of prediction, it was just as accurate as the Brock scoring system; however, the AI-based approach outperformed it when the radio mic scores were extremely high (Álvarez-Machancoses et al. 2020).

More research is being done to determine whether AI can identify neurological conditions (Olivier et al. 2019). To predict Auxiliary Diagnostic (AD,) the authors [34]@@ examined four ML algorithms: ANN, SVM, KNN, and naïve Bayes (NB), separately and in combination. According to the study, combined models with fewer features better predict AD in its early stages (Beckmann and Lew 2016). To identify neurological diseases, including Parkinson's disease (PD), major depressive disorder (MDD), Attention-Deficit Hyperactivity Disorder (ADHD), Autism Spectrum Disorder (ASD), and mental illnesses like schizophrenia, a variety of ML and DL algorithms are being researched (Ziad and Ezekiel 2016). Currently, polysomnography (PSG) uses the Apnea–Hypopnea Index (AHI) to diagnose obstructive sleep apnea (OSA). A CNN-based AI technique diagnoses OSA using a single-lead ECG signal with good sensitivity, specificity, and accuracy. This procedure is easy and effective, but PSG must ensure the calculated risk is significant (Meskó et al. 2017). AI aids in the monitoring of diabetic care components. The self-monitoring system warns the patient of hypoglycemia an hour early with the help of AI. AI is also applied in risk classification, diabetic diagnostic assistance, and retinal screening (Caudai et al. 2021). AI-powered imaging and diagnostics are becoming increasingly commonplace and used in all areas of medicine. Examining social norms and ethics in these algorithms' further research can improve patient care (Alipanahi et al. 2015).

## 3.2 Personalised Treatment and Precision Medicine

"Personalised medicine" considers an individual's genetic composition, lifestyle, and environment while treating and preventing disease (Zhou and Troyanskaya 2015). Accurate, customisation and prevention transform healthcare in a way that breaks from conventional diagnostic techniques that rely on shared patient features (Caudai et al. 2021). Numerous data sources are required, including transcriptomics, proteomics, genomes, and biological data (Liu et al. 2017). Large amounts of data are necessary for this procedure, and researchers are looking at how AI might analyse data more reliably (He et al. 2021; Vural et al. 2016). These cutting-edge instruments allow us to pinpoint the precise intervention targets and design customised treatment plans for every patient (Cleophas and Zwinderman 2013). ML builds models that can learn from experience and enhance performance using datasets and mathematical approaches (Dong et al. 2015).

To identify patients who would benefit from receiving early treatment or from participating in randomised control trials of innovative medicines, ML algorithms can analyse multimodal data from EMR and other carefully chosen sources (Xu et al. 2021). For example, thorough pharmacogenomics analyses were carried out across over 1000 cell lines in a work by Dong et al. Gaining an understanding of the mechanisms behind anticancer medication responses was the goal. This study accurately predicted the sensitivity of anticancer treatments by developing and evaluating a complex support vector machine (SVM) model with genomic data (Zhuang and Ibrahim 2021).

AI in healthcare faces several challenges that must be overcome for safety and efficacy. The main challenges are (Ali et al. 2023; Nadella et al. 2023; Nasr et al. 2021):

1. **Data Privacy and Security**: AI systems use large amounts of sensitive patient data, raising issues about data privacy, ownership, and security breaches. Healthcare data is sensitive and could harm people if accessed without permission. GDPR in Europe and HIPAA in the U.S. must be followed. AI models trained on sensitive data must anonymise and encrypt to protect patient privacy.
2. **Algorithm Transparency and Interpretability**: AI systems, especially DL models, are often "black boxes," making decisions without human input. Insufficient openness can erode patient and practitioner confidence, especially when

AI-driven suggestions contradict clinical intuition or protocols. Effective healthcare application of AI models requires optimising their comprehensibility using Explainable AI (XAI).

3. **Bias and Fairness**: AI algorithms may accidentally amplify data biases. If the training data is lacking in diversity or biased toward certain populations, AI may make biased or inaccurate predictions, adding to health inequities. This is especially concerning in healthcare, where prejudiced forecasts can lead to poor treatment or unequal care for vulnerable populations. Employing more representative datasets, rigorous model validation, and fairness metrics can mitigate this problems.
4. **Integration with Healthcare Systems**: Many healthcare systems use outdated infrastructure, making AI integration difficult. Incompatibility between AI systems and EHRs may slow adoption and reduce the efficiency of AI-driven technology. However, integrating AI into clinical operations without disrupting protocols requires careful planning and training for healthcare workers.
5. **Regulatory and Ethical Challenges**: Due to the exponential growth of AI technology, present regulatory frameworks cannot ensure the safety and effectiveness of AI-powered mechanisms. Guidelines are needed to control AI in diagnosis, treatment planning, and patient monitoring. Ethical issues, including patient permission for AI decision-making and AI replacing human judgment, should also be carefully considered.
6. **Training and Workforce Adaptation**: Healthcare professionals must be trained to use AI. Clinicians may overuse or avoid AI systems due to a lack of understanding of their pros and downsides. Long-term deployment requires healthcare professionals to understand and incorporate AI outputs into patient care.

To handle these challenges, some emerging AI technologies and techniques influencing the future of healthcare (Alhashmi et al. 2020; Roppelt et al. 2024; Saxena et al. 2024) are discussed below:

1. **Explainable AI (XAI)**: AI models' lack of transparency is a serious issue. Explainable Artificial Intelligence (XAI) studies models humans can understand, allowing doctors to trust AI system conclusions. This new technology is critical for clinical decision support systems since understanding AI suggestions is as important as advice.
2. **Edge AI**: Edge AI implements AI algorithms locally on devices like wearables and smartphones, reducing the need to send sensitive data to the cloud. In healthcare, this technology could provide fast monitoring and diagnosis while protecting privacy. It's ideal for remote patient monitoring since it lets patients digest data locally and share only relevant data with doctors.
3. **Federated Learning**: Federated learning trains AI models on several decentralised devices or organisations while keeping data local. This reduces privacy issues because patient data stays with the institution, and the AI model obtains insights from a larger sample. This approach could improve healthcare AI models while meeting strict privacy regulations.

4. **Reinforcement Learning (RL) in Clinical Decision Making**: Reinforcement learning, an AI method that iteratively learns to achieve a goal, may be useful in therapy optimisation. RL could generate individualised treatment regimens that change patient reactions over time, improving chronic illness control and cancer therapy results.
5. **AI-Driven Robotics**: The da Vinci robot for minimally invasive surgery exemplifies how AI-driven robots change surgery. New robotic technologies incorporating AI are projected to improve accuracy and efficiency, enabling autonomous surgical support or more complex treatments with less human intervention.
6. **AI in Genomics and Precision Medicine**: AI's ability to analyse genomic data enables precision medicine, which tailors treatments to specific genetic profiles. Deep learning and NLP are used to detect genetic changes and predict drug effects. AI will influence personalised treatment options as genome sequencing costs drop.
7. **AI for Early Disease Detection**: AI's ability to evaluate medical images, genetic data, and patient history is improving early disease detection. AI can detect early-stage cancers or neurological illnesses like Alzheimer's through imaging or biomarker analysis, possibly before symptoms appear.

In short, AI is transforming healthcare diagnostics, monitoring, and personalised medicines. AI-driven technologies like deep learning and ML models improve illness detection, enabling earlier diagnosis and greater precision in stroke detection, cancer diagnostics, and pandemic outbreak tracking. These technologies improve diagnostic procedures, reduce human error, and optimise treatment choices with unmatched accuracy. AI-enabled personalised medicine tailors treatments to each patient's genetic and lifestyle traits, improving efficacy and outcomes. These domains have great potential for AI, but more research is needed to address ethics, data reliability, and clinical practice.

## 4 Future Perspective of AI in the Health Sector

AI can transform pharmaceutical development, patient surveillance, and telemedicine. In numerous sectors, ML, DL, and natural language processing (NLP) are being utilised to enhance operations, reduce costs, and improve diagnostic precision. AI-powered chatbots and other digital technologies are improving personalised medication, remote care, and mental health therapies. Besides these benefits, ethical issues, including data privacy, biases, and integration issues, are addressed, emphasising the need for careful regulation and responsible AI use in medical care.

## 4.1 Drug Development and Discovery with Artificial Intelligence Support

AI is widespread in practically every facet of modern life, and its improvements are rapid. Thus, its importance in pharmaceutical research and development is expanding (Pu et al. 2019). Drug development is laborious and requires many trial-and-error experiments and powerful screening techniques to examine many molecules. NLP and ML can speed up the process, but they need a lot of data. Simplifying the procedure can lower errors and increase the calibre of analytical results (Strous et al. 2021). The DL approach facilitates determining a substance's efficacy. We can also ascertain whether a medication is hazardous by using AI methods (Beneke and Mackenrodt 2019). Studies on the potential for AI to produce and synthesise novel compounds have opened the possibility of employing AI to develop new medications (MacLennan 1996).

AI employs various methods, including an ML model, reasoning, problem-solving, and knowledge representation. The SVM classification model will be the main subject of this analysis. A hyperplane is used by the supervised ML module SVM to connect two sets of problematic subgroups, categorise them, and generate an analytical output. Thus, a decision boundary or hyperplane can be used to arrange the two sets into useful data (Khakharia et al. 2021). The application of ANNs extends to the ML field known as DL. These can be seen as sophisticated computers that use perceptions similar to biological neurons to simulate how the human brain sends and receives electrical impulses. An artificial neural network (ANN) comprises several nodes that use algorithms to solve problems and translate different inputs into outputs singly or multi-link structures (Zeng et al. 2020).

Using AI-based monitoring models to forecast the risks of infectious diseases worldwide could be highly beneficial (Khan et al. 2012). An integrated modelling approach combining many individual data model types, including travel, cell phone location tracking, epidemiology, and behavioural pattern data, is essential to developing an efficient surveillance system (De Farias et al. 2020). To effectively prevent the spread of infection, this modelling technique requires awareness of the targeted population to be gathered on a large scale and at the level of a single person (Mantena and Keshavjee 2021). Applications such as anticipatory surveillance and the detection of global infectious disease problems could be made with an integrated model-based platform.

## 4.2 Disease Treatment and Patient Monitoring by Artificial Intelligence

For public health monitoring to reliably identify signs of abnormal health and disease outbreaks from data sources, it must overcome data sourcing and analytics obstacles. DL, reinforcement learning, knowledge graphs, Bayesian networks, and multiagent systems are some of the techniques and methodologies that can be used to enable and improve public health surveillance to address these issues. Using traditional surveillance techniques is challenging. AI technologies facilitate monitoring a wider variety of data, including unstructured and semi-structured data, open, novel, or unexplored data, and complicated spatiotemporal data (Kitsiou et al. 2017).

By using digitally transmitted health-related data, Remote Patient Monitoring (RPM), sometimes called telemonitoring, aims to improve patient care by promoting early diagnosis, management, education, and patient-physician relationships (Shaik et al. 2023). Digital health platforms are utilised by telemonitoring technology to assess patients in addition to the standard clinical visit (Taylor et al. 2016). Antunes et al.'s retrospective cohort study (Antunes et al. 2017) found that the innovative Chronic Liver Failure Consortium Acute Decompensation Score (CLIF-C ADs) machine reliably predicted 30-day mortality better than earlier models. The COVID-19 epidemic has increased RPM use (Hohma 2021).

Edge AI and IoT technology enable a healthcare monitoring system to schedule patients in real time and assign resources based on their situation. This technology monitors and assists by collecting, communicating, and triggering actions on other linked devices. It helps seniors, disabled people, and pandemic victims (Klingler et al. 2017). Phenotypic Personalised Medicine (PPM) quantifies a patient's health by determining the best drug combination and dosage using data from prior treatments. The Quadratic Phenotypic Optimisation Platform (QPOP) applies this information to clinical treatment. PPM-based methods outperform traditional therapy regarding outcomes, side effects, and potential (Dey et al. 2022). Valvular heart disease therapy programs can be created using AI. AI-driven clinical decision systems improve aortic valve replacement planning by using anatomical data to swiftly and precisely assess valve size and type. This method is helpful in busy places where manual picture analysis is time-consuming and inconsistent (Pozza et al. 2024).

## *4.3 Artificial Intelligence in Telemedicine and Remote Patient Care*

Using information and communications technology to provide medical care while taking accessibility and distance into account is known as telemedicine. Telemedicine was first intended for tech-savvy patients, but the COVID-19 epidemic forced it into the global mainstream of healthcare (Mirmomeni et al. 2020). Its scalability enabled the implementation of crucial safety and health measures during the emergency and proved that it is a viable strategy after the pandemic. The ability of DL to learn and adapt on its own makes it the preferred supporting tool (Blasiak et al. 2020). AI algorithms have been thoroughly researched and applied in several medical domains to improve diagnostic accuracy, expedite exam interpretations, and save time and labour costs (Kalra et al. 2024).

Patient data in modern medicine includes large amounts of visual data and binary and numerical information. Such comprehensive reports are often too complex for traditional computer programs to understand. Multiple processing layers allow the network to extract functional visual patterns and information (Collins Oguine and Jane Oguine 2022). Diabetic Retinopathy (DR) is a common ophthalmic complication of diabetes mellitus (DM) and a primary cause of preventable blindness. The WHO recommends annual dilated fundoscopy screening for diabetics to prevent vision loss. However, human assessors and long-term financial sustainability make screening, referral, and treatment of many DR patients difficult. DR management should be optimised with digital technology and DL (Kropp et al. 2023). Age-related Macular Degeneration (AMD) causes considerable visual impairment in the elderly worldwide. Given the rising number of AMD patients and the necessity for frequent monitoring, a reliable automated solution is needed. DL and telemedicine can make AMD screening, follow-up, and therapy prediction easier. DL algorithms can analyse VA and OCT data to help determine AMD therapy methods.

Due to rising demand, high-resolution image requirements, and cutting-edge technology, AI in telemedicine, particularly dermatology, has grown rapidly. Teledermatology improves dermatological care by lowering wait times, offering diagnostic aid, and improving convenience, but in-person diagnosis is more accurate (Coustasse et al. 2019). By offering rapid support during in-person exams, AI improves dermatologists' skin condition diagnosis accuracy. Non-dermatologists with diverse skin condition experience can benefit from AI-assisted diagnosis, especially for melanoma. Applications like telehealth stroke triage and ML-based neuro-trauma prognostics show telemedicine's promise in neurosurgery (Iqbal et al. 2023).

A paediatric neurosurgery team has used semiautomated image analysis software to analyse digital photos to screen for craniosynostosis (Paro et al. 2022). By correctly identifying prevalent kinds of craniosynostosis, the ML algorithms demonstrated the efficacy of an AI-augmented telemedicine technique. This novel approach facilitates early diagnosis, lessens the strain of ambulatory appointments for families, and improves access to screening for head shape abnormalities. Non-specialists and primary care physicians can also use this method to identify individuals who might need to be referred for surgery. Numerous initiatives have evolved in the field of diabetes care that use Web 2.0 technologies, new information and communication technologies, and automatic data transmission for remote patient interpretation. A noteworthy feature is the extensive use of ML, which greatly expands these telemedicine programmes' capacities and efficacy. AI algorithms facilitate more accurate and efficient data analysis, enabling medical practitioners to treat diabetes patients individually and make well-informed decisions (Barrett et al. 2019).

### 4.4 Behavioural Health Therapies Powered by Artificial Intelligence

The proliferation of AI, the enhanced availability of digital mental health therapies, and the surging need for psychological services have all recently contributed to the emergence of Digital Mental Health Interventions (DMHI). AI-powered chatbots have several applications in DMHI, including information exchange, behaviour modification, symptom evaluation, and patient evaluation. A chatbot can range from a rudimentary program following predefined rules, such as ELIZA, to sophisticated AI models utilising ML and language processing techniques (Singh and Thakur 2020). DMHIs incorporate chatbot-based services that encompass assistance, screening, training, intervention, surveillance, and recurrence prevention. Chatbots capable of diagnosing and prioritising mental health conditions can alleviate the burden on healthcare professionals and accelerate treatment for individuals in urgent need. Chatbots have been employed to assess and identify various mental health illnesses, such as dementia, substance abuse, stress, depression, suicidal ideation, anxiety disorders, and symptoms related to posttraumatic stress disorder.

The chatbot serves as a conversational companion in such situations. The user will receive a medical diagnosis, prognosis, and treatment plan (along with information regarding the diagnosis) from the chatbot once they input their symptoms. In general, 51% of mental health experts who responded to the survey stated that it was challenging to use chatbots for diagnosis (Sweeney et al. 2021). Conversely, AI-assisted diagnostics may help detect individuals who are at risk early on, enabling more successful intervention and problem avoidance. Other forms of AI may also be able to examine behavioural patterns and data that the individual provides to anticipate when specific mental health disorders will manifest accurately.

Particular NLP chatbots may mimic a therapeutic dialogue to introduce and inform users about different therapies. According to a meta-analysis, the most studied and widely utilised kind of chatbot was cognitive behavioural therapy (CBT) in 10 out of 17 cases. One such chatbot is Woebot, which imitates social discourse using natural language processing and instant messaging to give consumers CBT from real clinicians. This year, a chatbot was launched by Happify Health, a digital mental health startup supported by ads. AI-driven chatbot Anna imitates a therapist's job and provides a variety of Happify activities (Khawaja and Bélisle-Pipon 2023).

### 4.5 Utilising Artificial Intelligence in Medical Training and Education

AI integration in healthcare enables the creative exploitation of EHRs for tasks like data processing and improved decision-making. Physicians are motivated to accurately enter data into EHRs to gain from the advantages offered by AI. Augmenting the expertise and understanding of healthcare professionals can result in enhanced clinical proficiency and superior health results for patients during the transition to the era of AI. Widespread use of AI can potentially harm several patients substantially due to a solitary mistake. This emphasises the critical need to gather precise data and integrate input from physicians to educate the AI efficiently. Hence, aspiring medical professionals must thoroughly comprehend data management, AI utilisation, and its associated limitations.

According to a scoping assessment, medical students' practical skills have improved due to AI-based support. These applications include ML, virtual reality (VR), and robotics training in many areas of medical education (Sapci and Sapci 2020). Virtual reality helps students make safe decisions during surgery and medicine. Even with expensive hardware infrastructure, training is cheaper than manikin or actor-based training over time. AI could enhance low-risk microsurgical education, ophthalmic surgeon training simulations, and video content analytics. Virtual reality head-mounted displays provide multiple options for simulating an operating theatre. To better study the attitudes and experiences of twelve Swedish medical students utilising a multiplayer virtual world, Creutzfeldt et al. (2016) educated them in cardiopulmonary resuscitation through avatars. Cardiopulmonary resuscitation training has benefits that have been validated by data-driven qualitative research. Additional clinical performance should be examined to exclude the possibility of mistaken self-belief bias. Participants believed that their lack of stress and inadequate psychomotor abilities were unrealistic, which could impact their performance in real life.

Even though ChatGPT has the potential to transform medical education, mentorship and practical clinical practice are still essential. Furthermore, algorithms use big databases, which might be biased, and provide biased system outputs that make it less fair to treat marginalised or minority populations. The output of AI systems should be checked to guarantee that there is no bias and to remove any that may be there. More studies can greatly alleviate the problem of prejudice. AI adoption is not without its difficulties, including liability concerns, data privacy concerns, and black box issues. Therefore, doctors must know the benefits and drawbacks of AI tools and their proper application.

## *4.6 Ethics in Artificial Intelligence*

In high-risk healthcare, errors can have serious consequences. Patients who seek medical assistance while their health is poor are vulnerable. AI must be thoroughly tested and researched before playing a major role in clinical care. AI in medicine's main ethical problems are safety and transparency, algorithmic equity and prejudices, informed consent for using personal data and health records, and data privacy. The FDA has approved machine-learning-based autonomous AI diagnostic systems. ML-HCAs use machine learning to develop algorithms utilising large datasets to forecast without programming to reduce biases and errors. This new technology could alter healthcare. It must be carefully organised and secured to protect user privacy and prevent mistakes and security breaches.

For ML, privacy protection poses challenges. It requires large datasets for ML to perform well during training and generate highly accurate results. However, using large amounts of data from several sources to train ML algorithms ultimately undermines privacy efforts. Patient data ownership and rights are complex based on jurisdiction, anonymisation, and data de-identification. Even though North American doctors may have evidence of patient data, European patients own and can use their data. Due to strict EU rules like the General Data Protection Regulation (GDPR), patients may be reluctant to consent to training their data in AI models. This may induce firms to choose less regulated environments with weak data ownership requirements. The GDPR may discourage US firms from operating in specific areas, and EU citizens may not exercise their right to privacy outside.

Accountability for wrong diagnostics and decisions by AI solutions is still unclear, with no specific person or organisation being held accountable. AI makes it difficult to assign decision-making authority and ownership. Due to accountability issues, untrusted AI systems in therapy raise questions about professional obligations and patient well-being. When using AI to treat problems, physicians must validate and verify. Supervisors of AI users should emphasise that physicians cannot hide behind AI technology. Automation bias can occur when healthcare personnel become complacent due to an inaccurate AI system. This can make patients and doctors accept the system's results without questioning its limits.

The Buolamwini and Gebru (2018) study's automated facial recognition and related datasets exhibited blatant selection bias, which reduced the system's accuracy in identifying individuals with darker skin tones, particularly women. MLs use datasets that come from specific populations. As a result, these AI systems are more likely to generate false positives when used on underprivileged or underrepresented patient groups. Owing to the abundance of large datasets that concentrate on common illnesses, AI may fail to identify rare or uncommon ailments that a radiologist could identify from X-rays or CT scans. The performance and trustworthiness of its application may be questioned if models are built on training data that are insufficiently representative of the population, case mix, modalities, and acquisition techniques.

The EU's GDPR, which aims to proactively address morally difficult circumstances that may occur from using AI in healthcare, was spurred by the acrimonious debate around the legality of AI systems. AI integration with medicine necessitates a solid governance structure that protects the patients. To protect patients' welfare, AI must always abide by moral standards and be able to adjust to constantly shifting conditions that are frequently interrupted. AIS and ML-HCAs must adhere to all ethical standards and avoid unfair prejudices as AI becomes more integrated into healthcare delivery. In several healthcare domains, AI is expected to outperform humans eventually; therefore, avoiding its usage would be unethical and counter-scientific. Table 1 summarises the advantages and constraints of different AI technologies implemented in the healthcare sector.

**Table 1** AI technology's advantages and flaws

| Technology | Advantages | Flaws |
|---|---|---|
| AI-driven diagnostics and imaging | Early disease detection, improved diagnostic accuracy, reduced human error, and faster processing | Concerns around data privacy ethics, reliance on high-quality datasets, and potential biases in AI calculations |
| AI in personalized medicine | Individualised therapy based on hereditary and lifestyle factors, improved treatment outcomes, illness prognosis | Massive dataset integration, huge computing costs, and privacy concerns |
| AI in telemedicine | Optimises healthcare accessibility, particularly in geographically isolated regions, minimises waiting periods and improves patient results through real-time monitoring | Suboptimal diagnostic precision in comparison to face-to-face assessments, difficulties in expanding and guaranteeing data protection |
| AI in drug discovery | Optimises medication development by forecasting drug efficacy, minimises expenses and time, and discovers novel therapeutic substances | Prerequisites extensive, top-notch databases, ethical concerns about the use of AI in clinical trials |
| AI-powered chatbots for mental health | Delivers scalable mental health assistance, alleviates the workload on healthcare practitioners, provides prompt support in behavioural health treatments | Constrained by predetermined norms and reactions, possible errors in diagnosis, and apprehensions over user confidence |
| AI in surgical assistance (e.g., da Vinci) | Optimises accuracy in minimally invasive procedures, decreases recovery duration, and enhances surgical results | Expensive technology, dependence on AI systems, and demand for proficient personnel to oversee the systems' operations |
| AI for disease surveillance | Accelerated identification of epidemics, enhanced public health intervention, and anticipatory analytics for managing viral outbreaks | Potential for data abuse, privacy issues, and susceptibility to errors in intricate predictive models |
| Remote patient monitoring (RPM) | Enhances patient care by identifying health problems early, maintaining ongoing surveillance, and promoting optimal patient involvement | Analysis of data security vulnerabilities, possible biases in AI forecasts, and the necessity for dependable infrastructure |

# 5 Conclusion and Future Directions

Advanced AI technologies, such as ML and DL, profoundly affect drug discovery, telemedicine, diagnostics, and patient monitoring. Disease detection and early intervention have significantly improved thanks to algorithms created using AI techniques for clinical decision support systems and disease prediction models. AI's groundbreaking advances in diagnosis, treatment planning, and patient monitoring have drastically changed the healthcare sector. AI has made great strides in clinical drug trials, personalised treatment plans, and telemedicine. However, ethical questions of data security and openness of AI-driven algorithms should still be asked. Despite its enormous promise in virtual training, medical education, and mental health therapies, AI must be supported by human knowledge.

## *5.1 Challenges and Future Directions*

The field of AI is advancing rapidly in its role in medicine. However, before AI can be widely implemented, various hurdles like constraints, ethical considerations, and safety concerns must be tackled. The integration of AI technology into diagnosis and treatment holds promise. However, it is essential to focus on ensuring safety within this realm. One major obstacle medical AI applications face is the scarcity of labelled data, particularly amplified by the large-scale nature of images and diverse data types involved in the pathology (Khanna et al. 2020). The application of AI in cancer research has yielded promising outcomes by addressing issues that challenge physicians in providing treatments and ultimately eliminating the disease. AI provides methods and platforms that enhance our understanding and approach to managing this life-threatening illness. By aiding pathologists with cancer diagnoses AI, AI-driven technologies significantly reduce error rates. Furthermore, AI has demonstrated advantages across drug discovery and development; however, challenges, such as acquiring top-tier data quality, still demand attention.

Rule-based systems have drawbacks in intelligence despite numerous benefits and technological advancements. These downsides encompass development costs, limitations stemming from linkages and the necessity for extensive medical knowledge. While AI is not widely utilised in patient care, it is expected to play a significant role in intricate clinical decision-support processes soon (Vasey et al. 2022). Establishing principles for AI applications is crucial, as engaging in discussions with patients for their informed consent and conducting thorough evaluations of AI implementation to develop a global standard. This underscores the significance of cybersecurity measures, safeguarding privacy, considering stakeholder concerns and ensuring algorithm transparency to address shortcomings (Familoni 2024). Proactive leadership from organisations can significantly enhance trust in the security and effectiveness of medical AI, thus fostering further exploration in this promising field.

The Internet of Medical Things allows AI to enhance processes and facilitate consultations using other tech advances such as telemedicine. AI can effectively determine the treatment approach and categorise cancer stages based on risk levels. Common characteristics are revealed across levels in data analysis, enabling AI algorithms to identify connections between variables. By merging this information, AI studies can analyse events related to cancer precisely. In time, comprehensive databases will emerge, producing datasets encompassing all health aspects. These records will enable the creation of models that could customise treatment options, accurate dosage calculations, monitoring tactics, timelines and other relevant factors.

The following key points have been identified from this survey:

1. Machine learning and deep learning have enhanced medication discovery, telemedicine, diagnostics, and patient health monitoring.
2. AI-driven clinical decision support systems improve early disease identification, intervention, and therapy.
3. AI-powered clinical medication trials and personalised treatment regimens are showing promising results.
4. Data protection, algorithm clarity, and AI system integrity are crucial ethical issues.
5. AI is used in virtual training, medical education, and mental health therapies, but with human supervision.
6. Pathologists benefit from AI systems' low cancer research and diagnosis error rates.

Furthermore, the potential areas for improvement are:

1. **Data Quality and Availability**: AI needs high-quality, labelled data, especially for medical imaging and different data kinds. Data gathering and annotation require improvement.
2. **Algorithm Transparency**: Making AI systems transparent and intelligible is a major challenge.
3. **An ethical perspective**: When using AI, patient privacy, data security, and informed consent must be addressed.
4. **Healthcare AI Integration**: Research is needed to integrate AI into clinical workflows and decision-making.
5. **Security**: Protecting sensitive medical data from breaches and misuse requires strong cybersecurity.
6. **Stakeholder Engagement**: In AI implementation talks, patients, healthcare professionals, and other stakeholders build trust and adoption.
7. **AI in Cancer Research**: AI offers promise in cancer research; however, data processing and stage categorisation need refinement.

## References

Akkus, Z., Kostandy, P., Philbrick, K. A., & Erickson, B. J. (2018). *Extraction of brain tissue from CT head images using fully convolutional neural networks*. 71. https://doi.org/10.1117/12.2293423

Alhashmi, S. F. S., Alshurideh, M., Al Kurdi, B., & Salloum, S. A. (2020). A Systematic Review of the Factors Affecting the Artificial Intelligence Implementation in the Health Care Sector. *Advances in Intelligent Systems and Computing, 1153 AISC*, 37–49. https://doi.org/10.1007/978-3-030-44289-7_4

Ali, O., Abdelbaki, W., Shrestha, A., Elbasi, E., Alryalat, M. A. A., & Dwivedi, Y. K. (2023). A systematic literature review of artificial intelligence in the healthcare sector: Benefits, challenges, methodologies, and functionalities. *Journal of Innovation and Knowledge, 8*(1). https://doi.org/10.1016/j.jik.2023.100333

Alipanahi, B., Delong, A., Weirauch, M. T., & Frey, B. J. (2015). Predicting the sequence specificities of DNA- and RNA-binding proteins by deep learning. *Nature Biotechnology, 33*(8), 831–838. https://doi.org/10.1038/nbt.3300

Álvarez-Machancoses, Ó., Galiana, E. J. D., Cernea, A., de la Viña, J. F., & Fernández-Martínez, J. L. (2020). On the role of artificial intelligence in genomics to enhance precision medicine. *Pharmacogenomics and Personalized Medicine, 13*, 105–119. https://doi.org/10.2147/PGPM.S205082

Andreeva, V., Aksamentova, E., Muhachev, A., Solovey, A., Litvinov, I., Gusarov, A., Shevtsova, N. N., Kushkin, D., & Litvinova, K. (2022). Preoperative AI-Driven Fluorescence Diagnosis of Non-Melanoma Skin Cancer. *Diagnostics, 12*(1), 72. https://www.mdpi.com/2075-4418/12/1/72/pdf%0A,http://ovidsp.ovid.com/ovidweb.cgi?T=JS&PAGE=reference&D=emed25&NEWS=N&AN=2015275603

Antunes, A. G., Teixeira, C., Vaz, A. M., Martins, C., Queirós, P., Alves, A., Velasco, F., Peixe, B., Oliveira, A. P., & Guerreiro, H. (2017). Comparison of the prognostic value of Chronic Liver Failure Consortium scores and traditional models for predicting mortality in patients with cirrhosis. *Gastroenterología y Hepatología (English Edition), 40*(4), 276–285. https://doi.org/10.1016/j.gastre.2017.03.012

Barrett, M., Boyne, J., Brandts, J., Brunner-La Rocca, H. P., De Maesschalck, L., De Wit, K., Dixon, L., Eurlings, C., Fitzsimons, D., Golubnitschaja, O., Hageman, A., Heemskerk, F., Hintzen, A., Helms, T. M., Hill, L., Hoedemakers, T., Marx, N., McDonald, K., Mertens, M., … Zippel-Schultz, B. (2019). Artificial intelligence supported patient self-care in chronic heart failure: a paradigm shift from reactive to predictive, preventive and personalised care. *EPMA Journal, 10*(4), 445–464. https://doi.org/10.1007/s13167-019-00188-9

Battineni, G., Chintalapudi, N., Amenta, F., & Traini, E. (2020). A comprehensive machine-learning model applied to magnetic resonance imaging (MRI) to predict Alzheimer's disease (ad) in older subjects. *Journal of Clinical Medicine, 9*(7), 1–14. https://doi.org/10.3390/jcm9072146

Beckmann, J. S., & Lew, D. (2016). Reconciling evidence-based medicine and precision medicine in the era of big data: Challenges and opportunities. *Genome Medicine, 8*(1). https://doi.org/10.1186/s13073-016-0388-7

Beneke, F., & Mackenrodt, M. O. (2019). Artificial Intelligence and Collusion. In *IIC International Review of Intellectual Property and Competition Law* (Vol. 50, Issue 1). https://doi.org/10.1007/s40319-018-00773-x

Blasiak, A., Khong, J., & Kee, T. (2020). CURATE.AI: Optimizing Personalized Medicine with Artificial Intelligence. *SLAS Technology, 25*(2), 95–105. https://doi.org/10.1177/2472630319890316

Bokhari, Y., Alhareeri, A., Aljouie, A., Alkhaldi, A., Rashid, M., Alawad, M., Alhassnan, R., Samargandy, S., Panahi, A., Heidrich, W., & Arodz, T. (2022). ChromoEnhancer: An Artificial-Intelligence-Based Tool to Enhance Neoplastic Karyograms as an Aid for Effective Analysis. *Cells, 11*(14). https://doi.org/10.3390/cells11142244

Brinker, T. J., Hekler, A., Enk, A. H., Berking, C., Haferkamp, S., Hauschild, A., Weichenthal, M., Klode, J., Schadendorf, D., Holland-Letz, T., Kalle, C. von, Frohling, S., Schilling, B., & Utikal, J. S. (2019). Deep neural networks are superior to dermatologists in melanoma image classification. *European Journal of Cancer*

Buolamwini, J., & Gebru, T. (2018). Gender Shades: Intersectional Accuracy Disparities in Commercial Gender Classification. *Proceedings of Machine Learning Research, 81*, 77–91.

Caudai, C., Galizia, A., Geraci, F., Le Pera, L., Morea, V., Salerno, E., Via, A., & Colombo, T. (2021). "It will change everything": DeepMind's AI makes gigantic leap in solving protein structures. *Computational and Structural Biotechnology Journal, 19*, 5762–5790. https://doi.org/10.1016/j.csbj.2021.10.009

Chandrashekar, A., Shivakumar, N., Lapolla, P., Handa, A., Grau, V., & Lee, R. (2020). A deep learning approach to generate contrast-enhanced computerised tomography angiograms without the use of intravenous contrast agents. *European Heart Journal, 41*(Supplement_2). https://doi.org/10.1093/ehjci/ehaa946.0156

Chang, H. Y., Yeh, C. Y., Lee, C. Te, & Lin, C. C. (2020). A sleep apnea detection system based on a one-dimensional deep convolution neural network model using single-lead electrocardiogram. *Sensors (Switzerland), 20*(15), 1–15. https://doi.org/10.3390/s20154157

Chen, P., Chen Xu, R., Chen, N., Zhang, L., Zhang, L., Zhu, J., Pan, B., Wang, B., & Guo, W. (2021). Detection of Metastatic Tumor Cells in the Bone Marrow Aspirate Smears by Artificial Intelligence (AI)-Based Morphogo System. *Frontiers in Oncology, 11*. https://doi.org/10.3389/fonc.2021.742395

Cleophas, T. J., & Zwinderman, A. H. (2013). Machine learning in medicine. In *Machine Learning in Medicine* (Vol. 9789400758). https://doi.org/10.1007/978-94-007-5824-7

Collins Oguine, O., & Jane Oguine, K. (2022). *AI in Telemedicine: An Appraisal on Deep Learning-based Approaches to Virtual Diagnostic Solutions (VDS)*. 229–243. https://doi.org/10.5121/csit.2022.121320

Comito, C., Falcone, D., & Forestiero, A. (2022). AI-Driven Clinical Decision Support: Enhancing Disease Diagnosis Exploiting Patients Similarity. *IEEE Access, 10*, 6878–6888. https://doi.org/10.1109/ACCESS.2022.3142100

Coustasse, A., Sarkar, R., Abodunde, B., Metzger, B. J., & Slater, C. M. (2019). Use of Teledermatology to Improve Dermatological Access in Rural Areas. *Telemedicine and E-Health, 25*(11), 1022–1032. https://doi.org/10.1089/tmj.2018.0130

Creutzfeldt, J., Hedman, L., & Felländer-Tsai, L. (2016). Cardiopulmonary resuscitation training by avatars: A qualitative study of medical students' experiences using a multiplayer virtual world. *JMIR Serious Games, 4*(2). https://doi.org/10.2196/games.6448

Currie, G., Hawk, K. E., Rohren, E., Vial, A., & Klein, R. (2019). Machine Learning and Deep Learning in Medical Imaging: Intelligent Imaging. *Journal of Medical Imaging and Radiation Sciences, 50*(4), 477–487. https://doi.org/10.1016/j.jmir.2019.09.005

De Farias, F. A. C. D., Dagostini, C. M., Bicca, Y. D. A., Falavigna, V. F., & Falavigna, A. (2020). Remote patient monitoring: A systematic review. *Telemedicine and E-Health, 26*(5), 576–583. https://doi.org/10.1089/tmj.2019.0066

Dey, A., Biswal, S., & Sundarapandian, S. (2022). Nanotechnology Applications in Nanomedicine: Prospects and Challenges. *Diverse Applications of Nanotechnology in the Biological Sciences*, 1–37. https://doi.org/10.1201/9781003277255-1

Dileep, G., & Gianchandani Gyani, S. G. (2022). Artificial Intelligence in Breast Cancer Screening and Diagnosis. *Cureus*. https://doi.org/10.7759/cureus.30318

Dong, Z., Zhang, N., Li, C., Wang, H., Fang, Y., Wang, J., & Zheng, X. (2015). Anticancer drug sensitivity prediction in cell lines from baseline gene expression through recursive feature selection. *BMC Cancer, 15*(1). https://doi.org/10.1186/s12885-015-1492-6

Fabrizio, C., Termine, A., Caltagirone, C., & Sancesario, G. (2021). Artificial intelligence for Alzheimer's disease: Promise or challenge? *Diagnostics, 11*(8). https://doi.org/10.3390/diagnostics11081473

Familoni, B. T. (2024). Cybersecurity Challenges in the Age of Ai: Theoretical Approaches and Practical Solutions. *Computer Science & IT Research Journal, 5*(3), 703–724. https://doi.org/10.51594/csitrj.v5i3.930

Gedefaw, L., Liu, C. F., Ip, R. K. L., Tse, H. F., Yeung, M. H. Y., Yip, S. P., & Huang, C. L. (2023). Artificial Intelligence-Assisted Diagnostic Cytology and Genomic Testing for Hematologic Disorders. *Cells, 12*(13). https://doi.org/10.3390/cells12131755

Hameed, B. S., & Krishnan, U. M. (2022). Artificial Intelligence-Driven Diagnosis of Pancreatic Cancer. *Cancers, 14*(21). https://doi.org/10.3390/cancers14215382

Hawkins, S., Wang, H., Liu, Y., Garcia, A., Stringfield, O., Krewer, H., Li, Q., Cherezov, D., Gatenby, R. A., Balagurunathan, Y., Goldgof, D., Schabath, M. B., Hall, L., & Gillies, R. J. (2016). Predicting Malignant Nodules from Screening CT Scans. *Journal of Thoracic Oncology, 11*(12), 2120–2128. https://doi.org/10.1016/j.jtho.2016.07.002

He, L., Bulanova, D., Oikkonen, J., Häkkinen, A., Zhang, K., Zheng, S., Wang, W., Erkan, E. P., CarpCrossed Dsign©n, O., Joutsiniemi, T., Hietanen, S., Hynninen, J., Huhtinen, K., Hautaniemi, S., Vähärautio, A., Tang, J., Wennerberg, K., & Aittokallio, T. (2021). Network-guided identification of cancer-selective combinatorial therapies in ovarian cancer. *Briefings in Bioinformatics, 22*(6). https://doi.org/10.1093/bib/bbab272

Hohma, B. E. (2021). *Assessing Fairness in AI-enabled Public Health Surveillance. October*

Iqbal, J., Cortés Jaimes, D. C., Makineni, P., Subramani, S., Hemaida, S., Thugu, T. R., Butt, A. N., Sikto, J. T., Kaur, P., Lak, M. A., Augustine, M., Shahzad, R., & Arain, M. (2023). Reimagining Healthcare: Unleashing the Power of Artificial Intelligence in Medicine. *Cureus*. https://doi.org/10.7759/cureus.44658

Iqbal, J., Jahangir, K., Mashkoor, Y., Sultana, N., Mehmood, D., Ashraf, M., Iqbal, A., & Hafeez, M. H. (2022). The future of artificial intelligence in neurosurgery: A narrative review. *Surgical Neurology International, 13*. https://doi.org/10.25259/SNI_877_2022

Kalra, N., Verma, P., & Verma, S. (2024). Advancements in AI based healthcare techniques with FOCUS ON diagnostic techniques. *Computers in Biology and Medicine, 179*. https://doi.org/10.1016/j.compbiomed.2024.108917

Khakharia, A., Shah, V., Jain, S., Shah, J., Tiwari, A., Daphal, P., Warang, M., & Mehendale, N. (2021). Outbreak Prediction of COVID-19 for Dense and Populated Countries Using Machine Learning. *Annals of Data Science, 8*(1), 1–19. https://doi.org/10.1007/s40745-020-00314-9

Khan, B., Fatima, H., Qureshi, A., Kumar, S., Hanan, A., Hussain, J., & Abdullah, S. (2023). Drawbacks of Artificial Intelligence and Their Potential Solutions in the Healthcare Sector. *Biomedical Materials and Devices, 1*(2), 731–738. https://doi.org/10.1007/s44174-023-00063-2

Khan, K., McNabb, S. J. N., Memish, Z. A., Eckhardt, R., Hu, W., Kossowsky, D., Sears, J., Arino, J., Johansson, A., Barbeschi, M., McCloskey, B., Henry, B., Cetron, M., & Brownstein, J. S. (2012). Infectious disease surveillance and modelling across geographic frontiers and scientific specialties. *The Lancet Infectious Diseases, 12*(3), 222–230. https://doi.org/10.1016/S1473-3099(11)70313-9

Khanna, S., Srivastava, S., Khanna, I., & Pandey, V. (2020). Current Challenges and Opportunities in Implementing AI/ML in Cancer Imaging: Integration, Development, and Adoption Perspectives. *Journal of Advanced Analytics in Healthcare Management, 4*(10), 1–25. https://research.tensorgate.org/index.php/JAAHM/article/view/104

Khawaja, Z., & Bélisle-Pipon, J. C. (2023). Your robot therapist is not your therapist: understanding the role of AI-powered mental health chatbots. *Frontiers in Digital Health, 5*. https://doi.org/10.3389/fdgth.2023.1278186

Kitsiou, S., Paré, G., Jaana, M., & Gerber, B. (2017). Effectiveness of mHealth interventions for patients with diabetes: An overview of systematic reviews. *PLoS ONE, 12*(3). https://doi.org/10.1371/journal.pone.0173160

Klingler, C., Silva, D. S., Schuermann, C., Reis, A. A., Saxena, A., & Strech, D. (2017). Ethical issues in public health surveillance: A systematic qualitative review. *BMC Public Health, 17*(1). https://doi.org/10.1186/s12889-017-4200-4

Kratz, A., Bengtsson, H. I., Casey, J. E., Keefe, J. M., Beatrice, G. H., Grzybek, D. Y., Lewandrowski, K. B., & Van Cott, E. M. (2005). Performance evaluation of the CellaVision DM96 system: WBC differentials by automated digital image analysis supported by an artificial neural network. *American Journal of Clinical Pathology, 124*(5), 770–781. https://doi.org/10.1309/XMB9K0J41LHLATAY

Kropp, M., Golubnitschaja, O., Mazurakova, A., Koklesova, L., Sargheini, N., Vo, T. T. K. S., de Clerck, E., Polivka, J., Potuznik, P., Polivka, J., Stetkarova, I., Kubatka, P., & Thumann, G. (2023). Diabetic retinopathy as the leading cause of blindness and early predictor of cascading complications—risks and mitigation. *EPMA Journal, 14*(1), 21–42. https://doi.org/10.1007/s13167-023-00314-8

Li, L., Qin, L., Xu, Z., Yin, Y., Wang, X., Kong, B., Bai, J., Lu, Y., Fang, Z., Song, Q., Cao, K., Liu, D., Wang, G., Xu, Q., Fang, X., Zhang, S., Xia, J., & Xia, J. (2020). Using Artificial Intelligence to Detect COVID-19 and Community-acquired Pneumonia Based on Pulmonary CT: Evaluation of the Diagnostic Accuracy. *Radiology, 296*(2), E65–E71. https://doi.org/10.1148/radiol.2020200905

Lin, K., Liu, J., & Gao, J. (2022). AI-Driven Decision Making for Auxiliary Diagnosis of Epidemic Diseases. *IEEE Transactions on Molecular, Biological, and Multi-Scale Communications, 8*(1), 9–16. https://doi.org/10.1109/TMBMC.2021.3120646

Liu, P., Lassén, E., Nair, V., Berthier, C. C., Suguro, M., Sihlbom, C., Kretzler, M., Betsholtz, C., Haraldsson, B., Ju, W., Ebefors, K., & Nyström, J. (2017). Transcriptomic and proteomic profiling provides insight into mesangial cell function in IgA nephropathy. *Journal of the American Society of Nephrology, 28*(10), 2961–2972. https://doi.org/10.1681/ASN.2016101103

MacLennan, A. (1996). The artificial life route to artificial intelligence: Building embodied, situated agents. *Journal of the American Society for Information Science, 47*(6), 482–483. https://doi.org/10.1002/(sici)1097-4571(199606)47:6<482::aid-asi14>3.0.co;2-0

Mantena, S., & Keshavjee, S. (2021). Strengthening healthcare delivery with remote patient monitoring in the time of COVID-19. *BMJ Health and Care Informatics, 28*(1). https://doi.org/10.1136/bmjhci-2020-100302

Meskó, B., Drobni, Z., Bényei, É., Gergely, B., & Győrffy, Z. (2017). Digital health is a cultural transformation of traditional healthcare. *MHealth, 3*, 38–38. https://doi.org/10.21037/mhealth.2017.08.07

Mirmomeni, M., Fazio, T., Von Cavallar, S., & Harrer, S. (2020). From wearables to THINKables: Artificial intelligence-enabled sensors for health monitoring. *Wearable Sensors: Fundamentals, Implementation and Applications*, 339–356. https://doi.org/10.1016/B978-0-12-819246-7.00012-7

Mostafa, F. A., Elrefaei, L. A., Fouda, M. M., & Hossam, A. (2022). A Survey on AI Techniques for Thoracic Diseases Diagnosis Using Medical Images. *Diagnostics, 12*(12). https://doi.org/10.3390/diagnostics12123034

Nadella, G. S., Nadella, G. S., Satish, S., Meduri, K., & Meduri, S. S. (2023). A Systematic Literature Review of Advancements, Challenges and Future Directions of AI And ML in Healthcare. *International Journal of Machine Learning for Sustainable Development, 5*(3), 115–130. https://www.ijsdcs.com/index.php/IJMLSD/article/view/519

Nasr, M., Islam, M. M., Shehata, S., Karray, F., & Quintana, Y. (2021). Smart Healthcare in the Age of AI: Recent Advances, Challenges, and Future Prospects. *IEEE Access, 9*, 145248–145270. https://doi.org/10.1109/ACCESS.2021.3118960

Nguyen, M. T., Nguyen, B. Van, & Kim, K. (2018). Deep Feature Learning for Sudden Cardiac Arrest Detection in Automated External Defibrillators. *Scientific Reports, 8*(1), 17196. http://www.embase.com/search/results?subaction=viewrecord&from=export&id=L625118467%0Ahttps://doi.org/10.1038/s41598-018-33424-9

Nomura, A., Noguchi, M., Kometani, M., Furukawa, K., & Yoneda, T. (2021). Artificial Intelligence in Current Diabetes Management and Prediction. *Current Diabetes Reports, 21*(12). https://doi.org/10.1007/s11892-021-01423-2

Olive-Gadea, M., Crespo, C., Granes, C., Hernandez-Perez, M., Pérez De La Ossa, N., Laredo, C., Urra, X., Carlos Soler, J., Soler, A., Puyalto, P., Cuadras, P., Marti, C., & Ribo, M. (2020). Deep Learning Based Software to Identify Large Vessel Occlusion on Noncontrast Computed Tomography. *Stroke, 51*(10), 3133–3137. https://doi.org/10.1161/STROKEAHA.120.030326

Olivier, M., Asmis, R., Hawkins, G. A., Howard, T. D., & Cox, L. A. (2019). The need for multi-omics biomarker signatures in precision medicine. *International Journal of Molecular Sciences, 20*(19). https://doi.org/10.3390/ijms20194781

Paro, M., Lambert, W. A., Leclair, N. K., Romano, R., Stoltz, P., Martin, J. E., Hersh, D. S., & Bookland, M. J. (2022). Machine Learning–Driven Clinical Image Analysis to Identify Craniosynostosis: A Pilot Study of Telemedicine and Clinic Patients. *Neurosurgery, 90*(5), 613–618. https://doi.org/10.1227/neu.0000000000001890

Pozza, A., Zanella, L., Castaldi, B., & Di Salvo, G. (2024). How Will Artificial Intelligence Shape the Future of Decision-Making in Congenital Heart Disease? *Journal of Clinical Medicine, 13*(10). https://doi.org/10.3390/jcm13102996

Pu, L., Naderi, M., Liu, T., Wu, H.-C., Mukhopadhyay, S., & Brylinski, M. (2019). eToxPred: a machine learning-based approach to estimate the toxicity of drug candidates. *BMC Pharmacology and Toxicology, 20*(1). https://doi.org/10.1186/s40360-018-0282-6

Raddo, T. R., Rommel, S., Cimoli, B., Vagionas, C., Perez-Galacho, D., Pikasis, E., Grivas, E., Ntontin, K., Katsikis, M., Kritharidis, D., Ruggeri, E., Spaleniak, I., Dubov, M., Klonidis, D., Kalfas, G., Sales, S., Pleros, N., & Tafur Monroy, I. (2021). Transition technologies towards 6G networks. *Eurasip Journal on Wireless Communications and Networking, 2021*(1). https://doi.org/10.1186/s13638-021-01973-9

Roppelt, J. S., Kanbach, D. K., & Kraus, S. (2024). Artificial intelligence in healthcare institutions: A systematic literature review on influencing factors. *Technology in Society, 76*. https://doi.org/10.1016/j.techsoc.2023.102443

Santosh, K. C., & Gaur, L. (2021). Artificial Intelligence and Machine Learning in Public Healthcare. *SpringerBriefs in Applied Sciences and Technology*. https://doi.org/10.1007/978-981-16-6768-8

Sapci, A. H., & Sapci, H. A. (2020). Artificial intelligence education and tools for medical and health informatics students: systematic review. *JMIR Medical Education, Query date: 2023–09–08 20:45:26*. https://mededu.jmir.org/2020/1/e19285/

Saxena, A. K., Ness, S., & Khinvasara, T. (2024). The Influence of AI: The Revolutionary Effects of Artificial Intelligence in Healthcare Sector. *Journal of Engineering Research and Reports, 26*(3), 49–62. https://doi.org/10.9734/jerr/2024/v26i31092

Shaik, T., Tao, X., Higgins, N., Li, L., Gururajan, R., Zhou, X., & Acharya, U. R. (2023). Remote patient monitoring using artificial intelligence: Current state, applications, and challenges. *Wiley Interdisciplinary Reviews: Data Mining and Knowledge Discovery, 13*(2). https://doi.org/10.1002/widm.1485

Shiwlani, A., Khan, M., Mannan, A., Sherani, K., Qayyum, M. U., & Hussain, H. K. (2024). Revolutionizing Healthcare: the Impact of Artificial Intelligence on Patient Care, Diagnosis, and Treatment. *Jurnalmahasiswa.Com, 1*(5), 779–790. http://jurnalmahasiswa.com/index.php/Jurihum/article/view/845

Singh, S., & Thakur, H. K. (2020). Survey of Various AI Chatbots Based on Technology Used. *ICRITO 2020 - IEEE 8th International Conference on Reliability, Infocom Technologies and Optimization (Trends and Future Directions)*, 1074–1079. https://doi.org/10.1109/ICRITO48877.2020.9197943

Strous, L., Johnson, R. (Computer historian), Grier, D. A., & Swade, D. (2021). *Unimagined futures - ICT opportunities and challenges*

Sweeney, C., Potts, C., Ennis, E., Bond, R., Mulvenna, M. D., O'Neill, S., Malcolm, M., Kuosmanen, L., Kostenius, C., Vakaloudis, A., McConvey, G., Turkington, R., Hanna, D., Nieminen, H., Vartiainen, A. K., Robertson, A., & McTear, M. F. (2021). Can Chatbots Help Support a Person's Mental Health? Perceptions and Views from Mental Healthcare Professionals and Experts. *ACM Transactions on Computing for Healthcare, 2*(3). https://doi.org/10.1145/3453175

Taylor, R. A., Pare, J. R., Venkatesh, A. K., Mowafi, H., Melnick, E. R., Fleischman, W., & Hall, M. K. (2016). Prediction of In-hospital Mortality in Emergency Department Patients with Sepsis: A Local Big Data-Driven, Machine Learning Approach. *Academic Emergency Medicine, 23*(3), 269–278. https://doi.org/10.1111/acem.12876

Topol, E. J. (2019). High-performance medicine: the convergence of human and artificial intelligence. *Nature Medicine*, 44–56

Tran, W. T., Sadeghi-Naini, A., Lu, F. I., Gandhi, S., Meti, N., Brackstone, M., Rakovitch, E., & Curpen, B. (2021). Computational Radiology in Breast Cancer Screening and Diagnosis Using Artificial Intelligence. *Canadian Association of Radiologists Journal, 72*(1), 98–108. https://doi.org/10.1177/0846537120949974

Tripathi, M. K., Nath, A., Singh, T. P., Ethayathulla, A. S., & Kaur, P. (2021). Evolving scenario of big data and Artificial Intelligence (AI) in drug discovery. *Molecular Diversity, 25*(3), 1439–1460. https://doi.org/10.1007/s11030-021-10256-w

Vasey, B., Nagendran, M., Campbell, B., Clifton, D. A., Collins, G. S., Denaxas, S., Denniston, A. K., Faes, L., Geerts, B., Ibrahim, M., Liu, X., Mateen, B. A., Mathur, P., Mccradden, M. D., Morgan, L., Ordish, J., Rogers, C., Saria, S., Ting, D. S. W., … Mcculloch, P. (2022). Reporting guideline for the early stage clinical evaluation of decision support systems driven by artificial intelligence: DECIDE-AI. *The BMJ*. https://doi.org/10.1136/bmj-2022-070904

Vural, S., Wang, X., & Guda, C. (2016). Classification of breast cancer patients using somatic mutation profiles and machine learning approaches. *BMC Systems Biology, 10*. https://doi.org/10.1186/s12918-016-0306-z

Xu, Y., Liu, X., Cao, X., Huang, C., Liu, E., Qian, S., Liu, X., Wu, Y., Dong, F., Qiu, C. W., Qiu, J., Hua, K., Su, W., Wu, J., Xu, H., Han, Y., Fu, C., Yin, Z., Liu, M., … Zhang, J. (2021). Artificial intelligence: A powerful paradigm for scientific research. *Innovation, 2*(4). https://doi.org/10.1016/j.xinn.2021.100179

Zeng, D., Cao, Z., & Neill, D. B. (2020). Artificial intelligence–enabled public health surveillance—from local detection to global epidemic monitoring and control. *Artificial Intelligence in Medicine: Technical Basis and Clinical Applications*, 437–453. https://doi.org/10.1016/B978-0-12-821259-2.00022-3

Zhang, Z., Li, G., Xu, Y., & Tang, X. (2021). Application of artificial intelligence in the mri classification task of human brain neurological and psychiatric diseases: A scoping review. *Diagnostics, 11*(8). https://doi.org/10.3390/diagnostics11081402

Zhou, J., & Troyanskaya, O. G. (2015). Predicting effects of noncoding variants with deep learning-based sequence model. *Nature Methods, 12*(10), 931–934. https://doi.org/10.1038/nmeth.3547

Zhuang, D., & Ibrahim, A. K. (2021). Deep learning for drug discovery: A study of identifying high efficacy drug compounds using a cascade transfer learning approach. *Applied Sciences (Switzerland), 11*(17). https://doi.org/10.3390/app11177772

Ziad, O., & Ezekiel, J. E. (2016). Predicting the Future — Big Data, Machine Learning, and Clinical Medicine. *New England Journal of Medicine, 375*(13), 1216–1219.

# Strategies for AI in Healthcare Implementation

**Jyotismita Talukdar, Deba Kanta Gogoi, and Rosy Sharma**

**Abstract** This chapter aims to explore various strategies for implementing artificial intelligence (AI) in healthcare implementation. AI in healthcare innovations have transformed healthcare by enhancing diagnostics, personalizing treatment plans, and enabling informed decision-making for both patients and healthcare providers. With increasing demand, key AI technologies such as machine learning (ML), deep learning (DL), natural language processing (NLP), and rule-based expert systems are discussed. The chapter also discuss how healthcare organizations can effectively implement AI technologies while navigating the complexities of healthcare environments. By integrating multiple AI modalities, healthcare delivery can be significantly improved, from operational efficiency to patient outcomes. Key contributions and challenges future directions associated with AI in healthcare are highlighted.

**Keywords** Artificial intelligence · Healthcare · Machine learning

## 1 Introduction

Artificial intelligence (AI) entails machines replicating human intelligence functions, including learning, reasoning, and self-correction. It incorporates diverse methodologies such as machine learning (ML), deep learning (DL), natural language processing (NLP), computer vision (CV), and robotics, with the objective of enabling machines to perform tasks often related to human intelligence. AI, showcasing machine intelligence, has a profound influence across diverse industries. Tedros A. Ghebreyesus, Director General, WHO said, "AI widely used in diagnosis and clinical care, drug development, disease surveillance, outbreak response, and health systems management. The future of healthcare is digital, and we must do what we can to encourage

J. Talukdar (✉) · D. K. Gogoi · R. Sharma
Department of CSE, Tezpur University, Tezpur, Assam, India
e-mail: jyoti4@tezu.ernet.in

R. Sharma
e-mail: rosy8@tezu.ernet.in

 
T. P. Singh et al. (eds.), *Revolutionizing Healthcare: Impact of Artificial Intelligence on Diagnosis, Treatment, and Patient Care*, Studies in Computational Intelligence 1182,
https://doi.org/10.1007/978-3-031-80813-5_21

equitable access to these innovations and prevent them from becoming another driver for inequity". The adoption of AI in the healthcare industry is accelerating, emerging as a crucial trend to enhance communication between patients and healthcare providers.

It markedly enhances diagnostics, treatment, and enables rapid, well-informed medical decisions for both patients and providers. Healthcare AI is expected to be a USD 20.9 billion worldwide market by 2024. And by 2029, it's projected to soar to USD 148.4 billion, with a forecasted compound annual growth rate (CAGR) of 48.1%. As a result, investments in AI-driven healthcare solutions have notably surged in recent years, mirroring the increasing demand. The growth of the AI healthcare market is propelled by the rise of large-scale healthcare datasets, need to reduce healthcare expenses, improved computational ability with cheaper hardware, increasing partnerships and collaborations over varied healthcare field, and a field for better health support to address the workforce-patient disparities.

## *1.1 AI Technologies Relevant to Healthcare*

Certain AI technologies hold significant importance in the domain of healthcare like ML, especially neural networks (NN) and DL, NLP, Rule based Expert systems (RBES), Physical Robots and Robotic process Automation (RPA) (Davenport and Kalakota 2019). ML, a statistical technique for modelling data and learning through training, is a prevalent aspect of AI. In healthcare, it's commonly utilized in precision medicine to foresee effective treatment protocols based on patient features and environment (MacEachern and Forkert 2021). NN, a more complex form of ML, have long been utilized in healthcare for categorization tasks, like disease prediction. DL, the most complex variant, employs NN with numerous layers to predict outcomes, often applied in tasks like cancerous lesion recognition in radiology images (Zafar et al. 2023). Its combination with radiomics shows promise in enhancing diagnosis accuracy in oncology (Vial et al. 2018). DL is also utilized for various omics data such as genomics, methylation, transcriptomics, and building genomic inference based on histopathology (Tran et al. 2021). However, DL models can be challenging to interpret due to the complexity of their features. NLP incorporates speech recognition, textual analysis, translation, and other language-based tasks. Among the two approaches of NLP, viz., statistical and semantic, statistical NLP is reliant on ML and DL thereby boosting recognition accuracies. In healthcare, NLP is crucial for interpreting and categorizing clinical record and research (Sheikhalishahi et al. 2019). It analyses unstructured clinical notes, produce reports (e.g., radiology findings) thereby supporting clinical decision-making, transcribes patient interactions, and enables patient engagement with NLP-powered chatbots and virtual assistants.

Rule based Expert systems in healthcare are computer-based systems that mimic the decision-making capabilities of human experts. They use 'if–then' rules to provide diagnostic or treatment recommendations based on patient data. While useful for decision support, they require frequent updates and have limitations in handling complex situations. Adapting to changes in the knowledge domain can also be cumbersome. Consequently, they will replace in healthcare by data-driven techniques and ML algorithms (Davenport and Kalakota 2019). Physical robots are quite familiar to us with over 3.4 million industrial robots installed worldwide executing jobs such as lifting, welding, and delivering supplies in places like factories and hospitals. Recent advancements include collaborative capabilities with humans and easier training methods.

With the integration of AI features into their operating systems these robots are becoming more intelligent. The review in Morgan et al. (2022) identified ten primary functions for robotic healthcare systems: surgical, rehabilitation and mobility, radiotherapy, socially assistive, telepresence, pharmacy, disinfection, delivery and transport, interventional, and imaging assistance.

Robotic process automation (RPA) involves computer programs that mimic human actions within information systems performing structured digital tasks typically done by humans, following scripts or rules. In healthcare, RPA automates repetitive tasks like prior authorization and billing. The technologies such as image recognition (IR), RPA can derive data from faxed images and feed it into systems. As these technologies evolve, they become more integrated, with robots gaining AI potentiality and IR merging with RPA. In the future, it's likely that integrated solutions merging these technologies will become increasingly prevalent.

## *1.2 Applications of AI in Healthcare*

In the healthcare field, various types of AI are relevant, each serving different purposes and applications:

1. Diagnostics AI: These systems scrutinize medical data such as images, test results, and patient symptoms to help in diagnosing diseases and conditions. For instance, AI-driven medical imaging tools assist in identifying abnormalities in X-rays, MRIs, or CT scans.
2. Predictive AI: Employing ML algorithms, predictive AI anticipates outcomes based on historical data. In healthcare, it forecasts disease progression, patient outcomes, and potential complications, informing personalized treatment plans and preventive care approaches.
3. Personalized AI: Tailored to individual patients, personalized AI systems leverage patient-specific data, including genetics, medical history, and lifestyle characteristics, to deliver personalized remedy recommendations and interventions.
4. Remote Monitoring AI: AI-powered remote monitoring facilitates healthcare providers to track patients' health condition beyond conventional clinical settings. By collecting and analyzing data from wearables, sensors, and other connected health devices, these systems monitor vital signs, medication adherence, and disease progression in real-time (Al-Antari 2023).

5. Administrative AI: Administrative AI applications streamline healthcare operations and administrative tasks such as appointment scheduling, billing, and electronic health record (EHR) management. NLP algorithms simplify clinical documentation and coding by automatically transcribing physician–patient interactions into structured data.
6. Drug Discovery AI: AI algorithms expedite drug discovery and development by analyzing extensive biological, chemical, and clinical data. These systems detect promising drug candidates, assess their effectiveness and safety profiles, and maximize treatment protocols, fostering more efficient drug development processes and innovative therapeutic solutions.

These examples underscore the diverse roles AI plays in healthcare, each contributing to enhanced patient care, clinical decision-making, and operational efficacy across the healthcare landscape. The above applications of AI have been extensively used in clinical and healthcare settings. While we've detailed these technologies separately, there's a rising inclination to merge them. Robots are acquiring AI-driven functionalities, and IR is being integrated into robotic process automation (RPA). Likewise, the fusion of CV and robotics can elevate surgical procedures by offering live guidance and heightened precision during operations. Furthermore, merging predictive analytics with remote monitoring technologies allows for early identification of health concerns and tailored interventions, resulting in better patient outcomes and lowered healthcare expenses. Additionally, incorporating AI into wearable gadgets and mobile apps empowers patients to take charge of their health and embrace preventive measures. Therefore, combining different AI methods in healthcare can revolutionize healthcare delivery, making it more personalized, efficient, and accessible for both patients and healthcare providers. In the future, these technologies may merge further, leading to more widespread and feasible composite solutions (Talukdar et al. 2023b).

## *1.3 Artificial Intelligence in Clinical Care*

### 1.3.1 Clinical Decision Tools

AI excels in clinical decision support by analyzing extensive patient data, surpassing traditional tools like Modified Early Warning Score (MEWS), commonly used by medical professionals to anticipate a patient's risk of clinical degradation within the next couple of hours, highlighting AI's superior accuracy over bedside calculations (Jitani et al. 2023). The efficacy of AI tools will soon hinge on their seamless integration into healthcare systems. These advanced tools require expert oversight for safety, strong IT infrastructure backing, and frontline users' readiness to embrace these models (Talukdar et al. 2023d).

### 1.3.2 Medical Diagnostics

Medical diagnostics involves analyzing symptoms, history, and test results to accurately diagnose conditions and provide effective treatment (Zafar et al. 2023). Diagnostic tests like imaging tests (for e.g., X-rays, MRI, CT, PET, fMRI scans), blood tests, and biopsies help guide treatment decisions, monitor progress, assess treatment effectiveness, and identify potential health issues early. With the AI revolution, there's potential to improve prediction accuracy, speed and efficiency in diagnostics, revolutionizing healthcare. AI can process vast quantities of patient data, including medical imaging, bio-signals (ECG, EMG, EEG, and EHR), vital signs (body temperature, pulse rate, respiration rate, blood pressure, blood oxygen, blood glucose etc.), medical history, demographic info, and laboratory results. This aids decision-making and enables precise predictions, empowering healthcare practitioners to make informed choices related to patient care.

Integrating multiple data sources provides a thorough understanding of a patient's health, lowering misdiagnosis risk and improving accuracy. Monitoring conditions over time with multimodal data allows for more efficient treatment of chronic diseases. Moreover, Explainable AI based on multimodal medical data can detect potential health issues prior, preventing severe complications (Ukwuoma et al. 2023). New AI technologies like quantum AI (QAI) are speeding up diagnostic processes by leveraging quantum computers' immense processing power. This allows for real-time analysis of large-scale medical data, improving diagnostic accuracy and decision-making in healthcare (Alonso Calafell et al. 2019). Various projects and companies are exploring general AI (GAI), such as OpenAI's DeepQA, IBM's Watson, and Google's DeepMind, to improve medical diagnoses. GAI analyses vast medical data, offering valuable insights to enhance patient outcomes and healthcare efficiency. Infact AI has penetrated traditional medical imaging, with deep-learning algorithms often outperforming radiologists in detecting anomalies in X-ray or MRI scans and now, it's advancing further (Savage 2024). Some researchers are investigating AI's ability to detect features overlooked by radiologists. Others are exploring the use of AI to diagnose neuro-developmental issues, like ADHD, using brain scans. AI has been extensively used in the diagnosis of different diseases viz., Alzheimer, cancer, anemia, diabetes, chronic heart disease, tuberculosis, viral diseases, stroke, cerebrovascular disease, hypertension, skin disease, liver disease and other rare diseases.

AI has revolutionized diagnostics, especially in imaging, like lung nodule detection on CT scans and breast imaging. It enhances clinical decisions for radiologists. With its ability to handle extensive data, AI has gained nearly 400 FDA approvals for radiology algorithms. Despite hospitals performing 3.6 billion imaging procedures yearly, 97% of this data remains unused (Pasquini et al. 2022). ML helps organize and leverage this data for more accurate diagnostics.

Imaging tools enhance diagnostics by leveraging DL to analyze medical data, improving radiology diagnoses and aiding in cancer aggressiveness assessment. They can even replace tissue samples with "virtual biopsies," identifying tumor phenotypes and genetic properties (Chang et al. 2018) These tools often yield more accurate conclusions than clinicians. For instance, the study in Bejnordi et al. (2017) revealed that 7 out of 32 DL algorithms diagnosed lymph node metastases in breast cancer more accurately than a panel of 11 pathologists.

Smartphones and other handheld devices could serve as effective diagnostic tools, especially in fields like dermatology and ophthalmology. AI in dermatology focuses on image analysis and classification to differentiate between benign and malignant skin lesions. Utilizing smartphones for image collection and sharing could broaden telehealth applications. In ophthalmology, companies such as Remidio utilize smartphone-based fundus cameras to detect diabetic retinopathy, enhancing diagnostic accessibility.

To grasp AI's role in disease diagnosis and prediction, it's crucial to explore a range of techniques such as Support Vector Machine, K-nearest neighbors, Naive Bayes, Decision Tree, Ada Boost, Random Forest, XGBoost, K-means and density based clustering, Convolutional Neural Networks (CNN), Recurrent Neural Network, Deep-CNN, Generative Adversarial Networks (GAN), and Long Short-Term Memory (LSTM), transfer learning among others, utilized in disease detection systems (Jitani et al. 2023; Owais et al. 2019; Nithya et al. 2020; Jitani et al. 2024).

### 1.3.3 Patient Safety

Integrating AI into clinical care presents notable opportunities to improve patient safety, but it also introduces new challenges that must be carefully managed. Some of the opportunities to enhance patient safety are:

1. Error Reduction: AI algorithms, particularly those using ML, can analyze vast datasets to detect patterns and predict outcomes more accurately than human clinicians in some cases. This capability can reduce diagnostic errors and enhance treatment plans.
2. Early Detection and Prevention: AI can help in the early detection of diseases by analyzing patient data from various sources (e.g., electronic health records, imaging, genomics) and identifying signs of conditions such as cancer, sepsis, or heart disease before they become clinically apparent.
3. Monitoring and Alerts: AI systems can monitor real-time patient vital signs and alert healthcare professionals to any significant changes or potential issues, enabling timely interventions (Silva et al. 2023).
4. Personalized Treatment: AI can help in personalizing treatment plans based on patient data, resulting in more efficient and safer treatments.
5. Operational Efficiency: AI can optimize workflows, reduce waiting times, and make sure that patient gets on time medical care, indirectly enhancing patient safety by reducing the risks associated with delayed treatment.

However, relying on AI heavily for ensuring patients safety also comes with several challenges and risks, some of which are:

1. Data Quality and Bias: AI systems rely massively on the data quality they are trained on. Poor data quality or biased datasets can lead to incorrect predictions or recommendations, posing significant safety risks.

2. Transparency and Explainability: Many AI algorithms, especially DL models, function as 'black boxes', making it complex for clinicians to understand how decisions are made. The absence of transparency can hinder trust and accountability.
3. Integration with Clinical Workflows: Successful integration of AI tools into existing clinical workflows is crucial. Poor integration can lead to workflow disruptions, increased clinician workload, and potential safety hazards (Talukdar et al. 2023c).
4. Over-reliance on AI: There is a risk that clinicians might over-rely on AI systems, potentially leading to complacency and a decrease in clinical skills. Ensuring that AI serves as an aid rather than a replacement for clinical judgment is essential.
5. Security and Privacy: AI systems often require access to huge amounts of patient data, bringing up concerns about data security and patients' privacy. Ensuring robust cybersecurity measures is critical to protect sensitive health information.

Implementing AI in healthcare requires thoughtful strategies to ensure that the technology enhances clinical outcomes, improves patient safety, and is integrated smoothly into existing workflows. Here are key strategies for successful AI implementation in healthcare:

1. Identify Clear Objectives

   - Clear and concise assessment of clinical needs: A clear assessment of the clinical need starts by identifying specific clinical problems or inefficiencies that AI can address, such as diagnostic accuracy, treatment planning, patient monitoring, or administrative tasks.
   - Finalize goals and objectives: Define clear, measurable goals for what the AI implementation should achieve, such as reducing diagnostic errors, improving patient outcomes, or increasing operational efficiency.

2. Data Management and Quality

   - Data Collection and Integration: Ensure access to high-quality, comprehensive data from numerous sources such as electronic health records (EHRs), medical imaging, laboratory results, and patient-generated data (Seo et al. 2020).
   - Data Cleaning and Preparation: Implement robust data cleaning processes to remove errors, inconsistencies, and biases that could affect AI performance.
   - Interoperability: Promote data interoperability standards to facilitate seamless data transfer between different healthcare systems and AI tools.

3. Algorithm Selection and Development

   - Appropriate selection of Algorithm: This process includes selection of an appropriate type of AI algorithms (e.g., ML, DL, NLP) based on the specific healthcare application to be implemented.
   - Customization: Customize algorithms to fit the unique needs of the healthcare organization, considering local population health data and clinical practices (Tajabadi et al. 2023).

- Validation and Testing: Conduct extensive validation and testing of AI models using diverse datasets to ensure reliability and accuracy across different patient populations.

4. Integration into Clinical Workflows

   - User-Centric Design: Design AI tools with input from end-users (clinicians, nurses, administrative staff) to ensure they fit seamlessly into existing workflows and address user needs (Talukdar et al. 2023a).
   - Pilot Programs: Start with pilot programs to test AI applications in real-world settings, gather feedback, and make necessary adjustments before full-scale deployment.
   - Interoperability with EHRs: Ensure AI systems can integrate smoothly with EHRs and other clinical systems to provide real-time insights and recommendations without disrupting workflow (Schork 2019).

5. Training and Education

   - Comprehensive Training: Provide thorough training for healthcare providers on how to use AI tools effectively, including understanding their capabilities, limitations, and best practices.
   - Continuous Education: It offers continuing education and support to keep staff updated on new AI developments and features.
   - Building Trust: Foster trust in AI tools by involving clinicians in the development and evaluation processes and ensuring transparency in how AI makes decisions.

6. Ethical and Regulatory Compliance

Implementing AI in healthcare requires adherence to ethical standards and regulatory compliance to ensure patient safety, trust, and the overall effectiveness of AI systems. Here are some key ethical standards for successful AI implementation in healthcare:

- Confidentiality: It ensures that patient data is kept confidential and is only accessible to authorized personnel and free from unauthorized access.
- Data Protection: Use encryption and other security precautions to safeguard patient data from breaches and unauthorized access (Love-Koh et al. 2018).
- Anonymization: Identity protection is a major characteristic to be considered when taking care of the ethical values of patients in accordance with AI methods.
- Transparency: Maintenance of transparency with the patients when AI is being used for diagnosis or treatment is very crucial while dealing with diagnosis and treatment.
- Understanding: Make sure that patients understand how their data will be used and the implications of AI on their care.
- Opt-In/Opt-Out: Provide patients with the option to consent to or decline the use of AI for treatments.

- Bias Mitigation: Identify and address any biases in AI algorithms that could result in unequal treatment of different patient groups.
- Equity: Ensure that AI systems provide equal treatment to all patients, regardless of ethnicity, gender, age, or socioeconomic status.
- Continuous Monitoring: Regularly evaluate AI systems to detect and correct any emerging biases.
- Validation: Rigorously test and validate AI systems to ensure their accuracy and reliability in different clinical settings.
- Transparency in Algorithms: Make the functioning of AI algorithms as transparent as possible to facilitate understanding and trust (Leung 2023).
- Responsibility: Clearly define who is responsible for the outcomes of AI systems, including developers, healthcare providers, and institutions.
- Liability: Establish clear guidelines for liability in situations when AI systems cause injury or fail to offer appropriate care.
- Auditability: Maintain detailed records of AI system decisions and actions to facilitate audits and investigations if needed.
- Interdisciplinary Collaboration: Involve ethicists, clinicians, and patient representatives in the development of AI systems.
- Continuous Improvement: Regularly update AI systems based on new medical research, clinical feedback, and ethical considerations.
- Sustainability: Ensure that AI systems are sustainable and can be maintained and updated over time without excessive resource consumption.
- Regulatory Compliance: Ensure compliance with relevant regulations and standards established by organizations such as the FDA, EMA, or local health authorities.
- Adherence to Standards: Comply with national and international regulations and standards for AI in healthcare.
- Ethical Guidelines: Follow established ethical guidelines from professional bodies and organizations.
- Bias Mitigation: Implement strategies to identify and mitigate biases in AI algorithms to ensure equitable care for all patient groups.

By adhering to these ethical standards, healthcare organizations can leverage AI technologies to improve patient well-being, enhance efficiency, and sustain the trust and confidence of patients and healthcare professionals.

7. Monitoring and Evaluation
    - Performance Monitoring: Continuously monitor AI performance and impact on clinical outcomes, patient safety, and operational efficiency.
    - Feedback Loops: Establish feedback mechanisms to collect user feedback and make iterative improvements to AI tools.
    - Outcome Measurement: Regularly measure the outcomes against the predefined goals and adjust strategies as necessary to achieve desired results.
8. Stakeholder Engagement

- Multidisciplinary Teams: Form multidisciplinary teams including clinicians, data scientists, IT professionals, and administrators to oversee AI implementation.
- Patient Involvement: Engage patients in the design and evaluation of AI tools to assure they meet patient needs and improve patient experiences.
- Communication: Maintain open communication channels with all stakeholders to ensure transparency and collaborative problem-solving (Klein et al. 2017).

9. Scalability and Sustainability

- Scalable Solutions: Design AI systems with scalability in mind, allowing for expansion across different departments or locations within the healthcare organization.
- Sustainability Planning: Develop plans for the long-term sustainability of AI systems, including maintenance, updates, and continuous learning to adapt to evolving healthcare needs and technological advancements.

By following these strategies, healthcare organizations can effectively implement AI technologies to enhance clinical care, improve patient well-being, and achieve operational excellence while navigating the complexities of healthcare environments. In conclusion, integrating various AI modalities in healthcare has significant potential to modernize healthcare delivery, rendering it more specific, effective, and accessible to both patients and healthcare providers. It's conceivable that in the future, these technologies will blend to the extent that composite solutions become more common and achievable.

AI's integration into medical diagnostics is still in early stages, facing technical, regulatory, and ethical hurdles. A key challenge is the quality and accessibility of medical data, critical for effective AI algorithms, which often suffer from biases due to non-representative datasets. Ethical concerns arise from using general AI (GAI) for diagnostics, particularly regarding data privacy and algorithm transparency. Although federated learning offers potential solutions, further investigation is warranted (Alowais et al. 2023). Interoperability standards are essential for seamless collaboration among AI-based diagnostic tools. Despite challenges, AI offers promise for personalized treatment planning, with ongoing research aimed at enhancing accuracy and efficiency. These advancements support healthcare staff and provide innovative solutions for combating global health crises.

## References


Al-Antari, M. A. (2023). Artificial intelligence for medical diagnostics—existing and future ai technology!

Alonso Calafell, I., Cox, J., Radonjic´, M., Saavedra, J., Garc´ıa de Abajo, F., Rozema, L., & Walther, P. (2019). Quantum computing with graphene plasmons. *npj Quantum Information*, *5*(1), 37.

Alowais, S. A., Alghamdi, S. S., Alsuhebany, N., Alqahtani, T., Alshaya, A. I., Almo-hareb, S. N., Aldairem, A., Alrashed, M., Bin Saleh, K., Badreldin, H. A., et al. (2023). Revolutionizing healthcare: The role of artificial intelligence in clinical practice. *BMC medical education, 23*(1), 689.

Bejnordi, B. E., Veta, M., Van Diest, P. J., Van Ginneken, B., Karssemeijer, N., Litjens, G., Van Der Laak, J. A., Hermsen, M., Manson, Q. F., Balkenhol, M., et al. (2017). Diagnostic assessment of deep learning algorithms for detection of lymph node metastases in women with breast cancer. *Jama, 318*(22), 2199–2210.

Chang, P., Grinband, J., Weinberg, B., Bardis, M., Khy, M., Cadena, G., Su, M.-Y., Cha, S., Filippi, C., Bota, D., et al. (2018). Deep-learning convolutional neural networks accurately classify genetic mutations in gliomas. *American Journal of Neuroradiology, 39*(7), 1201–1207.

Davenport, T., & Kalakota, R. (2019). The potential for artificial intelligence in healthcare. *Future healthcare journal, 6*(2), 94–98.

Jitani, N., Basumatary, A., & Sarmah, R. (2023). Deep learning-based tumor segmentation from ct images. *International Conference on Advanced Compu tational and Communication Paradigms*, 323–333.

Jitani, N., Singha, B. J., Barman, G., Talukdar, A., Sarmah, R., & Bhattacharyya, D. K. (2024). Medical image segmentation using automated rough density approach. *Multimedia Tools and Applications, 83*(13), 39677–39705.

Klein, M. E., Parvez, M. M., & Shin, J.-G. (2017). Clinical implementation of pharmacogenomics for personalized precision medicine: Barriers and solutions. *Journal of pharmaceutical sciences, 106*(9), 2368–2379.

Leung, R. (2023). Using ai–ml to augment the capabilities of social media for telehealth and remote patient monitoring. *Healthcare, 11*(12), 1704.

Love-Koh, J., Peel, A., Rejon-Parrilla, J. C., Ennis, K., Lovett, R., Manca, A., Chalkidou, A., Wood, H., & Taylor, M. (2018). The future of precision medicine: Potential impacts for health technology assessment. *Pharmacoeconomics, 36*, 1439–1451.

MacEachern, S. J., & Forkert, N. D. (2021). Machine learning for precision medicine. *Genome, 64*(4), 416–425.

Morgan, A. A., Abdi, J., Syed, M. A., Kohen, G. E., Barlow, P., & Vizcaychipi, M. P. (2022). Robots in healthcare: A scoping review. *Current robotics reports, 3*(4), 271–280.

Nithya, A., Appathurai, A., Venkatadri, N., Ramji, D., & Palagan, C. A. (2020). Kidney disease detection and segmentation using artificial neural network and multi-kernel k-means clustering for ultrasound images. *Measurement, 149*, 106952.

Owais, M., Arsalan, M., Choi, J., Mahmood, T., & Park, K. R. (2019). Artificial intelligence-based classification of multiple gastrointestinal diseases using endoscopy videos for clinical diagnosis. *Journal of clinical medicine, 8*(7), 986.

Pasquini, L., Napolitano, A., Pignatelli, M., Tagliente, E., Parrillo, C., Nasta, F., Romano, A., Bozzao, A., & Di Napoli, A. (2022). Synthetic post-contrast imaging through artificial intelligence: Clinical applications of virtual and augmented contrast media. *Pharmaceutics, 14*(11), 2378.

Savage, N. (2024). Ai's keen diagnostic eye. *Nature.*

Schork, N. J. (2019). Artificial intelligence and personalized medicine. *Precision medicine in Cancer therapy*, 265–283.

Seo, H., Badiei Khuzani, M., Vasudevan, V., Huang, C., Ren, H., Xiao, R., Jia, X., & Xing, L. (2020). Machine learning techniques for biomedical image segmentation: An overview of technical aspects and introduction to state- of-art applications. *Medical physics, 47*(5), e148–e167.

Sheikhalishahi, S., Miotto, R., Dudley, J. T., Lavelli, A., Rinaldi, F., Osmani, V., et al. (2019). Natural language processing of clinical notes on chronic diseases: Systematic review. *JMIR medical informatics, 7*(2), e12239.

Silva, B., Hak, F., Guimara˜es, T., Manuel, M., & Santos, M. F. (2023). Rule-based system for effective clinical decision support. *Procedia Computer Science, 220*, 880–885.

Tajabadi, M., Grabenhenrich, L., Ribeiro, A., Leyer, M., & Heider, D. (2023). Sharing data with shared benefits: Artificial intelligence perspective. *Journal of Medical Internet Research, 25*, e47540.

Talukdar, J., Singh, T. P., & Barman, B. (2023a). *Artificial intelligence in healthcare industry*. Springer.

Talukdar, J., Singh, T. P., & Barman, B. (2023b). Introduction to human and artificial intelligence. In *Artificial intelligence in healthcare industry* (pp. 1–16). Springer.

Talukdar, J., Singh, T. P., & Barman, B. (2023c). Robotic process automation: A path to intelligent healthcare. In *Artificial intelligence in healthcare industry* (pp. 159–168). Springer.

Talukdar, J., Singh, T. P., & Barman, B. (2023d). Rule-based expert systems. In *Artificial intelligence in healthcare industry* (pp. 145–158). Springer.

Tran, K. A., Kondrashova, O., Bradley, A., Williams, E. D., Pearson, J. V., & Waddell, N. (2021). Deep learning in cancer diagnosis, prognosis and treatment selection. *Genome Medicine, 13*, 1–17.

Ukwuoma, C. C., Qin, Z., Heyat, M. B. B., Akhtar, F., Bamisile, O., Muaad, A. Y., Addo, D., & Al-Antari, M. A. (2023). A hybrid explainable ensemble transformer encoder for pneumonia identification from chest x-ray images. *Journal of Advanced Research, 48*, 191–211.

Vial, A., Stirling, D., Field, M., Ros, M., Ritz, C., Carolan, M., Holloway, L., & Miller, A. A. (2018). The role of deep learning and radiomic feature extraction in cancer-specific predictive modelling: A review. *Translational Cancer Research, 7*(3).

Zafar, M., Sharif, M. I., Sharif, M. I., Kadry, S., Bukhari, S. A. C., & Rauf, H. T. (2023). Skin lesion analysis and cancer detection based on machine/deep learning techniques: A comprehensive survey. *Life, 13*(1), 146.

GPSR Compliance

*The European Union's (EU) General Product Safety Regulation (GPSR) is a set of rules that requires consumer products to be safe and our obligations to ensure this.*

*If you have any concerns about our products, you can contact us on ProductSafety@springernature.com*

In case Publisher is established outside the EU, the EU authorized representative is:

Springer Nature Customer Service Center GmbH
Europaplatz 3
69115 Heidelberg, Germany

**Batch number: 08262699**

Printed by Printforce, the Netherlands